Coherence
in Spectroscopy
and Modern Physics

NATO ADVANCED STUDY INSTITUTES SERIES

A series of edited volumes comprising multifaceted studies of contemporary scientific issues by some of the best scientific minds in the world, assembled in cooperation with NATO Scientific Affairs Division.

Series B: Physics

RECENT VOLUMES IN THIS SERIES

This series is published by an international board of publishers in conjunction with NATO Scientific Affairs Division

A Life Sciences	Plenum Publishing Corporation
B Physics	New York and London
C Mathematical and	D. Reidel Publishing Company
Physical Sciences	Dordrecht and Boston
D Behavioral and	Sijthoff International Publishing Company
Social Sciences	Leiden
E Applied Sciences	Noordhoff International Publishing
	Leiden

Coherence
in Spectroscopy
and Modern Physics

Edited by
F.T. Arecchi
*University of Florence
and National Institute of Optics
Florence, Italy*

R. Bonifacio
*University of Milan
Milan, Italy*

and
M.O. Scully
*University of Arizona
Tucson, Arizona*

Springer Science+Business Media, LLC

Library of Congress Cataloging in Publication Data

Nato Advanced Study Institute on Coherence in Spectroscopy and Modern Physics,
   Versilia, Italy, 1977.
   Coherence in spectroscopy and modern physics.

   (NATO advanced study institutes series: Series B, Physics; v. 37)
   "Proceedings of the NATO Advanced Study Institute on Coherence in Spectroscopy
and Modern Physics held at Villa LePianore, Versilia, Italy, July 17—30, 1977."
   Includes index.
   1. Coherence (Optics) — Congresses. 2. Spectrum analysis — Congresses. 3. Phys-
ics — Congresses. I. Arecchi, F. T. II. Bonifacio, R. III. Scully, Marlan Orvil, 1939-
      IV. Title. V. Series.
QC476.C6N37   1977   536   78-14474
   ISBN 978-1-4613-2873-5   ISBN 978-1-4613-2871-1 (eBook)
   DOI 10.1007/978-1-4613-2871-1

Proceedings of the NATO Advanced Study Institute on Coherence in
   Spectroscopy and Modern Physics held at Villa LePianore, Versilia,
   Italy, July 17—30, 1977

© 1978 Springer Science+Business Media New York
Originally published by Plenum Press, New York in 1978
Softcover reprint of the hardcover 1st edition 1978

PREFACE

This volume contains the lectures and seminars presented at the NATO Advanced Study Institute on "Coherence in Spectroscopy and Modern Physics," the seventh course of the International School of Quantum Electronics, affiliated with the "Ettore Majorana" Centre for Scientific Culture, Erice, Sicily. The Institute was held at Villa LePianore (Lucca), Versilia, Italy, July 17-30, 1977.

The International School of Quantum Electronics was started in 1970 with the aim of providing instruction for young researchers and advanced students already engaged in the area of quantum electronics or wishing to switch to this area from a different background. From the outset the School has been under the direction of Prof. F. T. Arecchi, then at the University of Pavia, now at the University of Florence, and Dr. D. Roess of Siemens, Munich. Each year the Directors choose a subject of particular interest, alternating fundamental topics with technological ones, and ask colleagues specifically competent in a given area to take the scientific responsibility for that course.

The past courses were devoted to the following themes:

1971: "Physical and Technical Measurements with Lasers"
1972: "Nonlinear Optics and Short Pulses"
1973: "Laser Frontiers: Short Wavelength and High Powers"
1974: "Cooperative phenomena in Multicomponent Systems"
1975: "Molecular Spectroscopy and Photochemistry with Lasers"
1976: "Coherent Optical Engineering"

The purpose of the present course was to provide instruction in the field of modern physics and spectroscopy as it is influenced by coherent quantum-mechanical phenomena. It is generally acknowledged that this rapidly developing field of research, besides offering an intellectual challenge, has had a significant impact on modern technology.

Prof. R. Bonifacio, University of Milan, and Prof. M. O. Scully, University of Arizona, undertook the scientific direction of the present course, selecting the specific topics and lectures.

The lectures collected in this volume cover the basic concepts and principles of coherence in quantum optics as well as in other areas of physics together with the most promising applications.

We wish to express our appreciation to the NATO Scientific Affairs Division, whose financial support made this Institute possible. We also acknowledge the contribution of the following:

> CISE
> IBM Italia
> Philips, Eindhoven
> Siemens, Munich

We finally thank the secretaries of Divisione Elettronica Quantistica, CISE, Mrs. G. Ravini and Miss M. Oriani, for their valuable assistance in the organization of the Institute and in the preparation of these proceedings, and Miss A. Camnasio of Servizio Documentazione, CISE, and again Mrs. G. Ravini for their assistance during the course itself.

F. T. Arecchi
R. Bonifacio
M. O. Scully

CONTENTS

A – Basic Tools

B – Coherence in Spin-like Systems

C – Coherence in Atomic Spectroscopy

D - Applications of Coherence to Physical Systems

E - New Laser Sources

COHERENCE IN SPECTROSCOPY

Alfred Kastler

Laboratoire de Spectroscopie Hertzienne de l'Ecole
Normale Supérieure – Université Paris VI
75231 PARIS Cedex 05, France

INTRODUCTION

The general subject of this conference is "coherence". The first object of our investigation must be the answer to the question: what is the meaning of the word "coherence". What is its definition? It may be used in every day language and in the language of science. Its opposite is "incoherence". I believe the word "coherence" means in general some correlation between things or events, some state of "order", while "incoherence" means "no relation" or "disorder".

By scientists the word "coherence" has been first used in the field of physical optics. We say that two light-beams, able to produce interference fringes by their superposition, must be coherent. In the wave-theory of light this means that their must be a definite relation, invariant in time, between the phases of their vibrations. If we note these vibrations by the expressions :

$$V_1 = a_1 \cos(\omega t - \varphi_1) \qquad , \qquad V_2 = a_2 \cos(\omega t - \varphi_2)$$

where ω is their commun circular frequency and where φ_1 and φ_2 are their phases, we say that these vibrations are coherent if their phase difference

$$\Delta\varphi = \varphi_1 - \varphi_2$$

remains constant in time during a time interval which is large compared to period $T = 2\pi/\omega$ of the vibrations.

The vibrations are incoherent if $\Delta\varphi$ changes rapidly in a random

1

manner during t i m e and takes all values between 0 and 2 π.

Such a definite relation between phases of vibrations is an element of "order" and can be expressed by a definite value of "neguentropy" while incoherence is an element of "disorder" and corresponds to a high entropy.

When, during the two last decades of the 19th century Ludwig Boltzmann[1], studying the behaviour of an ideal gas of molecules, established his famous H-theorem and established the relation between the thermodynamic concept of entropy S and the statistic concept of probability W, he had to make the hypothesis of "molekulare Unordnung", that is to say :he had to admit that in the initial state already there was same "disorder" among the molecules of the gas. This was necessary to guarantee the evolution of the gas to a higher state of disorder, to maximum entropy. When in 1900 Max Planck established his famous formula for black-body radiation, he had to make an analogous hypothesis which he called "Hypothesis der natürlichen Strahlung". In his mind the word "naturlich" was synonimous with "incoherent". He had to postulate the absence of any definite phase-relation between the vibration of the monochromatic "resonators" in equilibrium with cavity radiation. We know that instead of considering with Planck such material resonators, we can consider - as has been shown by Lord Rayleigh[3] in 1900 and used by Debye[4] in 1910 - the radiation itself in a cavity as composed of quantised "steady waves", still called "Eigenschwingungen" or " cavity-modes ". Their number per unit volume and unit-frequency range is independant of the shape of the cavity and of the nature of its walls. The vibrations of these cavity modes must be "incoherent". Only in this case the addition-theorem of their entropy is valid and can be applied. This incoherence is a necessary postulate for obtaining Planck's formula.

COHERENCE IN ATOMIC SPECTROSCOPY

We may now consider same special, but interesting cases of coherence in spectroscopy, limiting us to atomic spectroscopy.

MAGNETIC RESONANCE

As a first example we consider the simple case of paramagnetic atoms or nuclei in a steady magnetic field B. Each particle possesses a small magnetic moment $\vec{\mu}$ along an axis $\vec{OA}$. To this magnetic m oment is associated a mechanical angular momentum $\vec{P}$

along the same axis, and we write $\vec{\mu} = \gamma \vec{P}$, γ being called the gyro–magnetic factor. Let us call ϑ the angle between the field $\vec{B}$ and the momentum. It can be shown that the μ–axis will precess around the field $\vec{B}$ with a circular frequency $\omega = \gamma B$, the sens of precession betw een determined by the sign of γ. This is the well known Larmor precession.

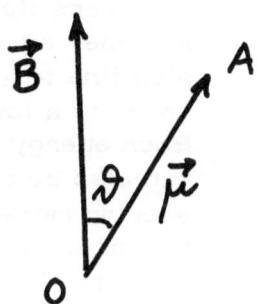

In a collection of N identical atoms, each one will precess with the same frequency ω, but the phases of precession of the different atoms will be distributed randomly, depending on the initial phase–angle φ_i^o of each atom: $\qquad \varphi_i = \omega t + \varphi_i^o$
The precession is incoherent. In the field B the collection of atoms gives rise to a "longitudinal" macroscopic moment $M_z = \sum_1^N \mu \cos\vartheta$ but there is no "transverse" m acroscopic moment.

If now we appl y to the atoms in the equatorial plane ($\theta = \frac{\pi}{2}$) a radiofrequency rotating field $\vec{\beta}$ of frequency ω_o, we get "magnetic resonance" if $\omega_o = \omega = \gamma B$. By this rotating field the precession of all atom s is "locked" to the field, and as a result of this the collection of atoms aquires a macroscopic rotating magnetic moment M_t w hich can be detected by sensitive instruments. The precession of all atoms has been made coherent. In general at resonance ($\omega_o = \omega$) the angle between the rotating field β and the rotating moment M_t is $\pi/2$.

COHERENC E IN THE OPTICAL RANGE

Let us now take an example of coherence in atomic spectroscopy in the optical range: in quantum theory the energetic properties of an atom are rep resented by an energy scale show ing its discrete energy levels numbered E_1, E_2, E_3 etc, E_1 being the lowest or graund state, the other states being excited states of short life –

time. In this scheme a spectral line of the atom is represented by a "transition" between two energy states whose circular frequency obeys Bohr's relation :

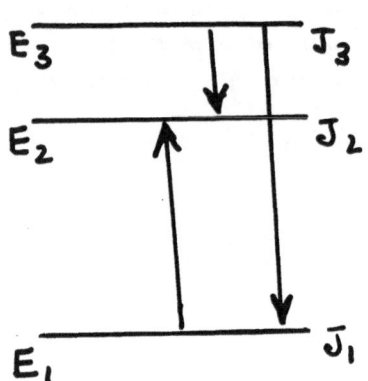

$$\hbar\omega_{12} = E_2 - E_1$$

An absorption line corresponds to a transition from a lower to an upper energy state, an emission line to a transition from an upper to a lower state.

Each energy state of the atom can also be characterized by its angular momentum, the quantum unit of angular momentum being equal to $\hbar$, the momentum of a state E_i being $P_i = J_i\hbar$, where J_i is called the inner quantum number, this number being integer or (integer $\pm$ 1/2). In a spectral transition energy and angular momentum must be conserved.

Let us now consider a magnetic field B. It is a consequence of the Zeeman splitting of the energy states, a state J_i being split into $2 J_i + 1$ equidistant magnetic substates characterized by a "magnetic quantum number" m which takes the values separated by unity between $- J_i$ and $+ J_i$, the equidistance between two neighbouring magnetic substates being, in the frequency scale

$$\omega_i = \gamma_i B$$

where γ_i is the gyromagnetic ratio of the energy state E_i. Let us consider the simple case of an atom whose ground state E_1 has the value $J_1 = 0$ and whose upper state E_2 has the value $J_2 = 1$. In a magnetic field B the ground state will remain unsplit, the upper state being split into three equidistant substates m = −1, m = 0 and m = +1.

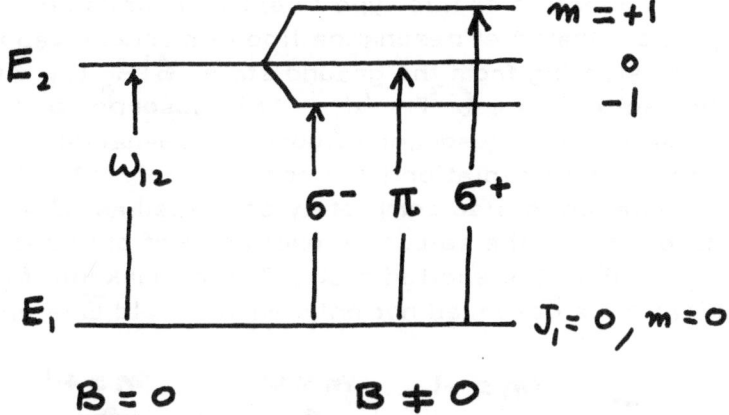

The spectral line between E_1 and E_2 with circular frequency in a 0-field will be split into three Zeeman-components having the frequencies

$$\omega_{12} - \gamma_2 B \quad , \quad \omega_{12} \quad , \quad \omega_{12} + \gamma_2 B \,.$$

A remarkable property of these Zeeman-components is their polarization. The central component, corresponding to a change $\Delta m = 0$ is linearly polarized with its electric vector $\vec{E}$ parallel to the field $\vec{B}$. It is called a π component. The two other components corresponding to $\Delta m = -1$ and $\Delta m = +1$ are circulalrly polarized in a plane perpendicular to the field-vector $\vec{B}$. Their circular polarizations are opposite and called $\sigma-$ for the $\Delta m = -1$ component and σ^+ for the $\Delta m = +1$ component. As Rulinanicz has shown, these polarizations are a consequence of conservation of angular momentum, the σ^- and σ^+ photons carrying angular momentum respectively equal to $-\hbar$ and to $+\hbar$, the π-photon carrying no angular momentum.

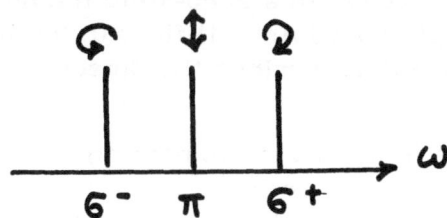

Scheme of the Zeeman splitting of a spectral line $J_1 = 0 \rightarrow J_2 = 1$.

An illustration of this type of splitting and polarization is given by the ultraviolet resonance line of mercury vapour Hg 2537 Å.
This line starting from the ground state 6^1S_0 ($J_1 = 0$) and reaching the excited state $6\,^3P_1$ ($J_2 = 1$) is absorbed by mercury vapour
and reemitted as " resonance fluorescence-radiation". If we pola-
rize the exciting radiation (if it contains only π - light), the resonance radiation is also completely polarized as π - light. This is
a consequence of the selective excitation of only one magnetic sublevel (m = 0) of the excited state. The remarkable fact is that this
polarization is observed not only when a field $\vec{B}$ is applied in the

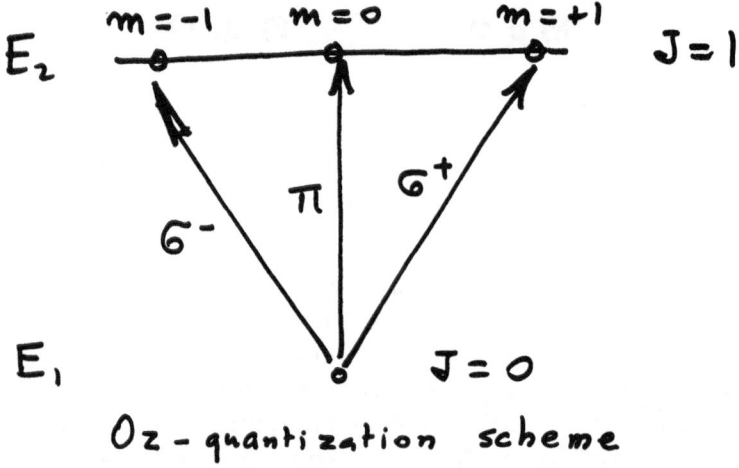

Oz - quantization scheme

direction of the $\vec{E}$ - vector of the exciting light (which we may take
as the 0z - direction of a 0xyz - referential), but also in the absence of any field. In fact the polarization of the fluorescence-
radiation is independent of the field-value if the field B has the same symmetry elements as the E - vector of the exciting radiation,
as was shown by Heisenberg [5] in 1926.

As he stated we observe in a zero-field the polarization we
would observe by applying a (virtual) field in the direction of the
E - vector. Such a field determines the direction of "space-quantization" of the atom.

But if we operate in a zero-field nothing obliges us to apply
the virtual field in the 0z-direction. We may apply it in any other
direction, for example the 0x-direction which is the direction of
observation of the fluorescence - radiation.

This will change nothing, of course, to the experimental facts, but it changes our theoretical interpretation :

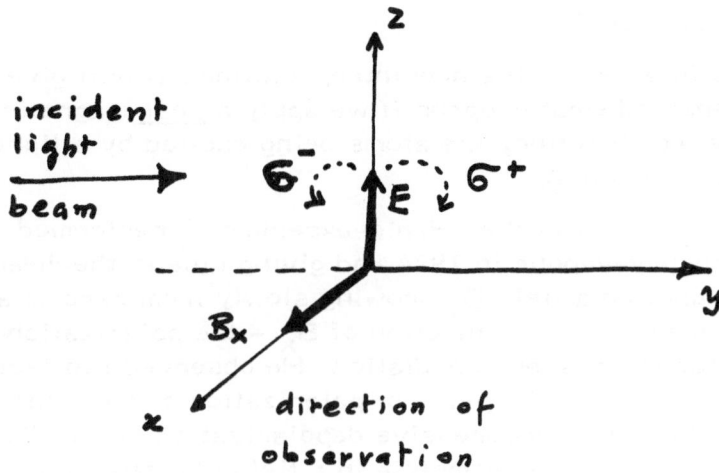

If we consider the space-quantization of the atom in the virtual field direction 0x, the $\vec{E}$-vector of the exciting light-beam, parallel to 0z, may be considered as a coherent superposition of two circular polarized light vectors rotating in the zy-plane.

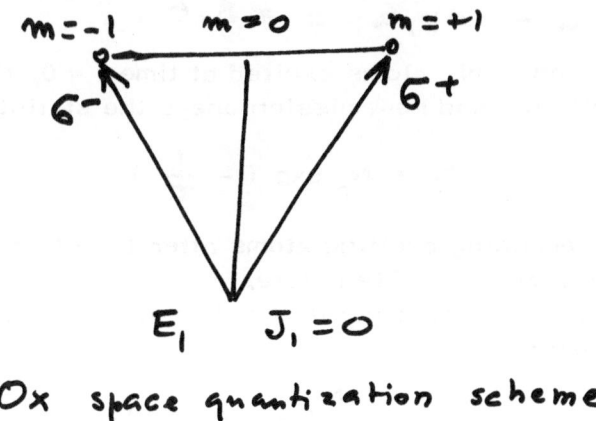

Ox space quantization scheme

From this new point of view, we may say that we obtain in this case a coherent excitation of the two substates m = - 1 and m = + 1 of the excited state. These two substates will re-emit "coherently" the fluorescent light vectors σ^- and σ^+ whose superposition will restore an E-vector in the Oz-direction.

THE HANLE-EFFECT

What is the interest of this new interpretation? It will give us the key to understand what happens if we apply a __real__ magnetic field B_x in the Ox-direction, the atoms being excited by a light vector E_z in the Oz-direction.

This new experiment is the "Hanle-experiment" performed by W. Hanle on mercury-vapour in 1924 and giving rise to the "Hanle-effect [6]. Hanle applied a field B_x growing slowly from zero to a high value. He measured as a function of B_x – the polarization-state of the re-emitted fluorescence radiation. He observed two facts: a progressive rotation of the plane of polarization of the emitted fluorescence light, and a progressive depolarization of this light, this depolarization becoming complete in a field of a few gauss only.

Let us explain Hanle's result:
The B_x field introduces a shift between the two substates m=-1 and m=+1 of the excited state. In the frequency scale this shift amounts to $\Delta = 2 \gamma B_x$. If the two coherently excited m-states emit time t after excitation, this frequency shift will introduce between the two emitted σ^+ and σ^- radiations a phase-shift $\Delta\varphi = t \Delta\omega = 2\gamma B_x t$, and the result will be a rotation of the plane of polarization by the angle

$$\alpha = \Delta\varphi/2 = \gamma B_x t$$

For a collection of N_o atoms excited at time t = 0, the time delay between excitation and re-emission obeys the statistical law

$$N = N_o \exp\left(-\frac{t}{\tau}\right)$$

N being the remaining exciting atoms after time t, and τ being the mean life-time of the excited state.
The number of atoms emitting in the time-interval between t and t + dt is given by

$$dN = \frac{N_o}{\tau} \exp\left(-\frac{t}{\tau}\right)dt$$

Integrating over all t-values from 0 to ∞ we get for the polarization P of the fluorescence light emitted :

$$P = \frac{P_o}{1 + (2\omega\tau)^2} = \frac{P_o}{1 + (2\gamma B_x \tau)^2}$$

In this relation the polarization is defined as

$$P = \frac{I_z - I_x}{I_z + I_x}$$

I_z and I_x being the intensities measured by an analyser of the light polarized in the z and in the x – directions.

Let us remark that in the small magnetic fields necessary to produce the Hanle effect, the Zeeman-splitting of the spectral line remains much smaller than the line-width which is the Doppler-width of the vapour at room-temperature, so that the exciting line covers at any field-value all the Zeeman-components of the line. In spite of the presence of the Doppler-width the Hanle experiment enables to measure τ . This is equivalent to a measurement of the "natural" line-width $\Delta\omega = \frac{1}{\tau}$.

To this relation corresponds a "Hanle-curve" of Lorentzian shape in the B_x-scale

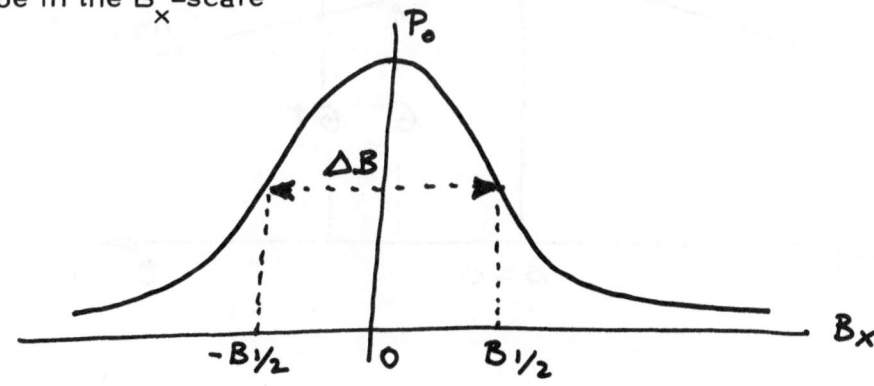

The "half-width" of this curve (the field value for which $P = \frac{1}{2}P_o$) is given by

$$2\gamma B_{\frac{1}{2}} \tau = 1 \quad \text{or} \quad B_{\frac{1}{2}} = \frac{1}{2\gamma\tau} \quad , \quad \Delta B = (\gamma\tau)^{-1}$$

For the 2537 Å resonance line of mercury-vapour Hanle found a $B_{1/2}$ values near 1 gauss. γ being known by the measurement of the Zeeman-splitting of the resonance line ($\omega = \gamma B$), Hanle showed that for the $6\,^3P_1$ excited state of the mercury-atom τ was of the order of 10^{-7} s.

At the time this method was performed, this result was one of the best measurements of life-times known in 1924.

We have seen that the "Hanle-experiment" is a typical "coherence" experiment. The Hanle-effect is a "decoherence-effect", the coherence observed in a zero-field being progressively destructed by a growing B_x-field.

Let us trace the energy scheme of the magnetic sublevels of the involved energy states of the mercury-atom as a function of the magnetic field B.

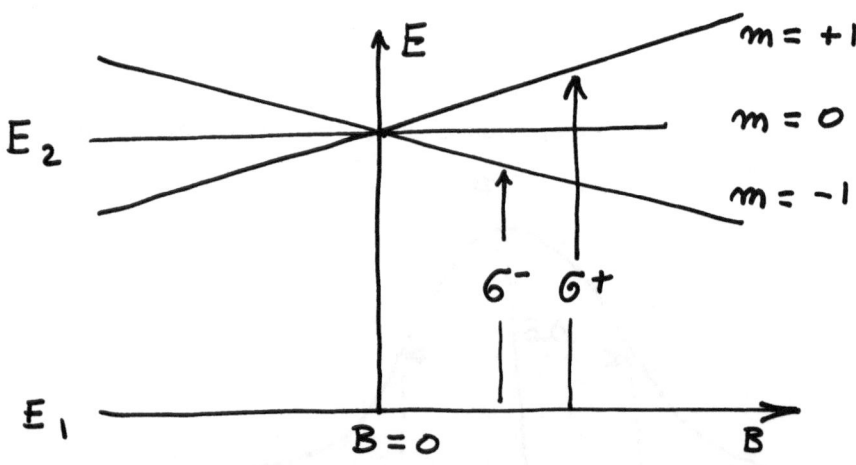

We notice that the three magnetic sublevels of the excited states are crossing each other at the field-value B = 0. At this value the frequencies of the σ^- and σ^+ components of the spectral line become equal, and these vibrations are able to interfere. The Hanle-curve can be interpreted as an interference-curve at "zero-field level crossing".

LEVEL-CROSSING SIGNALS

In more complicated Zeeman-diagrammes (E_m levels as a function of the B-field), especially in those of fine-structure and hyperfine structure levels of atoms, magnetic sublevels are crossing for field values B = 0. Peter Franken and his collaborators[7] have shown that for these field-values level-crossing signals can be observed which are analogous to the Hanle-curve, and Peter Franken[8] has shown that these signals are due also to a coherence effect of the emitted light vibrations. This level crossing technique has become very important for exploring fine-structure and hyperfine structure intervals. It has been used in our laboratory in Paris by J. P. Desconbes[9] who has determined the fine structure of the ^{4}He-atom and the fine structure and hyperfine structure of the ^{3}He-atom.

COHERENCE IN MULTIPLE SCATTERING OF RESONANCE LIGHT

A quite different coherence effect in atomic spectroscopy has been discovered in our Paris laboratory: the narrowing of the magnetic resonance-lines of the excited state $6\,^3P_1$ of the mercury atom by multiple scattering of the resonance radiation. This effect was observed by Brossel and Blamont[10] who studied the magnetic resonance signals by Brossel's double resonance method. Let us remember the principle of this method[11]. As we have shown already a selective excitation of the magnetic sublevel m = 0 of the excited state can be obtained by irradiation of the mercury vapour (containing only even isotopes) with linearly polarized 2537 $\overset{\circ}{A}$ light (π – light with its electric vector $\vec{E}$ in the 0z-direction). The resonance light observed at right angle of the incident light beam (in the 0x-direction) is also completely polarized and contains only π light, vibrating in the 0z-direction. Brossel applies to the vapour-vessel a steady magnetic field Bz in the oz direction. This field splits the magnetic sublevels of the upper state $6\,^3P_1$ the equidistance between the three levels being in the frequency scale : $\omega = \gamma B_z.$

With a pair of Helmholtz coils a radiofrequency field of circular frequency ω_o is applied creating and oscillating magnetic field of this frequency in a direction perpendicular to 0z.

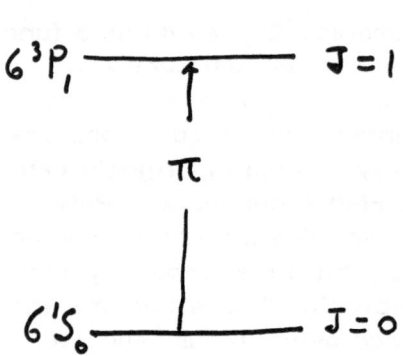

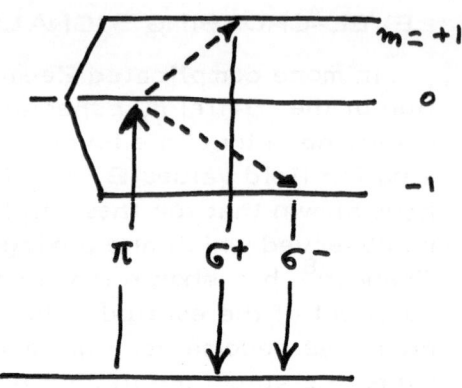

The frequency ω_o of the oscillating field (or the field value B_z) is slowly changed. At the point where ω_o becomes equal ω, magnetic resonance occurs. The oscillating field induces transitions from m = 0 substate to the two m = +1 substates. As a result of these transitions, the σ^+ and σ^- components appear in the emitted fluorescence light. In this way magnetic resonance can be monitored by observing the polarization of the resonance light emitted. It causes depolarization of this light. The resonance curve has a breadth which corresponds to the natural line-width.
For small values of the amplitude of the radiofrequency field, this width (for the points where the intensity of the σ-light is half of the intensity of the central peak at resonance) is given in the frequency scale by

$$\Delta \omega = 2/\tau$$

or in field scale (used by Brossel) by

$$\Delta B = \frac{\Delta \omega}{\gamma} = \frac{2}{\gamma \tau}$$

and its measure, obtained at very low vapour pressure, leads to a very precise value of the life-time τ of the excited state $\tau = 1,18 \cdot 10^{-7}$ s.

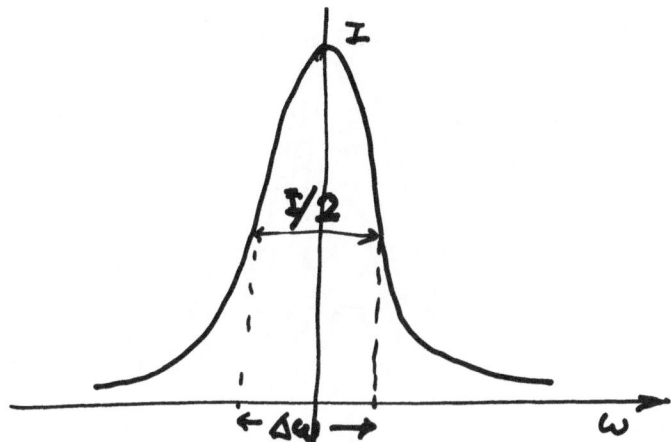

But the surprising effect, found by B lamont and Brossel was that, enhancing the density of the mercury vapour by heating the mercury drop in the tail of the vessel, and expecting to find a broadening of the resonance-curves by collisions, they found a remarkable narrowing of these curves when the density of the mercury vapour became higher and higher.

Fig. A shows three groups of curves (the different curves corresponding to raising amplitudes of the radio-frequency field) for three values of the temperature of the mercury drop : – 31°C, 0°C and 22°C. In this temperature range the density of the vapour is raised by a factor one hundred).

The explanation of this surprising result was given by Brossel and recognized by him to be a coherence effect: at very low vapour pressure the incident light photon is observed by one atom in the resonance vessel, and being spontaneously emitted, goes to the observer (the photodetector).

At higher vapour pressure the photon emitted by a first atom has a chance to be captured and absorbed by a second one before

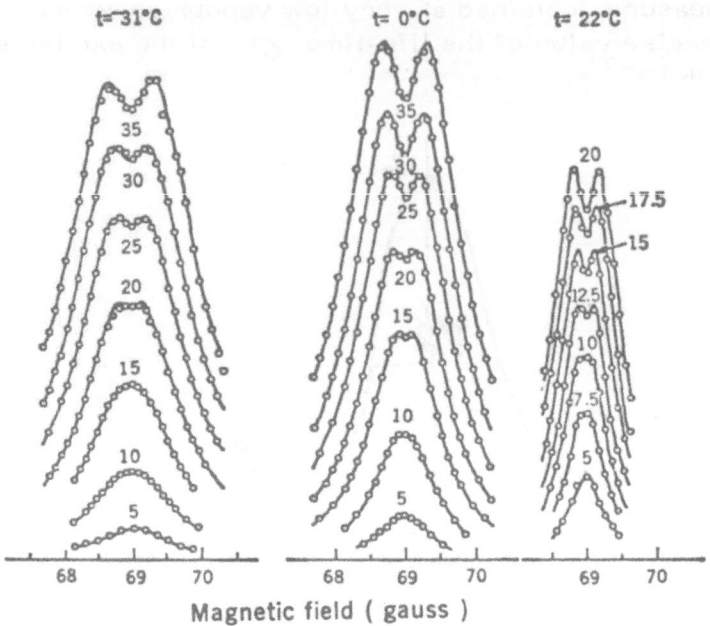

Fig. A

going to the observer, or even by a third one and so on. This phenomena is well known as "multiple scattering" or as " imprisonment of resonance radiation", and is illustrated by the figure.

At each vapour-pressure the width $\Delta\omega$ of the resonance curve is given by

$$B = \frac{1}{\gamma T_e}$$

where T_e is a "coherent time" ($T_e > \tau$). In the ground state $6\,{}^1S_0$ the mercury atom has no magnetic moment. When an atom, by absorbing a photon, is raised to the excited state $6\,{}^3P_1$ a magnetic moment is created inside the atom and begins its Larmor precession in the B-field. When the photon is jumping from atom 1 to atom 2 (and to atom 3 and so on), the phase of the precession angle of the magnetic moment of the excited state is conserved; and this phase-coherence produces the narrowing of the resonan-ce-curves. This effect was analyzed and experimentally studied in detail in Barrat's thesis[12]. Barrat showed that a similar effect can be observed also on the Hanle-curves.

LIGHT BEATS

In many other phenomena in atomic spectrscopy coherence plays a principal role. This is especially the case when by a pulse of laser light two or several levels of atomic fine structure are simultaneously excited, and when the emitted light is analyzed in time. In this case "light beats" are observed. They are due to the interference of the light emitted by the states which are "coherently excited". We refer to the publications on this topic[13].

ENTROPY OF INCOHERENT AND OF COHERENT LIGHT.
BLACK BODY RADIATION

Planck's principal aim, when he established his formula for black body radiation, was to find an expression for the entropy of this radiation[2]. He studied the mean energy of a material oscillator (resonator) of frequency ν_i in equilibrium with this radiation and he established two facts :

1) $$u_\nu = \frac{8\pi \nu_i}{c^3} u_i$$

u_ν = spectral density of the radiation,

2) he showed that Wien's displacement law takes the form

$$s_i = \varphi\left(u_i / \nu_i\right)$$

s_i and u_i being the mean values for resonator (or per cavity mode).

As a result of his postulate of "incoherent" radiation he could write also

$$S_\nu = g_i S_i = \frac{8\pi \nu_i^2}{c^3} S_i$$

All preceding results are valid either for Planck's resonators or for Lord Rayleigh's cavity-modes[3] which can be also identified to the "cells" of phase-space.

We know that to calculate the entropy S of the radiation, Planck divided the radiation energy into finite elements ξ_i and, using Boltzmann's method, introduced the relation :

$$S = k \ln W$$

If we consider inside the cavity the unity of volume and a frequency range between ν_i and $\nu_i + \Delta\nu_i$, putting $\Delta\nu_i = 1$, this domain contains $g_i = 8\pi\nu_i^2/c^3$ modes and we have to distribute the N_i energy elements between the g_i modes, taking into account the relations :

$$u_\nu = g_i u_i \quad , \quad u_\nu = N_i \xi_i$$

According to Planck the number of possible distribution is given by

$$W_\nu = \frac{(N_i + g_i - 1)!}{N_i! (g_i - 1)!}$$

This formula used by Planck which corresponds to undiscernable energy elements to be distributed among discernable modes, is identical with the formula of Bose - Einstein statistics.

Considering N_i and g_i as large numbers and using the Stirling approximation we can write :

$$S_\nu = k \ln W = (N_i + g_i) \ln(N_i + g_i) - N_i \ln N_i - g_i \ln g_i$$

Introducing the notation $n_i = N_i/g_i$, mean number of energy elements per mode, we obtain :

$$S_\nu = k g_i [(1 + n_i) \ln(1 + n_i) - n_i \ln n_i]$$

and

$$S_i = \frac{S_\nu}{g_i} = k [(1 + n_i) \ln(1 + n_i) - n_i \ln n_i]$$

Planck showed that for black-body radiation

$$n_i = \left[\exp\left(\frac{h\nu_i}{kT}\right) - 1\right]^{-1}$$

LIGHT IN FREE SPACE

Planck showed also that the entropy could be defined for the radiation of a monochromatic light beam in free space [14], and that the formula for S_ν can be used, n and g_i having in this case the values

$$n_i = \frac{B_\nu \lambda^2}{h\nu} = \frac{B_\nu c^2}{h\nu^3} \quad \text{and} \quad g_i = \frac{U}{\lambda^2} t \, \Delta\nu$$

In these formulas λ is the wave-length and ν the frequency, $\Delta\nu$ is the spectral width of the line and t the duration of the light beam. U is defined as the "étendue" of the beam,

$$U = \frac{S \times S'}{r^2}$$

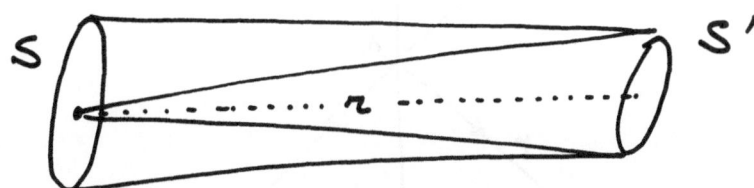

S and S' being the two apertures defining its extension and r their distance.
B_ν is the "brillancy" or luminosity of the beam, defined as its energy flux per unit time divided by U.

We remark that in case of "incoherence" the addition-theorem of entropy $S_\nu = g_i \, S_i$ is valid.

COHERENT LIGHT BEAMS

How have these formulaes to be changed in case of coherence?

This problem was solved by Max von Laue at a time (1905–1907) when he was a research student of Max Planck. He published his results in Annalen der Physik [14].

He started from the following question:
How can we produce two coherent light beams able to show interference?

In optics we start always from one single light beam and we split it into two parts.

The splitting can be obtained by the refraction—reflexion process. At the surface of separation of two media of different index of refraction an incoming beam is split into a reflected and a refracted beam which are coherent. Max van Laue considered a glass-plate (suppose for simplicity and are of its side is strongly reflecting, the other side covered by an antireflexion coating), for which for a given angle of incidence the coefficient of reflexion is r, the coefficient of transmission being 1—r, and be considered the following device which is analogous to a Michelson interferometer :

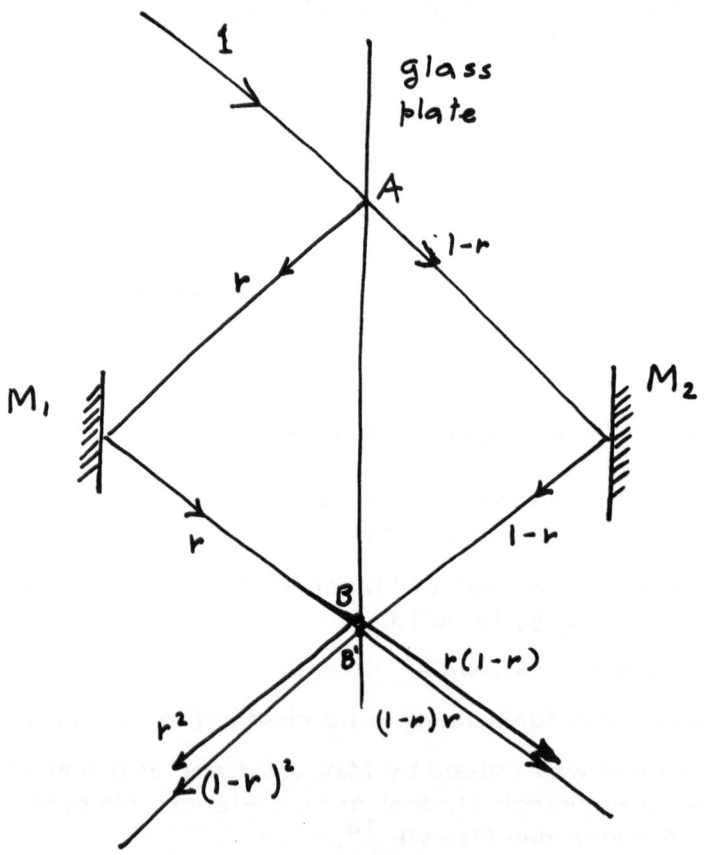

Let us consider an incident light beam of intensity 1 containing n photons per mode.

By the glass-plate in A this beam will be split into a reflected beam of intensity r and a refracted and transmitted beam of intensity 1—r.

The two beams, reflected by the mirrors M_1 and M_2 will join again at $B(^o)$ and give rise each one to a reflected and to a transmitted beam. The two beams emerging on the same side are coherent and will interfere. The result of the interference will depend on the optical path difference between AM_1B and AM_2B. If the two paths are made equal, the two beams emerging on the left will be in phase and their amplitudes r and 1-r will add to give amplitude 1. The two beams emerging on the right will be opposite in phase, resulting amplitude will be equal to the difference

$$\sqrt{r(1-r)} \quad - \sqrt{(1-r)r} \quad = 0$$

We see that in this special case the two separated beams will give rise again to the original beam, and this shows that the process of beam-splitting by reflection-refraction is a reversible process and must be an isentropic process. So the entropy of the two separated beams must be equal to the intensity of the original beam.

If we call n the number of photons per mode in the original beam, its entropy will be :

$$S_0 = \kappa g \left[(1+n) \ln (1+n) - n \ln n \right]$$

The corresponding number of photons per mode in the reflected and in the transmitted beams are

$$n_1 = rn \qquad , \quad n_2 = (1-r)n$$

If the beams would be <u>incoherent</u> their total entropy would be, according to the addition theorem :

$$S_{in} = S_1 + S_2 \quad = \quad \kappa g \left[(1+n_1) \ln (1+n_1) - n_1 \ln n_1 \right.$$
$$\left. + (1+n_2) \ln (1+n_2) - n_2 \ln n_2 \right]$$

But as these beams are <u>coherent</u>, their total entropy is equal S_0. We can say that the coherence of the beams introduces an entropy difference

$$\Delta S = S_0 - S_1 - S_2$$

and it can be shown that this difference is negative.

(^{o}) For reason of clarity we have separated in the figure the two points B and B' which in fact coincide.

The coherence corresponds to a neg-entropy.
For example, in the case of ordinary optical light sources, where
n ≪ 1, we find

$$\Delta s = k g n \ln r^r (1 - r)^{1-r}$$

which in the case of $r = \frac{1}{2}$ gives $\Delta S = - k g n \ln 2$.

Max van Laue has also given expressions for the entropy of
partially coherent light beams. We refer to his two papers in
Annalen der Physik.

I want to draw attention on this very important work of Max
van Laue which seems to have been forgotten and which should
be remembered.

REFERENCES

1) L. Boltzmann, Vorlesungen über Gastheorie, (1895)
2) M. Planck, Annalen der Physik 4 (1901), 553 and Theorie der
           Wärmestrahlung, 1906.
3) L. Rayleigh, Phil. Mag. 49 (1900), 539
   L. Rayleigh, Nature 71 (1905), 559
   L. Rayleigh, Nature 72 (1905), 54 and 243
4) P. Debye, Annalen der Physik 33 (1910), 1427
5) W. Heisenberg, Zeit. Phys. 31 (1926), 617
6) W. Hanle, Zeit Phys. 30 (1924), 93
   W. Hanle and R. Pepperl, Physikal Blätter 27 (1971), 19
   A. Kastler, Physikal Blätter 30 (1974), 394
   Mitchell and Zemansky, Resonance Radiation and Excited Atoms,
   Cambridge Un. Press 1971, p. 262
7) F. O. Colegrove, P. A. Franken, R. R. Lewis, R. A. Sands,
   Phys. Rev. Letters 3 (1959), 420
8) P. A. Franken, Phys. Rev. 121 (1961), 508
9) J. P. Descoubes, C. R. Ac. Sci. 259 (1964), 327 and 3733
10) M. A. Guichon, J. E. Blamont, J. Brossel, C. R. Ac. Sc. 243
    (1956), 1859 and Journal de Physique 18, (1957), 99
11) J. Brossel, Annales de Physique 7 (1952), 622
    J. Brossel and F. Bitter, Phys. Rev. 86 (1952), 311
12) J. P. Barrat, Journal de Physique 2° (1959) 541, 633 and 657
13) S. Haroche, Quantum Beats and Time Resolved Fluorescence
    Spectroscopy, in "High Resolution Laser Spectroscopy",
    K. Shimoda ed., Springer Verlag 1976.
14) M. Planck, Annalen der Physik 6 (1901), 818
15) M. von Laue, Annalen der Physik 20 (1906), 365 and 23 (1907), 1

COHERENCE IN SYNERGETICS

H. Haken

Institut für theoretische Physik

der Universität Stuttgart, West-Germany

1. Introduction

My lectures will be organized as follows. In chapter 2 I shall
recall some basic ideas about coherence to the reader. This
chapter will lead us in a natural way to the definition of correlation functions. In chapter 3 I shall remind the reader of some
basic properties of the laser and why it is such an interesting
subject to study, also with respect to other disciplines. In chapter
4 I shall then explain what synergetics means and shall give an outline of this new interdisciplinary field of research. In chapter 5
finally I shall again discuss the laser as an outstanding example of
a synergetic system.

Many fields, such as those listed in table 1 have contributed
to make the laser come into its existence.

electronics
optics
solid state physics
gas discharge physics
chemistry } ⟶ laser
nonlinear wave theory
quantum mechanics
quantum electrodynamics
quantum statistics
etc.

Table 1

Over the past years it has become more and more evident that the
laser pays back what has been invested into its development. Indeed
it finds numerous important applications in various fields ranging

from physics over chemistry to medicine. There is, however, another
aspect of the laser which I have stressed since many years. A
thorough interpretation of the laser process leads us to an understanding of many processes of selforganization including models of
biological processes. The title "Coherence in Synergetics" has a
double meaning. It may refer to the role coherence plays in synergetics and it may mean that synergetics is itself a coherent
discipline. In what follows I will elaborate on both aspects.

2. Coherence in Optics[+]

Since the topic of this summer school is coherence in physics
quite a number of lecturers deal with this subject. Therefore I remind the reader only of those aspects which are of direct relevance
to what will follow below[1]. In classical physics light is treated
as electromagnetic wave. Let us consider, for example, a plane wave
whose electric field strength is represented by

$$E = E_0 \exp(ikx) \exp(-i\omega t) + c.c. \tag{2.1}$$

In it E_0 is a constant amplitude and k the propagation vector. ω is
the frequency. Because the electric field strength is a real quantity the conjugate complex of the first expression has been added
in (2.1). In the following we shall write instead of (2.1)

$$E = E^{(+)} + E^{(-)} \tag{2.2}$$

where we keep the same sequence of the expressions as in (2.1). The
decomposition (2.2) holds also if we deal with a superposition of
waves. We now consider the intensity I which is proportional to the
square of E. Because the light frequency is rather high and we are
making measurements which are long compared to one period of light
oscillation we are led to define the intensity by an expression in
which E^2 is averaged over many periods of light oscillation. Introducing a suitable proportionality factor we define as intensity

$$I = \frac{1}{2} \overline{E^2} \tag{2.3}$$

When we insert (2.1) or, more generally, (2.2) into (2.3) we immediately find that expressions of the form $E^{(-)2}$ and $E^{(+)2}$ are va-
nishingly small on account of the quickly oscillating exponential
functions, so that we retain only

$$I = \overline{E^{(-)} E^{(+)}} \tag{2.4}$$

+ Readers acquainted with the concept of the complex degree of mutual
degree of coherence can immediately go on to chapter 3.

We now study the intensity I of the superposition of two waves E_1 and E_2

$$E = E_1 + E_2 \qquad (2.5)$$

Inserting (2.5) in (2.4) we obtain

$$I = E_1^{(-)}E_1^{(+)} + E_2^{(-)}E_2^{(+)} + (E_1^{(-)}E_2^{(+)} + c.c.) \qquad (2.6a)$$

$$\quad G(1,1) \quad + \quad G(2,2) \quad + \quad G(1,2) \quad + G(2,1) \qquad (2.6b)$$

where the second row stands just as abbreviation for the corresponding expressions above. We remind the reader of the classical experiment of Young [2] which allows us to create a superposition of two waves and which simultaneously measures the resulting intensity I. In it light radiated from a source is led through two slits of a screen. The intensity distribution on a second screen shows dark and bright fringes (compare fig. 1 r.h.s.). To analyze this result let us consider fig. 2.

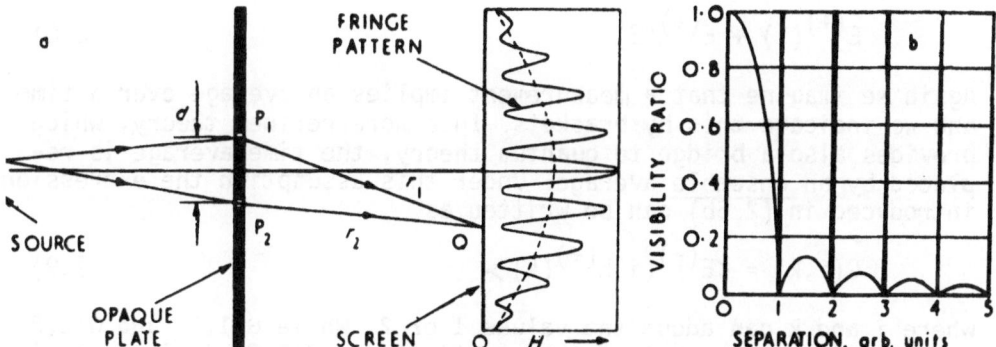

Fig. 1 *Young's double slit interference experiment.*
Left: typical experimental set-up.
Right: Fringe visibility (after T. Young, Phil. Trans.
Roy. Soc. 12, 387 (1802))

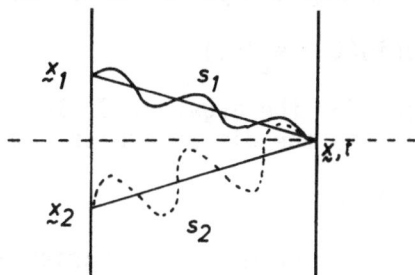

Fig. 2 *Schematic representation of path and phase-differences.*

We assume that equal lightfields are created at the points x_1 and x_2. According to Huygens' principle these excitations then propagate in the form of spherical waves. These two waves meet each other again at each point x of the screen. However, the paths from x_1 to x and from x_2 to x, which will be denoted by s_1 and s_2, respectively, are in general different. As is well known from wave theory such differences of paths result in a phase difference. Accordingly we have to replace the time t in the complex amplitude $E^+(x_1,t)$ by the new argument $t_1 = t - s_1/c$. Of course, we must do the corresponding with the second path. Ignoring factors which are practically constant and are not essential here the lightfield on the screen at point x at time t is represented by

$$E^{(+)}(x,t) = \underbrace{E^{(+)}(x_1,t - s_1/c)}_{t_1} + \underbrace{E^{(+)}(x_2,t - s_2/c)}_{t_2} \qquad (2.7)$$

For brevity we shall write (2.7) in the form

$$E^{(+)}(1) + E^{(+)}(2) \qquad (2.8)$$

Again we imagine that a measurement implies an average over a time and we indicate this by brackets. In a more refined theory, which provides also a bridge to quantum theory, the time average is replaced by an ensemble average. Under this assumption the expressions introduced in (2.6b) can be written as

$$G(i,k) = \langle E^{(-)}(i)E^{(+)}(k) \rangle \qquad (2.9)$$

where i and k can adopt the values 1 or 2. While $G(1,1)$ and $G(2,2)$ are positive on account of their definition, $G(1,2)$ is in general a complex quantity which we decompose according to

$$G(1,2) = |G(1,2)| \exp(i\phi) \qquad (2.1o)$$

In order to see in which way the phase varies we consider the example

$$E^{(-)}(1) \sim \exp(i\omega(t - s_1/c)) \qquad (2.11)$$

and a corresponding one for the argument 2. In the present case the phase ϕ is proportional to the difference of paths

$$\phi = -(s_1 - s_2)\omega/c \qquad (2.12)$$

When we vary the path difference, also ϕ varies and I depends on ϕ according to

$$I = \underbrace{G(1,1) + G(2,2)}_{A} + \underbrace{2|G(1,2)|\cos\phi}_{B} \qquad (2.13)$$

For physical reasons the intensity can never become negative. Two typical cases are presented in fig. 3.

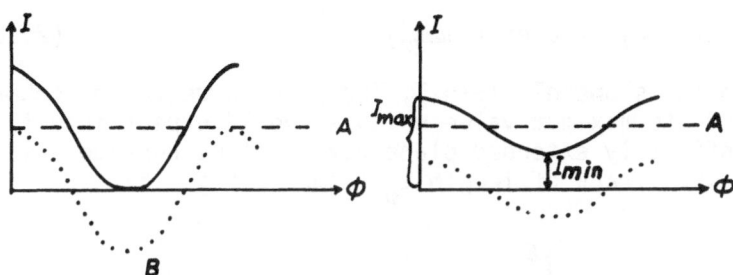

Fig. 3 *The total intensity is composed of the practically constant quantity A and the rapidly changing quantity B. Abscissa: angle ϕ. Left: $|B| = A$, right: $|B| < A$.*

Because the intensity distribution (2.13) means alternating darkness and brightness there are in principle two limiting cases. One in which the interference fringes are sharp (left side of fig. 3) and one in which they are washed out. Apparently the fringe visibility is maximum if

$$|G(1,2)| = |< E^{(-)}(1)E^{(+)}(2) >| \qquad (2.14)$$

acquires a maximum, while it is zero when (2.14) vanishes. For reasons which will transpire below, (2.14) is called mutual coherence function. As can be immediately seen from fig. 3 the relative differences in brightness are not so much determined by the absolute size of (2.14) but rather determined by its relative size compared to the first two expressions in (2.13). This leads us to define a normalized function by

$$\gamma(1,2) = \frac{G(1,2)}{\sqrt{G(1,1)G(2,2)}} \qquad (2.15)$$

As can be shown mathematically

$$|\gamma| \leq 1. \qquad (2.16)$$

(2.15) is called the complex degree of mutual coherence. We now want to explain what the functions (2.14) or (2.15) and thus also Young's experiment have to do with coherence. Consider as a first example an

infinitely extended plane wave. We assume that this wave is now
split into two waves where a difference of paths generates a phase
difference (fig. 4). The wave amplitude occurring in (2.14) now
reads

$$E^{(-)}(1) \;\; = a \, \exp(i\omega t_1) \tag{2.17}$$

$$E^{(+)}(2) \;\; = a \, \exp(-i\omega t_2) \tag{2.18}$$

As one convinces oneself $|\gamma| = 1$. The complex degree of coherence
has adopted its maximum value which we would expect intuitively
from an infinitely extended plane wave. Let us consider as a second
example a wave track of length t_0 (compare fig. 4b)

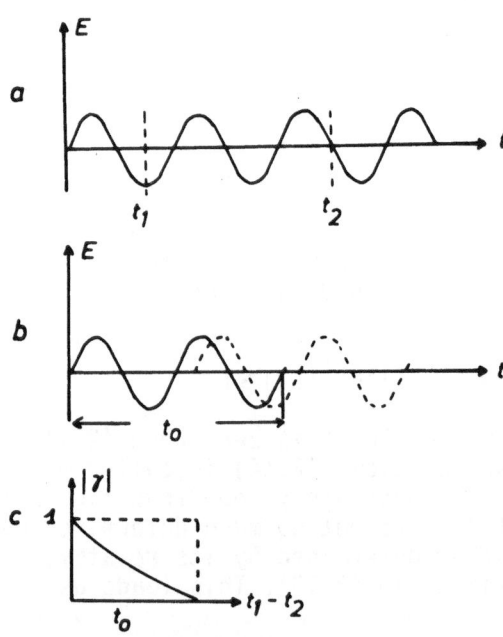

Fig. 4   a) *Electrical field strength of infinitely extended wave.*
By splitting the wave into two partial waves a time-
difference t_2-t_1 is created.
b) *Wavetrack of length t_0.*
c) *Complex degree of coherence as function of time-*
difference in case b).

The solid and the dashed lines represent the original wave and a
second wave to which we have added a phase difference. If the phase
lag between the two wave tracks is zero we obtain 1 for the degree
of mutual coherence. If the time difference between the two waves
is bigger than t_0 we obtain $\gamma = 0$. If the time difference lies

between 0 and t_0 we obtain a partial overlap between the two waves which leads to $|\gamma|$ lying inbetween 0 and 1. This variation of $|\gamma|$ is exhibited in fig. 4c. The complex degree of mutual coherence is thus a function of time lag t_2-t_1. This example shows clearly that studying γ we get a measure for how long a wave track is coherent, i.e. uninterrupted. Expressed in more general terms we can say that γ is a measure for how long a process persists.

Besides Young's experiment which allows us to measure γ another famous experiment is that of Hanbury-Brown-Twiss which measures [3] the correlation of intensities. When we denote the photonnumber by $n \propto E^{(-)}E^{(+)}$ the photon correlation function is defined by

$$\bar{K}(1,2) = \langle E^{(-)}(1)E^{(-)}(2)E^{(+)}(2)E^{(+)}(1) \rangle \qquad (2.19)$$

In a quantum mechanical treatment of the lightfield care with respect of the sequence of $E^{(-)}$, $E^{(+)}$ must be exercised. We refer the reader to the classical papers of Glauber [4], some of which have been published in previous Italian summer schools. It is often advantageous to subtract from (2.19) the product of the mean values of the photon numbers:

$$K(1,2) = \bar{K}(1,2) - \langle E^{(-)}(1)E^{(+)}(1)\rangle\langle E^{(-)}(2)E^{(+)}(2)\rangle \qquad (2.2o)$$

In a similar way one can define a whole series of higher order correlation functions.

3. Some Important Properties of Laser Light

A typical experimental set up of the laser [5] is shown in fig. 5.

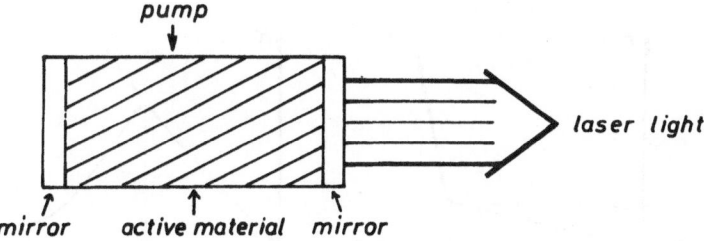

Fig. 5 Typical set-up of a laser

The active material, for instance a ruby crystal or a gas discharge tube, carries at its endfaces two mirrors. These mirrors serve for a selection of waves according to their wavelengths (see fig. 6) Furthermore they favor modes propagating in axial direction. The laser-active atoms are energetically pumped from the outside. Eventually laser light is emitted through one of the semitransparent mirrors. When the laser is pumped only weakly it acts as a usual

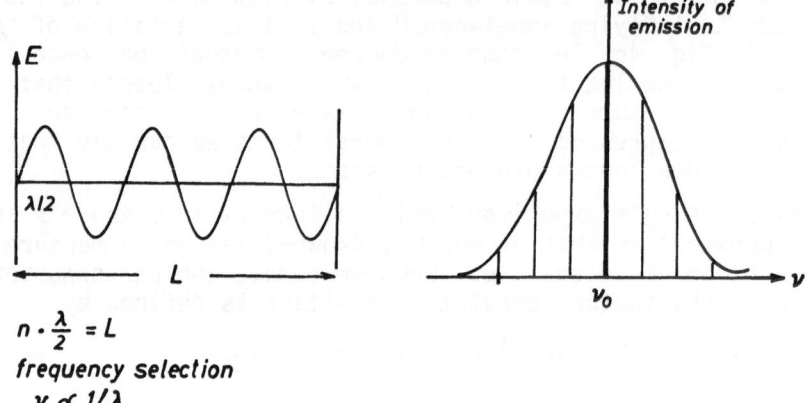

$$n \cdot \frac{\lambda}{2} = L$$

frequency selection
$$\nu \propto 1/\lambda$$

Fig. 6   The mirrors act as Perot-Fabry-interferometer (left),
line profile of spontaneous emission (right)

lamp. Its light consists of individual wave tracks each of a few
meters in length. This coherence property can be easily measured
by the complex degree of coherence. With increasing pump strength
the linewidth becomes narrower and narrower indicating that the
wave tracks become longer and longer. For some time most physicists
believed that this ever increasing length of wave tracks and this
ever decreasing linewidth is the only typical feature of laser
light. However, we now know that the basic difference between laser
light and light from thermal sources rests in different features. To
this end consider fig. 7 left hand side, which represents the
light intensity as a function of pump power.

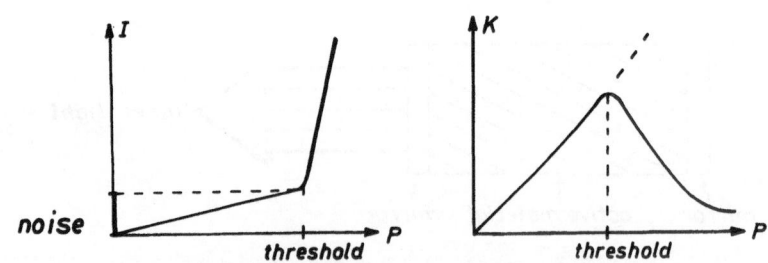

Fig. 7   Left: Light intensity vs. pump power.
Right: Intensity fluctuations over output intensity
vs. pump power.

First we obtain a rather slow increase of intensity when pump power
is increased. But suddenly beyond a certain threshold the intensity
increases very steeply. This behavior is explained in a good
approximation by semiclassical laser theory [6]. Semiclassical theory

neglects, as is well known, fluctuations. This theory predicts that below a certain threshold there is no light emission at all, whereas above it a coherent wave imerges so that the light intensity increases steeply. The occurrence of light even below threshold is, of course, due to the fact that light from a lamp is mere noise. We get a still much deeper insight into the laser process when we consider the intensity fluctuations as a function of pump power. More precisely we consider the expression (2.2o) divided by the output power and taken at equal times and plot it as a function of pump power (compare fig. 7, right hand side. The quantity K first increases with pump power. At the same threshold as mentioned before some qualitatively completely new phenomenon occurs. The relative intensity (or amplitude fluctuations) decreases. This effect, which most clearly exhibits the fundamental difference between laser light and light from normal lamps was first predicted in my 1964 paper [7] and has since been verified experimentally. A thorough experimental study of a laser close to and at its threshold has been performed in particular by Arecchi and coworkers [8]. While the Langevin equation approach which I had used did not allow me to treat the immediate vicinity at threshold, this problem could be solved by Risken [9] who interpreted my quantum mechanical Langevin equation as a classical one and established its corresponding Fokker-Planck equation which he then solved. Results equivalent to Risken's have been obtained by Hempstead and Lax [1o]. Quantum mechanical treatment of the photon distribution close to laser threshold have been performed by Scully and Lamb [11] and Weidlich, Risken and Haken [12]. The relation between these results has been more recently demonstrated by Casagrande and Lugiato [13]. Furthermore a very elegant technique developed more recently by Arecchi and Ricca [14] allows for treating the Langevin equations even at threshold. What is of immediate relevance to us here is the apparent change of the statistical properties of laser light at threshold. Obviously the internal state of the laser must have changed dramatically. Indeed we can easily visualize what happens at threshold when we look at Fig. 8.

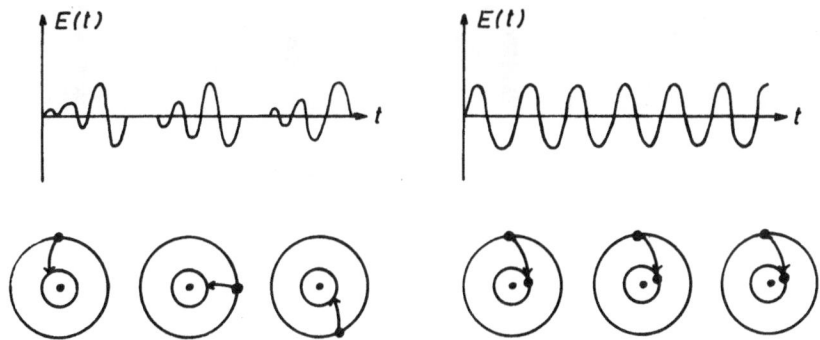

Fig. 8 Electric field strength of light as function of time.
        Left: lamp, right: laser

The upper left part shows the random wave tracks of a usual lamp.
The right hand side shows the amplitude stabilized coherent wave of
the laser. While the atoms in the lamp act individual it seems that
in the laser the atoms are told to behave in an entirely organized
fashion. To visualize the situation consider a channel filled with
water (fig. 9). Let us assume that at its border several men are
pushing sticks into the water (the men represent the individual
atoms, the water the lightfield). When the men push their sticks
at random a completely chaotic motion of the water surface evolves.
On the other hand a coherent motion evolves if the men push their
sticks in a well regulated fashion into the water. Who is the little
demon who tells the men or the atoms to behave in such a well
organized manner? In the realm of laser physics we now know how to
cope with this problem. First, randomly emitted wave tracks hit excited atoms which can perform stimulated emission and thus enhance
the wave tracks. Those wavetracks, whose center frequency is closer
to the center frequency of the atomic emission, are enhanced more
than the others. Since all wave tracks "live" on the same atoms, a
competition between different wave tracks sets in and only the one
which has the longest lifetime survives. During this build-up
process more and more atoms are dragged into the oscillation determined by this wave. Thus this wave "slaves" the oscillations of all

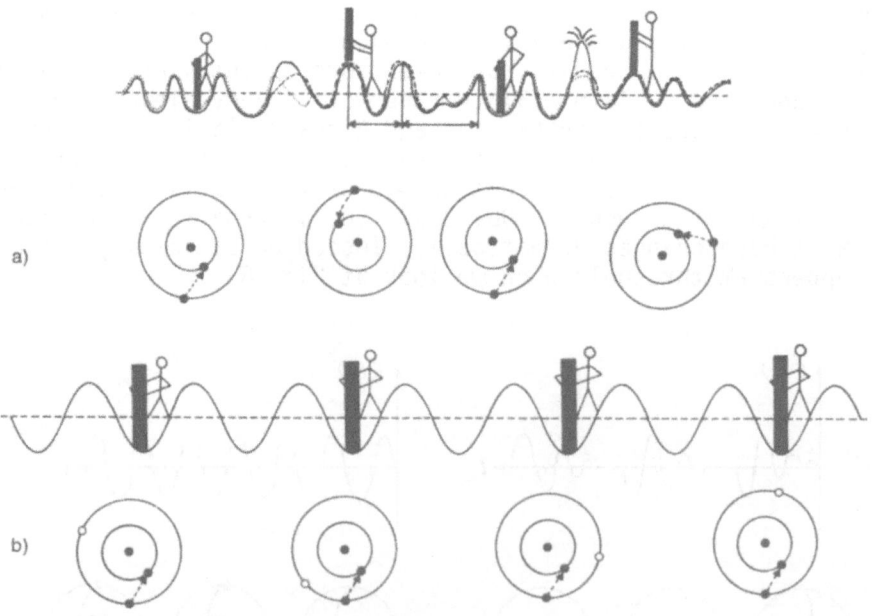

Fig. 9

atoms. On the other hand this wave now describes the order in the
laser. We shall call its amplitude "order parameter". The concept of
slaving and order parameters has turned out to be extremely useful in
other sciences. Since the mathematical theory is rather extensive
and was presented elsewhere we must refer the reader to the literature
(cf. my textbook on Synergetics [15]). The sharp transition at laser
threshold has many features in common with a second order phase
transition of systems in thermal equilibrium. This analogy including
fluctuations has been fully recognized in independent papers by Graham
and Haken [16] and DeGiorgio and Scully [17]. As we now know such
dramatic changes of properties are shared by many other systems in
physics and other fields. One of the tasks of synergetics is just the
exploration of such phase-transition analogies.

4. Synergetics [15], [18]

4.1 Examples. I first give a number of typical examples of such
disorder-order transitions which are described in figs. 1o, 11 and 12.

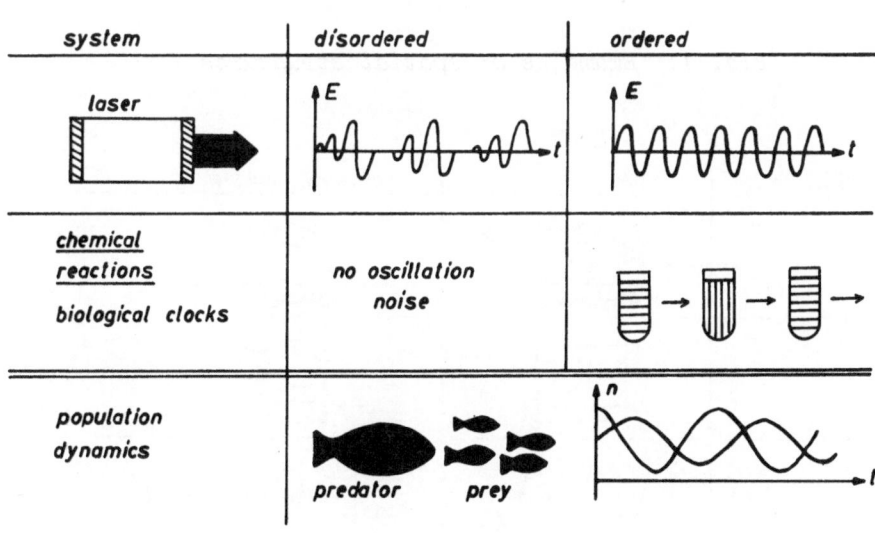

Fig. 1o   Example of oscillations

Fig. 1o shows again the laser and as a second example the onset of
oscillations in chemical reactions for which now numerous examples
are well known. Also the number of predator and prey fish (and other
animals) can show an oscillatory behavior. Fig. 11 exhibits a number
of suddenly occurring spatial structures. The first row shows a fluid
layer heated from below. When the temperature gradient is small, heat

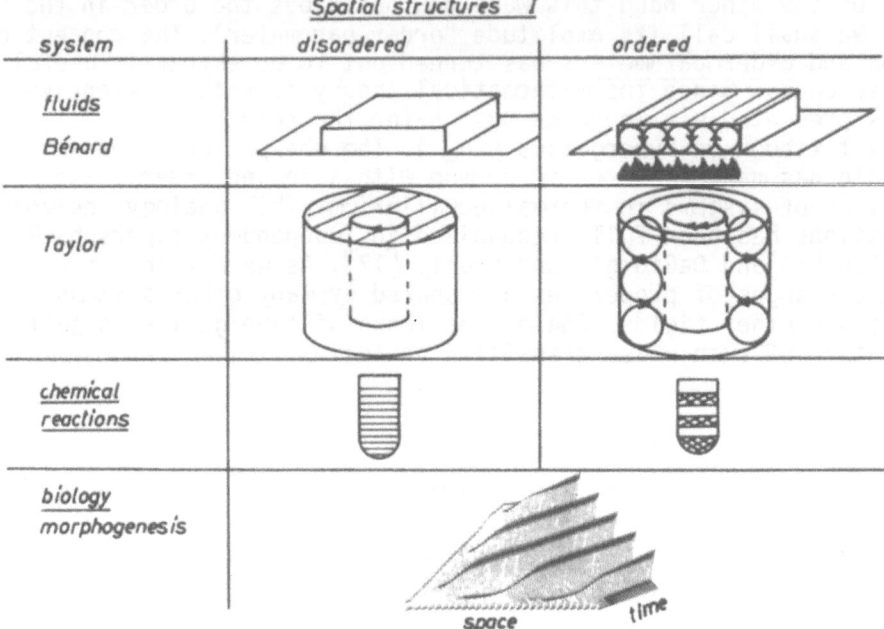

Fig. 11   Examples of spatial structures I

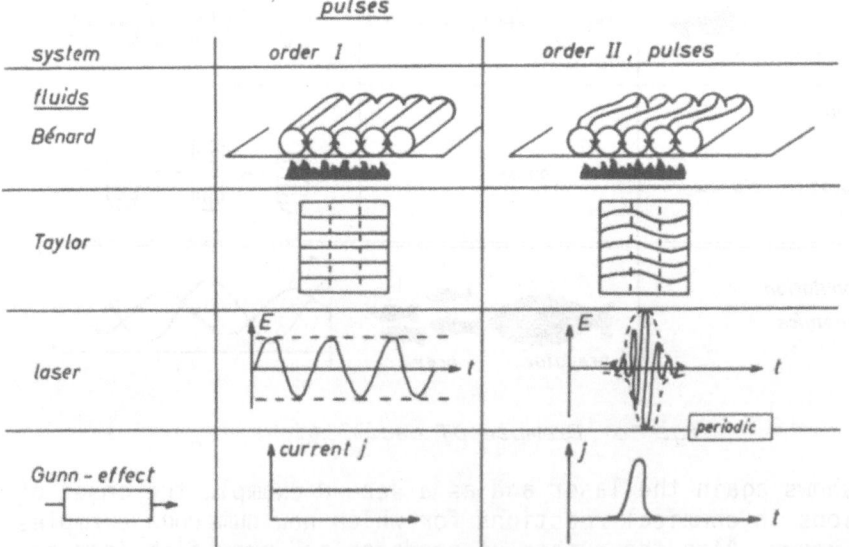

Fig. 12   Examples of pulses

is transported by conduction and the fluid remains macroscopically
at rest. At a critical temperature gradient the liquid starts a
macroscopic motion in the form of rolls. In the Taylor instability
the liquid moves within two coaxial cylinders, the inner one:rotating.
At a critical speed a macroscopic motion in form of rolls appears.
In the third row we represent an example of a chemical reaction giving
rise to a spatial pattern, for instance in the form of alternating red
and blue layers. Spatial structures arise also in mathematical models
of morphogenesis. In a number of cases further new patterns occur
when control parameters, i.e. input power of the laser, heat flux
etc. are changed. Examples are exhibited in fig. 12. The Bénard rolls
can start a wavy motion as do the Taylor vortices. The coherent
laser light emission can break into ultrashort pulses or the
constant current of a semiconductor can break into ultrashort pulses.

   4.2 Concept and mathematical tools. The mathematical tools to
cope with these above mentioned phenomena require a lot of space and
cannot be represented here. I refer the reader to an adequate representation in my book on synergetics [15]. However, it is worthwhile to describe these methods by means of the underlying concept.
We start from a given state, for instance a quiescent fluid. Then
we change external parameters and check whether that state of the
fluid is still stable. If it becomes unstable, a number of waves can
now be excited. They grow similar to the wave tracks of a laser. The
winning mode serves as order parameter and slaves the other kinds of
motion, i.e. all the other modes. Because these other modes are determined by the order parameter they can be eliminated from the
equations of motion and we deal with equations for the order parameter
alone. In contrast to the usual laser process in general systems
also several order parameters can coexist. For instance, in the
Bénard instability, in some chemical processes and in morphogenetic
models three waves forming a triangle may coexist. Their cooperation
leads to the formation of hexagon cells. By means of the slaving
principle the enormous number of degrees of freedom of each system
can be reduced to the small number of order parameters. The resulting
order parameter equations can then be classified using symmetry properties and,in many cases, the smallness of order parameters when the
new structure evolves. This leads to the formulation of "universality
classes" much in the same way as in phase transition theory. An important class of order parameter equations can be considered as
extensions of the famous Ginzburg-Landau equation of superconductivity. In any case the whole dynamics of the system, which initially
had very many degrees of freedom, is now governed by a few degrees
of freedom of the remaining order parameters.Fig. 13 shows the
build up of an ultrashort laser pulse, calculated by this method.
Fig. 14a - c shows the build-up of the hexagonal pattern of a morphogenetic model.

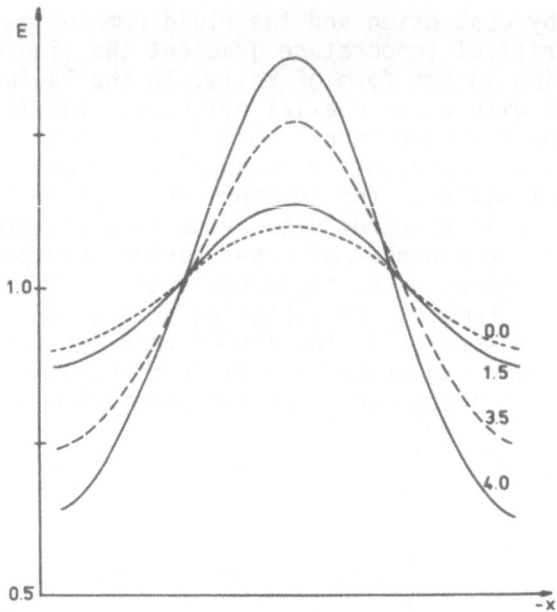

Fig. 13   Build-up of ultra-short laser pulse
          (after Ohno and Haken)

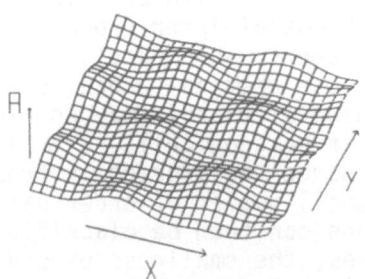

Fig. 14a-c   Growth of (pre-)morphogenetic pattern (after Haken
             and Olbrich).A : concentration of activator substance
             X,Y: spatial coordinates

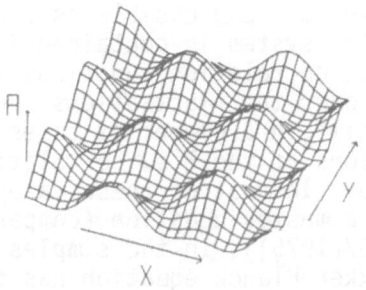

Fig. 14b

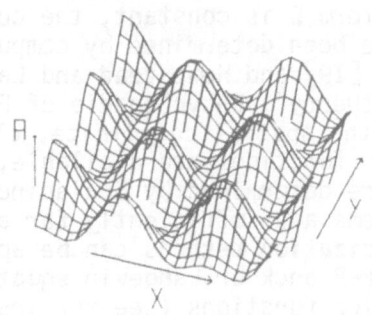

Fig. 14c

4.3 Coherence and correlation functions. A fundamental question concerning synergetic systems relates to their persistence in time. It is here where the coherence functions introduced in chapter 2 come in. Denoting the amplitude of the order parameter by $\xi(t)$ we have now to study correlation functions of the form $<\xi(t)\,\xi(t')>$ and those containing still more ξ's. Since the amplitudes of the socalled slaved modes are expressible as functions of ξ all the information about the system is contained in correlation functions of the ξ's. In these correlation functions the bracket means that we have to average over the fluctuations. Unfortunately this is a formidable task. While in a number of cases we know the stationary distribution functions of the ξ's, correlation functions are known only for a few cases. In most of these cases the ξ's obey a Fokker-Planck equation or a master equation (compare for instance Haken, (Rev.Mod.Phys. 47,67(1975)). In the simplest case of a single order parameter ξ the Fokker-Planck equation has the form

$$\dot{f}(\xi,t) = (- \frac{\partial}{\partial\xi} k(\xi) + D \frac{\partial^2}{\partial\xi^2}) \, f(\xi,t) \tag{4.1}$$

The difficulty in solving this Fokker-Planck equation arises because the drift coefficient K is a nonlinear function of ξ. Provided K has the form

$$K(\xi) = a\xi + b\,\xi^3 \tag{4.2}$$

and the diffusion term D is constant, the correlation functions $<\xi(t)\,\xi(t')>$ have been determined by computer calculations by Risken and Vollmer [19] and Hempstead and Lax [1o]. For a summary see Haken, "Laser theory", Encyclopedia of Physics, XXV/2c). If ξ depends also on the spatial coordinate, close to threshold only certain correlation functions are available, for instance $<\xi(x)\xi(x')>$ These functions were determined by Scalapino et al [2o] and others [21].When the systems are sufficiently far away from their instability points, linearization methods can be applied, which allow us to solve the Fokker-Planck or Langevin equations explicitly and to calculate correlation functions (see for instance Haken, Rev.Mod. Phys. 47, 67 (1975) or Z.Physik, B 22,73 (1975)). A method valid near threshold consists in the application of renormalization group techniques [22] where we can exploit the analogy between many of the nonequilibrium phase transitions with usual phase transitions. In this context a word ought to be said on the Landau theory of phase transition which is nowadays very often criticized and which is replaced by renormalization group techniques. It should be noted that the Landau theory consists of two parts, namely one which guesses the distribution function and one which then goes over to a mean field approach which amounts to neglecting fluctuations. The first step has been nowadaysreplaced by more rigorous methods, for instance by explicit solutions of the Fokker-Planck equation which yield in many cases exactly the same type of distribution function which had been guessed before. The second step, namely the mean field approach, is in many cases not correct and it is here where renormalization group techniques are so important for the evaluation of correlation functions.

5. Instability hierarchy and turbulence in the laser

In the new field of synergetics the laser is a very nice example for the occurrence of several subsequent instabilities and thus of the importance of new order parameters. In a model with waves running in a single direction the laser equations can be written in the form

$$\frac{\partial E}{\partial t} + c \frac{\partial E}{\partial x} + \varkappa E = \alpha P \tag{5.1}$$

$$\frac{\partial P}{\partial t} + \gamma P = \beta ED \tag{5.2}$$

$$\frac{\partial D}{\partial t} = \gamma_{\shortparallel} (D_0 - D) + \delta EP \tag{5.3}$$

E is the electric field strength, c the light velocity, $\varkappa$ the cavity damping constant, α a constant proportional to the dipole matrix element, P the atomic polarization. t is time and x the spatial coordinate of our one-dimensional laser model. γ is the inverse relaxation time of the atomic dipole moment, β a constant, D the atomic inversion, $\gamma_{\shortparallel}$ the damping time of the atomic inversion, D_0 the equilibrium inversion caused by pump and incoherent decay processes, δ a constant. The control parameter of primary importance is the prescribed inversion D_0. When $D_0 < D_c$ there is no light emission at all, because in this model we neglect fluctuations. With $D > D_{c_1}$ a c.w. wave arises which in the above laser equations is described by time and space independent E, P and D. When $D_0 > D_{c_2}$ the c.w. solution becomes unstable and is replaced by ultrashort pulses. The onset of these pulses can be either a first or second order phase transition depending on the length of the laser [23]. These pulses are regular provided $\varkappa < \gamma + \gamma_{\shortparallel}$. If $\gamma + \gamma_{\shortparallel} > \varkappa$ an entirely new kind of motion appears which is called chaotic. We could show that in this case the laser equations are equivalent [24] with those introduced by E.N. Lorenz [25] to describe turbulence in fluids. Therefore in a way we now get laser light turbulence (An interesting different kind of laser turbulence has been considered by Arecchi and Ricca [26] (and in this volume)). As has been remarked by Bonifacio [27] the condition $\varkappa > \gamma + \gamma_{\shortparallel}$ is the one used also in fluorescence [28], [29]. While in the above mentioned Lorenz model the pump continues steadily, in superfluorescence the atomic system is assumed to be initially in a totally inverted state and then no more pumped. In spite of this

difference one might expect that also in some cases of superfluorescence chaotic motion becomes visible, as was stressed by
Bonifacio [27].

6. Concluding Remarks

In conclusion I would like to say a few words about the
relations of synergetics to other disciplines and about some of
its future aspects. In numerous talks I recently gave on synergetics I have been very often asked what its relation is to the
work of Thom and the work of Prigogine. So let me make a few
comments about that.

The work of the mathematician Thom [3o] is based on equations
of the form

$$\dot{\underset{\sim}{q}} = \text{grad}_q \, V(\underset{\sim}{q}) \tag{6.1}$$

In many examples I have studied it turned out that kinetic equations
cannot be described by means of a potential V. This is quite natural because we cannot apply to systems far from equilibrium or to
many non-physical systems the principle of detailed balance which
would allow us to invoke a potential. Our approach of synergetics
is not confined to equations of that sort. In our approach the
order parameter concept allows for an enormous reduction of the
degrees of freedom in large classes of kinetic or stochastic
equations. Furthermore it should be noted that the interpretation
of the meaning of the q's and the control parameters is entirely
different in our theory and in that of Thom and Zeeman.

The work of Prigogine [31] starts from conceptions of irreversible
thermodynamics and culminates in the formulation (jointly with
Glansdorff) of the excess entropy production principle. While this
principle may be useful to find the onset of an instability, synergetics goes far beyond that because it gives us also means to
determine the newly evolving structures and to classify them.
Furthermore it should be noted that synergetics comprises both
systems exhibiting "dissipative structures" (an expression coined
by Prigogine) and systems to which the concept of dissipation is not
applicable, for instance sociology.

There are some interesting links between synergetics and other
disciplines. One becoming more and more evident is the relation between synergetics and bifurcation theory. Another one which has been
obvious from the very beginning is its profound relation to phase
transition theory. Here the concept of order parameters is still in
its place, though, of course, we must not fall into the trap of mean
field theory (at least close to threshold). As we have seen in
chapter 4.1 a system may not show only a single instability but a
whole series of subsequent instabilities. A systematic study of
higher order instabilities in various systems and their comparison

seems to be one of the next important steps to be undertaken in the field of synergetics.

I want to thank the Volkswagenwerk Foundation for its support of the project synergetics.

References

[1] For introductions into the subject of optical coherence see
Born, H. and E. Wolf, "Principle of Optics", Pergamon, MacMillan, Oxford, 2nd ed. 1964
Mandel, L. and E. Wolf, Rev.Mod.Phys. 37, 231 (1965)
An excellent collection of papers in
Mandel, L. and E. Wolf, eds. "Coherence and Fluctuations of Light", Vol. I + II, Dover Publ., N.Y. 1970

[2] Young, T., Phil.Trans.Roy.Soc. 12, 387 (18o2)

[3] Hanbury-Brown, R. and R.Q.Twiss, Nature 178, 1447 (1956)

[4] Glauber, R.J., Phys.Rev. 131, 2766 (1963)
Glauber, R.J., ed. "Proceedings of the Int.School of Physics E. Fermi", Course XLII, Quantum Optics, Varenna 1967, Academic Press, N.Y. 1969

[5] Since the first proposal of the laser by
Schawlow, A.L. and C.H. Townes, Phys.Rev.112, 194o (1958)
an enormous literature has evolved. An excellent survey of the whole field gives
Arecchi, F.T. and O.E. Schulz-Dubois, eds. "Laser Handbook", North-Holland,Amsterdam, 1972
For theoretical treatments see
Haken, H.,"Laser Theory" in Encyclopedia of Physics, Vol. XXV/2c, Springer, Berlin, 197o
Sargent III, M., M.O.Scully and W.E.Lamb jr. "Laser Physics" Addison-Wesley, Reading Mass., 1974
Louisell, W.H., "Quantum Statistical Properties of Radiation" Wiley, N.Y., 1973 (which includes work by M. Lax)
These books contain many further references.

[6] Haken, H., talk given at the Int.Conference on Optical Pumping, Heidelberg, 1962
Haken, H. and H. Sauermann, Z.Phys. 173, 261 (1963)
Lamb, W.E., Phys.Rev. 134, 1429 (1964)
Some historical aspects of semiclassical laser theory stressing its independent development by Haken and by Lamb are described by M.Sargent III, Opt. Comm. 1974

[7] Haken, H., Z.Phys. 181, 96 (1964)

[8] Arecchi, F.T. and V.DeGiorgio in "Laser Handbook" ed.F.T.Arecchi and O.E.Schulz-Dubois, North-Holland, Amsterdam, 1972, with further references

[9] Risken, H., Z.Phys. 186, 85 (1965)

[1o] Hempstead, R.D. and M.Lax, Phys.Rev. 161, 35o (1967)

[11] Scully, M.O. and W.E.Lamb jr., Phys.Rev. 159, 2o8 (1967),
     Phys.Rev. 166, 246 (1968)
[12] Weidlich, W., H.Risken and H.Haken, Z.Phys. 2o1, 396 (1967)
[13] Casagrande, F. and L.A. Lugiato, Phys.Rev. A15, 429 (1977)
[14] Arecchi, F.T., M.Asdente and A.M.Ricca,Phys.Rev.A14, 383 (1976)
[15] Haken, H. "Synergetics, An Introduction" Nonequilibrium Phase
     Transitions and Selforganization in Physics, Chemistry and
     Biology, Springer, Berlin, 1977
[16] Graham, R. and H.Haken, Z.Phys. 237, 31 (197o)
[17] DeGiorgio, V. and M.O.Scully,Phys.Rev.A2, 117o (197o)
[18] Haken, H. and R. Graham, Umschau 6, 191 (1971)
     Haken, H. ed. "Synergetics", Teubner, Stuttgart, 1973
     Haken, H., ed. "Cooperative Effects", North Holland, Amsterdam
     1974

     Haken, H., ed. "Synergetics, A Workshop", Springer, Berlin,
     1977
[19] Risken, H., and H.D.Vollmer, Z.Phys. 2o4, 24o (1967)
     see also in Haken, H. [5]
[20] Scalapino, D.J., M.Sears and R.A.Ferrell, Phys.Rev.B6, 34o9
     (1972)
[21] Gruenberg, L.W. and L.Gunther,Phys.Lett.38A, 463 (1972)
     Nauenberg, M., F.Kuttner and M.Furman,Phys.Rev.A13,1185 (1976)
[22] Wilson, K.G. and J. Kogut,Phys.Reports 12C, 75 (1974)
[23] Haken, H. and H.Ohno, Opt.Comm. 16, 2 (1976)
     and to be published
[24] Haken, H., Phys.Lett. 53A, 77 (1975)
[25] Lorenz, E.N., J.Atmos.Sci. 2o, 13o (1963)
[26] Arecchi, F.T. and A.M.Ricca,Phys.Rev.A15, 3o8 (1977)
[27] Bonifacio, R. discussion remark at this summerschool
[28] Bonifacio, R., P.Schwendimann and F.Haake,
     Phys.Rev.A4, 3o2 (1971), Phys.Rev. A4, 854 (1971)
     Bonifacio, R. in "Cooperative Effects", ed.H.Haken,
     North Holland, Amsterdam, 1974
[29] Bonifacio, R., this volume. Further references are given there.
[3o] Thom, R., "Structural Stability and Morphogenesis",
     Benjamin, Reading, Mass., 1975
[31] Glansdorff, P. and I. Prigogine, "Thermodynamic Theory of
     Structure, Stability and Fluctuations", Wiley Intersc., New York
     1971

COHERENT OPTICAL TRANSIENTS[*][†]

Richard G. Brewer

IBM Research Laboratory

San Jose, California 95193, U.S.A.

LECTURE   I: The Bloch equations

LECTURE   II: The Stark switching technique

LECTURE III: Photon echoes and molecular collisions

LECTURE   IV: Two-photon transients

LECTURE   V: The laser frequency switching technique

---

[*]Presented at the NATO Advanced Study Institute on Coherence in
Spectroscopy and Modern Physics, Villa le Pianore, Versilia, Italy,
July 17-30, 1977.

[†]Supported in part by the U.S. Office of Naval Research.

41

INTRODUCTION

The early practitioners of nuclear magnetic resonance [1,2] probably never dreamed that their sophisticated coherence techniques would one day be adapted to the optical region. Coherent radiation sources were needed and in those days, the early 1950's, lasers were not being discussed. We now know that the subject of *coherent optical transients* has developed steadily, beginning in 1964 with the initial photon echo measurement of Kurnit, Abella, and Hartmann [3]. Indeed, at the present time, the optical analogs of pulsed NMR transients have to some degree been realized.

One major objective in this area is to understand the coherent optical transient effects encountered. This is done in part by theoretical analysis using the Maxwell-Schrödinger equations and in part by carefully chosen experiments. Another objective is to use these new techniques to study time-dependent atomic or molecular interactions. As we shall see, these coherent transient methods are quite powerful. They allow one to study, for example, collision phenomena in gases and phonon or magnetic hyperfine interactions in solids. The methods are *selective* in that different dephasing mechanisms of coherently prepared systems can be examined independently. These methods are also highly *quantitative* particularly when the data handling capability of computers is utilized. And last, a wide variety of dynamic and spectroscopic studies of atoms, molecules and solids can now be carried out in the infrared and in the visible-ultraviolet regions. It appears possible in fact to measure in a highly controlled manner dynamic events of atomic and molecular processes on a time scale ranging from milliseconds to about 50 picoseconds.

Lectures I through IV resemble my 1975 summer school notes [4], which are now published. Therefore, there is no need to reproduce the derivations in detail here. Instead, I will sketch some physical arguments, state conclusions where appropriate, and review the recent progress. In Lecture IV, I discuss the recent two-photon transient experiments in light of our earlier theoretical work. Lecture V summarizes our understanding of the laser frequency switching technique, which is now one year old. Applications to the study of molecular collision phenomena and dephasing processes in low temperature solids are presented. Following the suggestion of the directors of this NATO Summer School on *Coherence in Spectroscopy and Modern Physics*, these notes are more of an outline than a detailed exposition of the subject.

LECTURE I: The Bloch Equations

1.1 Classical description

The classical description of nuclear magnetic resonance, first developed by F. Bloch [5], is the starting point for understanding coherent spin transient effects. The resulting Bloch equations describe the precessional motion of a magnetic dipole in an external magnetic field. The great advantage is that complicated time-dependent interactions are easily visualized geometrically as a series of precessional motions when the radio frequency field is pulsed in some prescribed sequence.

In the optical region, the Bloch formalism can be derived for electric as well as magnetic dipole transitions by a transformation of the Schrödinger equation. Indeed, the same precessional motion appears and one can derive all the optical analogs of spin transients.

The equation of motion

$$\frac{d\vec{J}}{dt} = \vec{L} \tag{1.1}$$

for a magnetic dipole moment μ in the presence of an external magnetic field H follows directly from the definition of angular momentum

$$\vec{J} = m\vec{r} \times \vec{v} \tag{1.2}$$

where the torque

$$\vec{L} = \vec{\mu} \times \vec{H} . \tag{1.3}$$

If μ behaves like a compass bar magnet with no internal angular momentum in the $\vec{\mu}$ direction, the dipole simply oscillates in the plane of $\vec{\mu}$ and $\vec{H}$.

On the other hand, when $\vec{\mu}$ possesses internal angular momentum $\vec{J}$ such that

$$\vec{\mu} = \gamma\vec{J} , \tag{1.4}$$

we may write (1.1) as

$$\frac{d\vec{\mu}}{dt} = \vec{\mu} \times \gamma\vec{H} . \tag{1.5}$$

This is the equation of motion for a gyroscope where $\vec{\mu}$ precesses about H at a frequency γH and through a phase angle $\phi(t) = \gamma H t$ in time t.

Note that if a static magnetic field

$$\vec{H} = \vec{k}H_o \tag{1.6a}$$

is directed along the z axis, μ_z will be a constant of the motion, i.e., μ precesses about the z axis in the laboratory reference frame. In spectroscopy, we are interested in applying resonant radiation fields

$$\vec{H}_x(t) = \vec{H}_1\cos\omega t , \tag{1.6b}$$

perhaps along the x direction, which complicates the precessional motion considerably. The earlier simplicity is regained, however, by transforming from the laboratory frame (L) to a rotating coordinate system (R) where the rotation frequency about the laboratory z axis is Ω and

$$\left(\frac{d\vec{\mu}}{dt}\right)_L =\left(\frac{d\vec{\mu}}{dt}\right)_R + \vec{\Omega}\times\vec{\mu} . \tag{1.7}$$

Using (1.5), we obtain

$$\left(\frac{d\vec{\mu}}{dt}\right)_R = \vec{\mu}\times(\gamma\vec{H}+\vec{\Omega}) , \tag{1.8}$$

$$= \vec{\mu}\times\gamma\vec{H}_e . \tag{1.9}$$

Thus, the effective field

$$\vec{H}_e = \vec{k}(H_o-\omega/\gamma)+\vec{i}H_1 \tag{1.10}$$

behaves like a static field in the rotating frame. In this rotating coordinate system, the dipole $\vec{\mu}$ precesses about the effective field $\vec{H}_e$. In (1.9), we have assumed that the coordinate system rotates about the laboratory $\vec{k}$ axis with a frequency equal to that of the radiation field ($\Omega=\omega$). The *rotating wave approximation* has been imposed also by retaining only the component of (1.6b) that is resonant with the (Larmor) precession frequency γH_e; the out of phase or second harmonic component at 2ω is ignored.

1.2 Quantum Description

The transformation of the Schrödinger equation into Bloch-like form, Eq. (1.9), was carried out initially by Feynman, Vernon and Hellwarth [6]. A derivation using the density matrix formalism has appeared elsewhere [4,7]. We assume a two-level quantum system, with lower and upper levels labeled 1 and 2 respectively, that interacts with an optical wave

$$E_x(z,t) = E_o \cos(\Omega t - kz) \qquad (1.11)$$

through an electric-dipole interaction

$$H_I = -\mu \cdot E_x(z,t) \; . \qquad (1.12)$$

The equation of motion of the density matrix

$$i\hbar \frac{d\rho}{dt} = [H,\rho] + i\hbar\lambda + \text{damping terms} \qquad (1.13)$$

includes the total Hamiltonian

$$H = H_o + H_I \; , \qquad (1.14)$$

the phenomenological damping times

$$T_2 \text{ and } T_1 \qquad (1.15)$$

for the off-diagonal and diagonal density matrix elements, and a population source term λ due to relaxation from other levels, which act as a thermal reservoir in contact with levels 1 and 2.

In the optical region, a laser beam induces a dipole.

$$\langle\mu\rangle = \text{Tr}(\mu\rho) \qquad (1.16)$$

in an atom and a polarization

$$P = N\mu_{12} \langle\langle\tilde{\rho}_{12}\rangle\rangle e^{-i(\Omega t - kz)} + c.c. \qquad (1.17)$$

in a sample of molecular number density N where $\mu_{12} = \langle 1|\mu_x|2\rangle$ is the electric dipole matrix element and ρ_{12} is the slowly varying amplitude of the off-diagonal density matrix, i.e., $\rho_{12} = \tilde{\rho}_{12} e^{-i(\Omega t - kz)}$. The first bracket in (1.16) and (1.17) is a quantum mechanical average and the second bracket $\langle \; \rangle$ in (1.17) designates an average over the inhomogeneous lineshape. The polarization in turn is the source of a transient signal field

$$E_x(z,t) = \tilde{E}_x(z,t) e^{-i(\Omega t - kz)} + c.c. \qquad (1.18)$$

which obeys the Maxwell wave equation

$$\frac{\partial E_x}{\partial z} = 2\pi i k N \mu_{12} \ll \tilde{\rho}_{12} \gg \tag{1.19}$$

in the slowly varying envelope approximation, i.e., when the sample is optically thin.

The object of these calculations, therefore, is to calculate $\ll \tilde{\rho}_{12} \gg$.

Linear combinations of the density matrix equations of motion yield the Bloch equations

$$\dot{u} + \Delta v + u/T_2 = 0$$

$$\dot{v} - \Delta u - \chi w + v/T_2 = 0$$

$$\dot{w} + \chi v + (w-w^o)/T_1 = 0 \tag{1.20}$$

where the three components of the Bloch vector $\vec{B}(u,v,w)$ are defined as

$$u = \tilde{\rho}_{12} + \tilde{\rho}_{21}$$

$$v = i(\tilde{\rho}_{21} - \tilde{\rho}_{12})$$

$$w = \rho_{22} - \rho_{11} . \tag{1.21}$$

In these equations, the dipole dephasing time T_2 and the population decay time T_1 have been introduced. Equations (1.20) can be written compactly in the absence of damping as

$$\frac{d\vec{B}}{dt} = \vec{\beta} \times \vec{B} . \tag{1.22}$$

We note that (1.22) is exactly of the same form as (1.9) where the Bloch vector

$$\vec{B} = \vec{i}u + \vec{j}v + \vec{k}w \tag{1.23}$$

precesses about the effective field

$$\vec{\beta} = -\vec{i}\chi + \vec{k}\Delta , \tag{1.24}$$

as in Figure 1. The Rabi frequency is defined as

$$\chi = \mu_{12} E_0 / \hbar \tag{1.25}$$

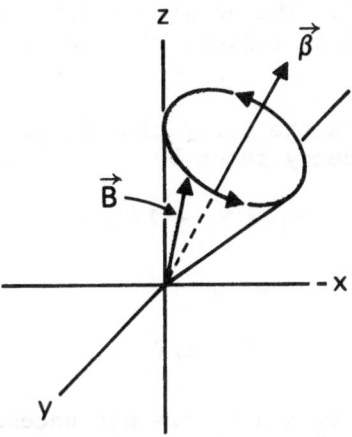

Fig. 1. The Bloch vector $\vec{B}$ in its precession about the effective field $\vec{\beta}$ provides a simple geometrical interpretation of the time-dependent interaction of an optical field with a two-level quantum system.

and the off-resonant tuning parameter as

$$\Delta = -\Omega + kv_z + \omega_{21} \qquad (1.26)$$

where kv_z is the Doppler shift associated with a molecule having longitudinal velocity v_z.

1.3 Comment on Decay Phenomena

It should be realized that the density matrix equations (1.13) are more general than the Bloch equations (1.20). In NMR, a two level spin system constitutes an isolated entity where population is conserved. The time-dependent population difference of the two spin levels is characterized by the single decay time T_1. However, in the optical region, population is not generally conserved among the two transition levels as the level structure is more varied and the decay channels more numerous. Therefore, T_1 is not defined in general and one must introduce different decay rates γ_1 and γ_2 for the two transition levels in the equations of motion [8]. The density matrix solutions, appropriate to the optical region, will therefore differ from the NMR case.

To summarize, in NMR the decay time T_1 is related to the individual population decay rates by

$$\gamma_1 = \gamma_2 = 1/T_1 \; , \qquad (1.27)$$

and the dipole dephasing rate is

$$\Gamma = 1/T_2 \; . \qquad (1.28)$$

In the optical region, γ_1 and γ_2 are not necessarily equal in which case T_1 is not defined in general, and the dipole dephasing rate, which follows from section 1.2, is

$$\Gamma = \frac{1}{2}(\gamma_1 + \gamma_2) \qquad (1.29)$$

If in addition there are phase interrupting interactions without population decay, then

$$\Gamma = \frac{1}{2}(\gamma_1 + \gamma_2) + \gamma_\phi \qquad (1.30)$$

where γ_ϕ is the rate of phase interruptions. Diffusion effects, such as elastic velocity-changing collisions produce additional dephasing effects with a characteristic nonexponential decay and will be considered in Lecture III.

References to Lecture I

1. A. Abragam, The Principles of Nuclear Magnetism, Oxford, London (1961).
2. C. P. Slichter, Principles of Magnetic Resonance, Harper and Row, New York (1963).
3. N. A. Kurnit, I. D. Abella and S. R. Hartmann, Phys. Rev. Lett. 13, 567 (1964).
4. R. G. Brewer, "Coherent Optical Spectroscopy" in Frontiers in Laser Spectroscopy, Volume 1, North-Holland, Amsterdam, p. 341, edited by R. Balian, S. Haroche and S. Liberman; see also R. G. Brewer, Physics Today, May 1977, p. 50.
5. F. Bloch, Phys. Rev. 70, 460 (1946).
6. R. P. Feynman, F. L. Vernon, and R. W. Hellwarth, J. Appl. Phys. 28, 49 (1957).
7. M. Sargent III, M. O. Scully and W. E. Lamb Jr., Laser Physics, Addison-Wesley, Reading, Mass. (1974), p. 91.
8. A. Schenzle and R. G. Brewer, Phys. Rev. A14, 1756 (1976).

LECTURE II: The Stark Switching Technique

Since the introduction of the Stark switching technique [1] in 1971, over ten different coherent optical transient effects have been observed. They include photon echoes involving two [1] or multiple pulses (stimulated [2] and Carr-Purcell [3] echoes), optical nutation [1], free induction decay (FID) [4], coherent Raman beats [5], adiabatic fast passage [6], FID interference pulses [7], quantum beats [8], and pulse Fourier transform spectroscopy [9]. Nearly all of these effects are the optical analogues of spin transients [10-14]. Most of these studies have been carried out in the infrared in NH_2D, $^{13}CH_3F$, and NH_3. However, the method has not been restricted to molecules and recently has been applied in the visible region by Liao et al. [15] to atomic sodium vapor and by Szabo [16] to ruby at liquid helium temperatures.

In our experiments (Fig. 2), a molecular gas sample that is Stark tunable is irradiated by a continuous wave CO_2 laser beam in the 9-10μm region. Hence, vibration rotation transitions are excited. Electronic dc field pulses are applied repetitively to the sample thereby switching the molecular level structure into or out of resonance with the fixed laser frequency. A particular coherent transient effect can be selected by simply varying the pulse sequence.

In Fig. 2, we consider a pair of levels which exhibit a change in their transition frequency when a Stark pulse is applied. Initially, molecules of velocity v are excited coherently in steady-state by laser light of frequency Ω. Hence, a set of phased dipoles oscillating at the optical frequency are induced and this constitutes the preparative step. When the Stark pulse appears, this initial velocity group is no longer in resonance, but because of its preparation will radiate, by analogy with NMR [13], a free induction decay signal (Fig. 3). At the same time, a second velocity group v' will be switched into resonance and will alternately absorb and emit radiation. This optical ringing or nutation effect [11] appears as a damped oscillation (Fig. 4). When the pulse terminates, the group v is suddenly excited and it too begins to nutate while the second group v' emits a FID signal. If two pulses are applied, the group v' emits the photon echo, the optical analogue of the spin echo [10].

Transient light signals that are emitted by molecules switched out of resonance propagate in the forward direction, because of the preparative step, and are monitored together with the transmitted laser beam by a photodetector. Heterodyne detection occurs automatically as in the Raman beat, photon echo and FID since the emission signal is colinear with the laser beam and is Stark-shifted from the laser frequency. Heterodyne detection

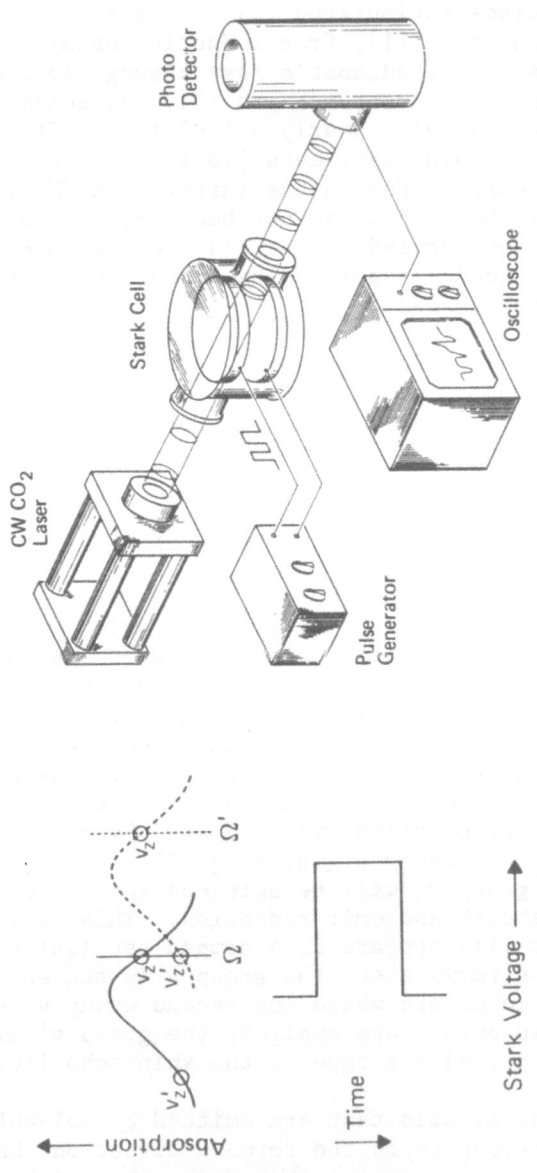

Fig. 2. Left: Stark switching principle applied to a Doppler-broadened optical transition. Right: Stark-switching apparatus.

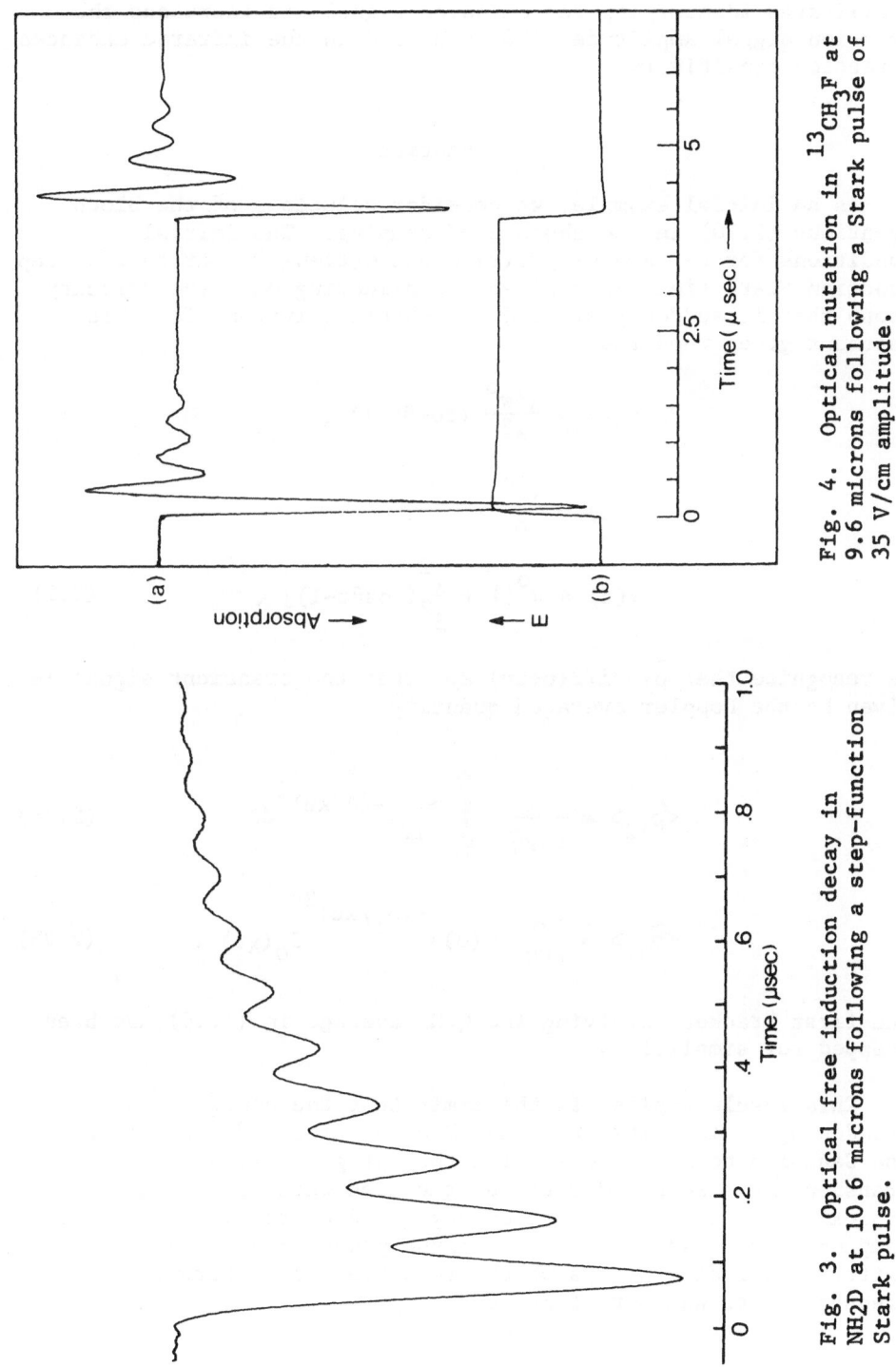

Fig. 4. Optical nutation in $^{13}CH_3F$ at 9.6 microns following a Stark pulse of 35 V/cm amplitude.

Fig. 3. Optical free induction decay in NH2D at 10.6 microns following a step-function Stark pulse.

facilitates identifying the emission signal; it increases the
emission signal amplitude 1000 fold; and in the infrared enhances
detection sensitivity.

2.1 Nutation

As an initial example, we consider solutions of the Bloch
equations (1.20) in the absence of damping. The initial
conditions for $t \leq 0$ are $u(t) = v(t) = 0$ and $w(t) = w(0)$. At $t = 0$, a step
function Stark field is applied. Considering only the velocity
group that is suddenly excited, we obtain solutions for this
resonant group when $t \geq 0$

$$u(t) = \frac{\chi \Delta w^o}{\beta^2} (\cos\beta t - 1) \, ,$$

$$v(t) = \frac{\chi w^o}{\beta} \sin \beta t \, ,$$

$$w(t) = w^o [1 + \frac{\chi^2}{\beta^2}(\cos\beta t - 1)] \, . \tag{2.1}$$

We recognize that $\tilde{\rho}_{12} = 1/2(u + iv)$ and that the transient signal is
given by the Doppler averaged quantity

$$\langle \tilde{\rho}_{12} \rangle = \frac{1}{ku\sqrt{\pi}} \int_{-\infty}^{\infty} \tilde{\rho}_{12} e^{-(\Delta/ku)^2} d\Delta \tag{2.2a}$$

$$\langle \tilde{\rho}_{12} \rangle \simeq \frac{i\sqrt{\pi}}{2ku} \chi w(0) e^{-(\Delta_1/ku)^2} J_0(\chi t) \, . \tag{2.2b}$$

The first bracket involving the Q.M. average in (1.16) has been
dropped for simplicity.

This result applies in the limit that the homogeneous
broadening is much less than the Doppler width, $\Gamma \ll ku$, allowing
the Gaussian to be factored from the integral (2.2a). The
transient is a zero order Bessel function which oscillates
approximately at the Rabi frequency χ. Even though we excluded
damping terms, damping occurs because power broadening causes
different velocity groups which are driven at different
frequencies to get out of phase.

2.2 Free Induction Decay

In this case, we focus on the coherent preparation of the sample during steady state conditions and the coherent emission (FID) which follows a step function Stark pulse. For $t \leq 0$, we set $\dot{u}=\dot{v}=\dot{w}=0$ to obtain the steady-state solutions

$$u(0) = - \Delta \chi w^o / (\chi^2 T_1/T_2 + \Delta^2 + 1/T_2^2)$$

$$v(0) = (\chi w^o/T_2)/(\chi^2 T_1/T_2 + \Delta^2 + 1/T_2^2)$$

$$w(0) = w^o[1-(\chi^2 T_1/T_2)/(\chi^2 T_1/T_2 + \Delta^2 + 1/T_2^2)] \qquad (2.3)$$

When the Stark switch is applied at $t=0$,

$$\Delta \rightarrow \Delta' = \Delta + \Delta\omega_{21} \ .$$

Furthermore, when the frequency jump is large enough so that $\Delta\omega_{21} \gg 1/T_2$, we can set $\chi=0$ in the Bloch equations (1.20) and obtain solutions for the regime $t \geq 0$

$$u(t) = [u(0)\cos\Delta't - v(0)\sin\Delta't]e^{-t/T_2} \ ,$$

$$v(t) = [u(0)\sin\Delta't + v(0)\cos\Delta't]e^{-t/T_2} \ ,$$

$$w(t) = w^o + [w(0)-w^o]e^{-t/T_1} \ . \qquad (2.4)$$

Proceeding as in the nutation calculation [4,17], the Doppler averaged result is

$$\langle\tilde{\rho}_{12}\rangle = \frac{i\sqrt{\pi}}{2ku} e^{-(\Delta_1'/ku)^2} \chi w^o \exp\left[- \frac{t}{T_2} \left(1 + \sqrt{\chi^2 T_1 T_2 + 1}\right)\right]$$

$$\times \left(\frac{1}{\sqrt{\chi^2 T_1 T_2 + 1}} - 1\right) \cos\Delta\omega_{21}t \ . \qquad (2.5)$$

Since the laser acts as a local oscillator, it produces with the signal field $E_s \propto \langle\tilde{\rho}_{12}\rangle$ a cross term in the intensity, i.e., a heterodyne beat signal essentially given by (2.5). The beat frequency $\Delta\omega_{21}$, which is the Stark shift, is evident in (2.5). In addition, the decay rate

$$\lim_{\chi \to 0} \frac{1}{T_2}\left(1+\sqrt{\chi^2 T_1 T_2 + 1}\right) \to \frac{2}{T_2} \qquad (2.6)$$

shows a power dependence due to power broadening in the preparative stage. In the zero power limit, the factor of 2 in (2.6) arises due to the preparative step which adds an additional $1/T_2$ decay rate. Hence, at low laser power the dephasing time T_2 can be obtained directly from a FID signal in a single burst, a procedure which is somewhat simpler than the echo technique.

In the visible-ultraviolet region, in contrast to the infrared and rf, the quantity T_1 may not be well defined. In that case, the Bloch equations are to be replaced by the density matrix equations of motion (1.13). The result of an FID calculation [18] when $\gamma_1 \neq \gamma_2$ is

$$\langle \tilde{\rho}_{12} \rangle = \frac{i\sqrt{\pi}\chi}{2ku} \rho_{11}^0 \left(1 - \frac{\Gamma}{(\Gamma^2 + \hat{\Gamma}^2)^{1/2}}\right) \exp\{-[\Gamma + (\Gamma^2 + \hat{\Gamma}^2)^{1/2}]t\} \qquad (2.7)$$

where

$$\hat{\Gamma}^2 = \frac{\chi^2 \Gamma}{2\Gamma_1 \Gamma_2}(\gamma_1 + \gamma_2 - \gamma) , \qquad (2.8)$$

and the decay rate is

$$\Gamma + (\Gamma^2 + \hat{\Gamma}^2)^{1/2} . \qquad (2.9)$$

The dipole dephasing time is $\Gamma \equiv 1/T_2$ and the spontaneous emission time of level 2 for the $2 \to 1$ transition is γ. We see that (2.9) resembles (2.6) and reduces to it when $\gamma=0$ and $\gamma_1 = \gamma_2$.

The nutation and FID calculations lead directly to the two pulse echo solutions which we consider in various applications in the next lecture.

References to Lecture II

1. R. G. Brewer and R. L. Shoemaker, Phys. Rev. Lett. 27, 631 (1971).
2. R. G. Brewer in Very High Resolution Spectroscopy, Academic Press, London (1976), p. 127, edited by R. A. Smith.
3. J. Schmidt, P. R. Berman, and R. G. Brewer, Phys. Rev. Lett. 31, 1103 (1973).
4. R. G. Brewer and R. L. Shoemaker, Phys. Rev. A6, 2001 (1972).
5. R. L. Shoemaker and R. G. Brewer, Phys. Rev. Lett. 28, 1430 (1972).
6. M. M. T. Loy, Phys. Rev. Lett. 32, 814 (1974).

7. K. L. Foster, S. Stenholm and R. G. Brewer, Phys. Rev. A10, 2318 (1974).

8. R. L. Shoemaker and F. A. Hopf, Phys. Rev. Lett. 33, 1527 (1974).

9. S. B. Grossman, A. Schenzle and R. G. Brewer, Phys. Rev. Lett. 38, 275 (1977).

10. E. L. Hahn, Phys. Rev. 80, 580 (1950).

11. H. C. Torrey, Phys. Rev. 76, 1059 (1949).

12. H. Y. Carr and E. M. Purcell, Phys. Rev. 94, 630 (1954).

13. E. L. Hahn, Phys. Rev. 77, 297 (1950).

14. A. Abragam, The Principles of Nuclear Magnetism, Oxford, London (1961).

15. P. F. Liao, J. E. Bjorkholm and J. P. Gordon, Phys. Rev. Lett. 39, 15 (1977).

16. A. Szabo (unpublished).

17. F. A. Hopf, R. F. Shea and M. O. Scully, Phys. Rev. A7, 2105 (1973).

18. A. Schenzle and R. G. Brewer, Phys. Rev. A14, 1756 (1976).

LECTURE III: Photon Echoes and Molecular Collisions

3.1 Photon Echoes

The spin echo [1] is an interference involving a coherent
set of oscillating dipoles that dephase in the first pulse
interval because of a spread in their frequencies (destructuve
interference) and rephase in the second pulse interval
(constructive interference). The Bloch vector model allows one
to understand this complicated time-dependent behavior as a
sequence of four precessional motions (see Fig. 5), and we will
return to it shortly.

In the time domain, homogeneous and inhomogeneous line
broadening assume different roles. Homogeneous broadening is
due to time-dependent interactions which are random in character
and exhibit irreversible time behavior. On the other hand,
inhomogeneous broadening often arises from static interactions
where time-dependent fluctuations do not appear. Therefore, in
the time domain inhomogeneous broadening can exhibit
reversibility. The echo effect occurs because of this
reversibility property. However, the homogeneous broadening
remains and damps the echo signal. In fact, the echo envelope
function provides a way of measuring the dipole dephasing time
T_2 without the complication of inhomogeneous dephasing effects.

The racetrack analogy of E. L. Hahn [2] while not exact
expresses the spirit of echo formation. Runners which start out
together at the firing of a gun soon fan out (the FID effest).
When a second firing occurs at time τ, the runners reverse
direction (the dipole phase changes sign). At time 2τ, all
runners come back to the starting line (echo formation).

An equivalent representation in terms of the dipole phase-time
diagram is shown in Fig. 6 for a two pulse sequence where the
first and second pulses appear at times 0 and τ and the echo at
2τ. The dephasing-rephasing behavior is evident from the figure.
A common question is why the second pulse reverses the sign of
the dipole phase. If we remember that the precession angle $\phi=\beta t$
of the Bloch vector $\vec{B}$ about the effective field $\vec{\beta}$ is the dipole
phase, then the duration of excitation t and the light intensity
determine the phase angle. For example, a perfect π pulse ($\pi=\beta t$)
for the resonant velocity group ($\beta=\chi$) tips the Bloch vector from
a positive to its negative value. Other pulse areas or tipping
angles give echoes with smaller amplitude.

In the first observation of the photon echo effect by
Kurnit, Abella and Hartmann [3], they irradiated a ruby crystal
at liquid Helium temperatures with pulsed ruby laser light. The

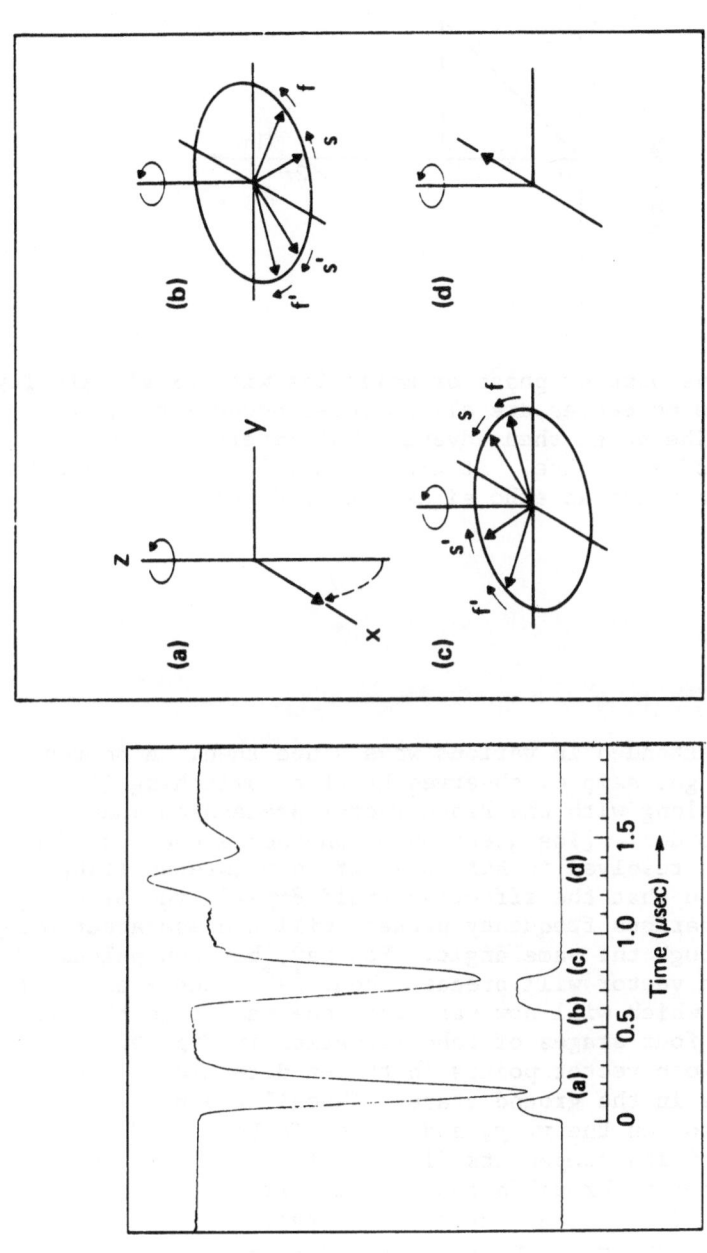

Fig. 5. The third pulse in the upper trace at the left is an infrared photon echo for a 9.7 micron vibration-rotation transition in 13CH3F. The four diagrams at the right indicate four stages in the precessional motion of the Bloch vector, corresponding to the times marked a, b, c and d on the pulse sequence.

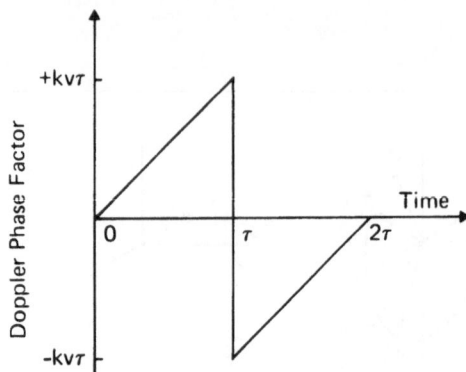

Fig. 6. Relative Doppler phase of molecules with axial velocity
v is shown. The molecules are all in phase because of a $\pi/2$
pulse at t=0. The phase then advances but reverses sign at t=τ
by application of a π pulse. At t=2τ, the molecular dipoles have
all rephased again and an echo signal is produced.

method has been extended in various ways since then. A photon
echo in a $^{13}CH_3F$ gas sample, observed by Stark switching [4], is
given in Fig. 5 along with the Bloch vector precession model
which we now consider. [The question of whether echoes could be
seen in a gas was resolved in Ref. 5.] If only intense light
pulses are used so that the effective field $\beta \sim \chi \gg \Delta$, the Bloch
vector $\vec{B}$ of the various frequency packets will precess about $\vec{\beta} = \vec{i}\chi$
(the x axis) through the same angle. However, between pulses
$\chi = 0$ and the Bloch vector will precess about $\vec{\beta} = \vec{k}\Delta$ (the z axis) at
the frequency Δ, which will now vary from one packet to the next.
Now consider the four stages of echo formation in Fig. 5.
Initially, the Bloch vector points in the −z direction because
the molecules are in the ground state. (Recall the projections
of the Bloch vector on the x, y, and z axes in Eq. (1.23) and
the definitions of its components (1.21).) (a) The first pulse
tips the Bloch vector through a phase angle $\pi/2 = \beta t_w$ where t_w is
the pulse width. (b) In the next time interval, the packets fan
out precessing in the x-y plane about the z axis. (c) The second
pulse tips the Bloch vector through a phase angle $\pi = \beta t_w'$, thereby
reversing the sign of the dipole phase; the fast packets now lag
behind the slow ones, in contrast to (b). (d) In the next
interval, the fast packets catch up with the slow ones precisely

at time 2τ where τ is the pulse delay time. At this point, the dipoles are all in phase and the sample can radiate the echo pulse.

To calculate the echo signal, we make use of the nutation and free induction decay solutions, (2.1) and (2.4) appropriate to the Stark switching experiments. Details of this calculation are given elsewhere [6]. The echo signal

$$[E_c^2(t=2\tau)]_{beat} = 2\pi\hbar kNL\chi^4 w(0)e^{-t/T_2}$$

$$\times < \frac{1}{\beta^3} \sin\theta_{10}\sin^2(\theta_{32}/2)> \qquad (3.1)$$

appears as a heterodyne beat signal with beat frequency $\Delta\omega_{21}$ equal to the Stark shift. Therefore, the echo field amplitude is observed and not the intensity. The beat is not evident in (3.1) because the modulation term $\cos\Delta\omega_{21}(t-2\tau)=1$ when the echo is formed at $t=2\tau$. The echo envelope has a decay time T_2 independent of Doppler inhomogeneous dephasing, as mentioned earlier.

We note that the Doppler integral, $< >$ of (3.1), differs from the nutation and FID expressions. At high light intensities when $\chi/ku \gg 1$, the echo signal is a maximum when the first and second pulse angles (areas) are $\theta_{10}=\pi/2$ and $\theta_{32}=\pi$. In that case, the Doppler integral of (3.1) is

$$<\tilde{P}_x(t=2\tau)> = \frac{1}{u\sqrt{\pi}} \int_{-\infty}^{\infty} \tilde{P}_x(t=2\tau)e^{-v^2/u^2}dv \qquad (3.2)$$

where

$$\tilde{P}_x = \frac{1}{\beta^3} \sin\beta t_1 \sin^2 \frac{1}{2}\beta t_2 , \qquad (3.3)$$

reduces to the NMR result [1]

$$<\tilde{P}_x> = \frac{1}{\chi^3} \sin\chi t_1 \sin^2 \frac{1}{2}\chi t_2 . \qquad (3.4)$$

At low light intensity when $\chi/ku \ll 1$, i.e., when a small fraction of the Doppler width is excited, we obtain

$$G(t_{1,2}) = \int_{-\infty}^{\infty} \frac{1}{\beta^3} \sin^2 \frac{1}{2}\beta t_2 \sin\beta t_1 d\Delta$$

$$2G(t_{1,2}) = F(t_1) - \frac{1}{2}[F(t_1-t_2)+F(t_1+t_2)] \qquad (3.5)$$

where

$$F(t) = \int_{-\infty}^{\infty} \frac{\sin\beta t}{\beta^3} d\Delta \; . \tag{3.6}$$

Equation (3.6) can be written [7] as

$$F(t) = \pi t\chi[1-1/2t\chi_1 F_2(\frac{1}{2};\frac{3}{2},2,-\frac{1}{4}t^2\chi^2)] \tag{3.7}$$

in terms of the hypergeometric function

$$_1F_2(a;b,c,y) = \sum_{n} \frac{(a)_n}{(b)_n(c)_n} \frac{y^n}{n!} \; ,$$

$$(\ell)_n = \ell(\ell+1)(\ell+2)...(\ell+n-1) \; . \tag{3.8}$$

A comparison of (3.7) with a direct numerical integration of (3.2) shows that only a few terms in the series are needed to obtain precise results. The intensity dependence of (3.7) is due to the slight dephasing which occurs during the second pulse; this affects the echo amplitude but not the decay behavior.

3.2 Molecular Collisions

Molecular collisions influence coherent optical transients in various ways [8]. We have discussed how the dipole dephasing time T_2 can be measured in a two pulse echo experiment, and this is sensitive to both elastic and inelastic collisions. The rate of inelastic collisions can be obtained also in the same echo experiment, simply from the envelope function of the second nutation signal. The first pulse redistributes the population among the transition levels while the second pulse monitors the rate of population recovery as equilibrium is approached in the pulse interval. The decay is exponential.

The effects of elastic collisions are more subtle and can modify the simple T_2 decay mentioned above. If an elastic collision between two molecules is not state-dependent and produces only a change in their relative velocity, the resulting phase change at time t in the coherently prepared collision partner is $\Delta\phi=k\Delta v_z t$. Here, $\vec{k}$ is the propagation vector of light and Δv_z the longitudinal jump in velocity due to the collision. Thus, the phase change freely evolves in time because of an earlier collision, and the dipoles eventually get out of phase. We may view the effect as Brownian motion in velocity space.

Elastic collisions of the velocity changing type influence echo signals in the following way [8]. In the absence of elastic collisions, the echo signal is of the form

$$S \sim e^{i\vec{k}\cdot\vec{v}(t_2-t_1)} \; e^{-i\vec{k}\cdot\vec{v}(t-t_3)} , \qquad (3.9)$$

and S=1 for perfect dephasing-rephasing behavior where t_2-t_1 is the first pulse interval and $t-t_3$ the second interval. Since elastic collisions reduce the echo amplitude, it is necessary to perform a collision average of (3.9)

$$S \sim \langle e^{ik\Delta v_z \tau} \rangle_{collision} . \qquad (3.10)$$

For long times when $k\Delta v_z\tau \gg 1$, the phase excursions are large and most of the molecules have dephased. The echo signal is due to those molecules that didn't collide, and since their survival probability is an exponential in time, we obtain as the limiting long-time behavior

$$S \sim e^{-\Gamma\tau} \qquad (3.11)$$

where Γ is the rate of elastic collisions. For short times when $k\Delta v_z\tau \ll 1$, Eq. (3.10) can be expanded as

$$S \sim 1-k^2\langle\Delta v^2\rangle\tau^2/2 . \qquad (3.12)$$

Since $\langle\Delta v^2\rangle = \Gamma\tau\Delta u^2/2$, we obtain

$$S \sim \exp[-k^2\Delta u^2\Gamma\tau^3/2] \qquad (3.13)$$

where Δu is a characteristic rms change in velocity per collision along the z direction (optic axis). Thus, the echo signal obeys a cubic decay law initially and eventually becomes linear. A more rigorous calculation [8] based on the Boltzmann transport equation gives the same results apart from numerical factors. Experimental results [8] confirm these predictions as seen in Fig. 7. The upper curve resulting from two pulse nutation described earlier yields the population decay time T_1. This curve coincides with Carr-Purcell echo measurements, which essentially eliminate the velocity changing collision dephasing effect, and yields the dipole dephasing time T_2. Hence, $T_2=T_1$. The two-pulse echo (lower curve) shows the effect of elastic collisions in agreement with the limiting forms of (3.11) and (3.13) and more detailed calculations. From measurements of this kind, Γ and Δu are obtained.

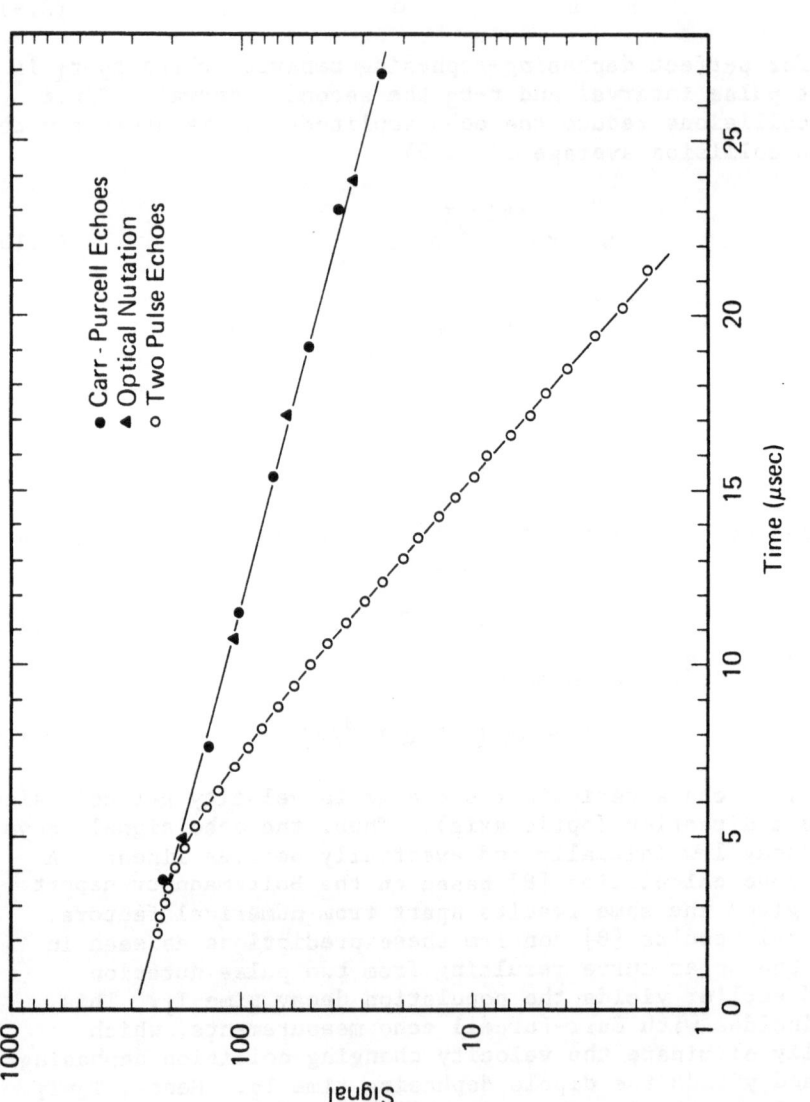

Fig. 7. Decay curves of three optical coherent transient effects in $^{13}CH_3F$.

3.3 Pulse Fourier Transform Spectroscopy

In these lectures, we have considered only one two level system interacting resonantly with a classical radiation field. In a more general case, several transitions of different frequency might be simultaneously excited and all would produce coherent transient signals, such as the echo, simultaneously. However, it is possible to separate the various spectral components independently in frequency, and this can be accomplished by means of Fourier transform methods using a digital computer. The method has been used extensively in NMR over the past several years. The great advantage is that closely spaced spectral lines can be resolved under high resolution and the decay behavior can be obtained simultaneously as a function of quantum state for any particular coherent transient effect.

An example of such a spectrum [9] is shown in the three dimensional computer plot of a two pulse echo signal versus frequency and pulse delay time (Fig. 8). The $^{13}CH_3F$ vibration-rotation lines are split by ~1 MHz due to the Stark effect, and the line widths are ~100 kHz due to the finite pulse width and the intensity of the laser beam. Since the Doppler

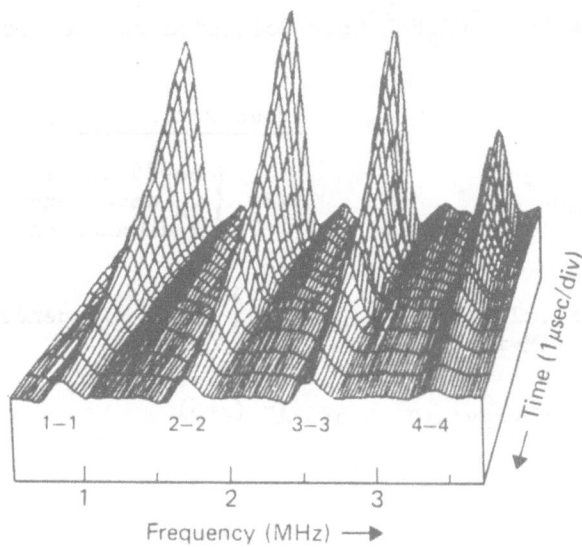

Fig. 8. Fourier-transform heterodyne beat spectrum of $^{13}CH_3F$ derived from two-pulse echoes as a function of pulse delay time. The infrared transitions 1-1,2-2,... designate the $|M|$ states involved.

width is 66 MHz, it is clear that a Doppler-free spectrum is
resolved. In addition, one obtains the decay behavior for each
line. By moving the detector's window in time, the amplitude of
the second nutation signal can be monitored also and the decay
rate of inelastic collisions can be determined. Clearly, other
coherent transient effects can be seen as well.

It is also possible by means of a Stark bias field to tune
different velocity groups v_z into resonance and examine how the
decay rate of elastic and inelastic collisions depend on the
velocity and hence deduce the molecular force laws with the aid
of scattering theory. Figure 9 shows the results of these
measurements. The solid lines are a modified form of a
Landau-Lifshitz elastic scattering theory where the interaction
is assumed to be of the form $V=-c/r^s$. Inelastic collisions are
velocity independent in agreement with other theories which
predict a first order dipole-dipole interaction (s=3). The
elastic collisions behave differently and seem to follow a second
order dipole-dipole interaction (s=6). Table I summarizes the
collision data obtained for $^{13}CH_3F$ by means of these coherent
transient effects, including the versatile and quantitative
Fourier transform method.

Table I: $^{13}CH_3F$-$^{13}CH_3F$ Collision Parameters

	Cross-section ($\AA^2$)
Elastic	
Velocity changes[*]	430 $\left\{ \begin{array}{l} \Delta u=85 \text{ cm/sec} \\ v_z \text{ dependence} \rightarrow s=6 \\ M \text{ dependence: none} \end{array} \right.$
Phase interruptions	~0
Inelastic	
Rotation ($\Delta J=\pm1,\pm2...$)	330 $\left. \begin{array}{l} v_z \text{ \& M dependence:} \\ \text{none} \end{array} \right\}$
Reorientation ($\Delta M=\pm1,\pm2...$)	140

[*]Landau-Lifshitz theory: $\sigma_e = 590$ $\AA^2$ (s=6)

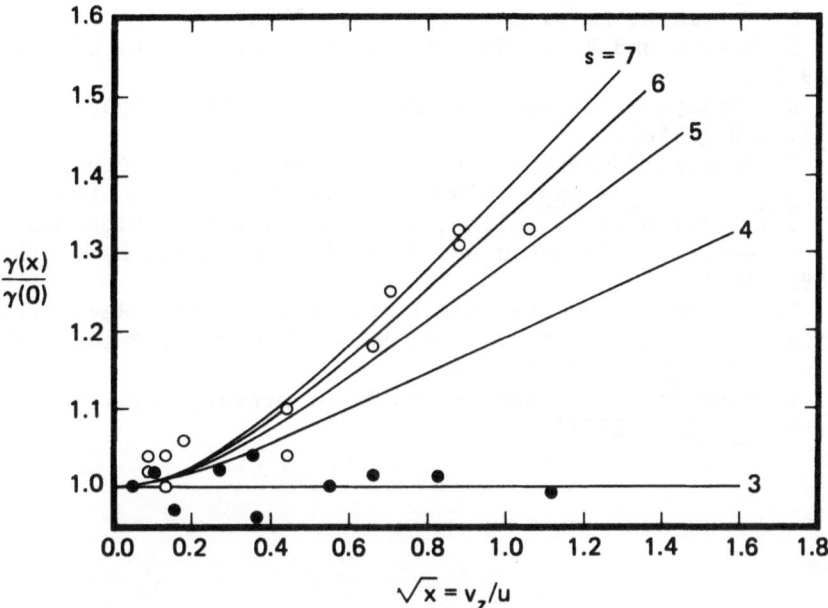

Fig. 9. Normalized decay rate $\gamma(\chi)/\gamma(0)$ as a function of the velocity ratio V_z/u. Experiment: elastic scattering (open circles); inelastic scattering (filled circles). Theory: elastic scattering and the s=3 curve for inelastic scattering (curves).

References to Lecture III

1. E. L. Hahn, Phys. Rev. 80, 580 (1950).
2. E. L. Hahn, Physics Today, November 1953, cover illustration.
3. N. A. Kurnit, I. D. Abella, and S. R. Hartmann, Phys. Rev.
    Lett. 13, 567 (1964); I. D. Abella, N. A. Kurnit, and
    S. R. Hartmann, Phys. Rev. 141, 391 (1966).
4. R. G. Brewer and R. L. Shoemaker, Phys. Rev. Lett. 27, 631
    (1971).
5. M. O. Scully, M. J. Stephen and D. C. Burnham, Phys. Rev.
    171, 213 (1968).
6. R. G. Brewer, "Coherent Optical Spectroscopy" in Frontiers
    in Laser Spectroscopy, Volume 1, North-Holland, Amsterdam,
    p. 341, edited by R. Balian, S. Haroche, and S. Liberman.
7. A. Schenzle, S. Grossman and R. G. Brewer, Phys. Rev. A13,
    1891 (1976).
8. P. R. Berman, J. M. Levy and R. G. Brewer, Phys. Rev. A11,
    1668 (1975); J. Schmidt, P. R. Berman and R. G. Brewer, Phys.
    Rev. Lett. 31, 1103 (1973).
9. S. B. Grossman, A. Schenzle and R. G. Brewer, Phys. Rev.
    Lett. 38, 275 (1977).

LECTURE IV: Two-photon Transients

Only one-photon processes have been considered thus far.   In
this lecture, coherent two-photon transients are treated.   The
subject has attracted interest recently not only in optical
problems but in NMR as well. We begin with the coherent Raman
beat [1-3] effect which was observed by the Stark switching
technique. Consider the three level system of Fig. 10 where an
incident laser beam of frequency Ω excites two coupled
transitions, 1→3 and 2→3.   Initially, levels 1 and 2 are
degenerate and excitation proceeds under steady-state conditions,
causing all three levels to be in coherent superposition.   Now
assume that a step-function Stark field is applied which breaks
the degeneracy.   This action causes the one-photon transitions
1→3 and 2→3 to be nonresonant with the laser field.   However,
two-photon forward scattering can now occur.   The laser again
acts as a local oscillator producing a heterodyne beat signal at
the photo detector where the beat frequency ω_{12} is the 1-2 level
splitting.

In this two-photon process, the Doppler shift kv_z associated
with the absorbed photon cancels that of the emitted photon, so
that the transient beat signal shown in Fig. 11 is independent
of Doppler dephasing.   Therefore, the decay is long-lived being
limited only by population decay.   For this reason, elastic
velocity changing collisions are ineffective also.

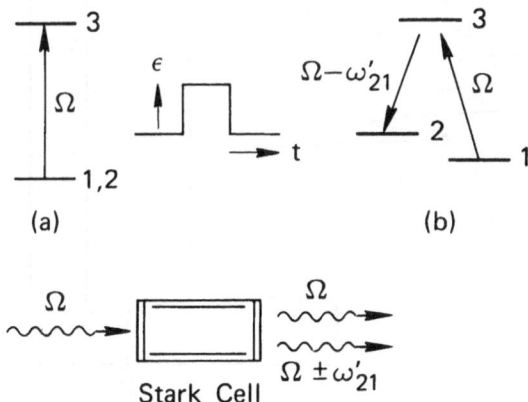

(a)   (b)

Stark Cell

Fig. 10. Molecular energy-level diagrams illustrate (a) coherent
preparation of a three-level quantum system during a Stark pulse
when the levels 1 and 2 are degenerate ($\omega_{21}=0$); and (b) transient
forward scattering following the pulse because of the coherent
Raman-beat effect.

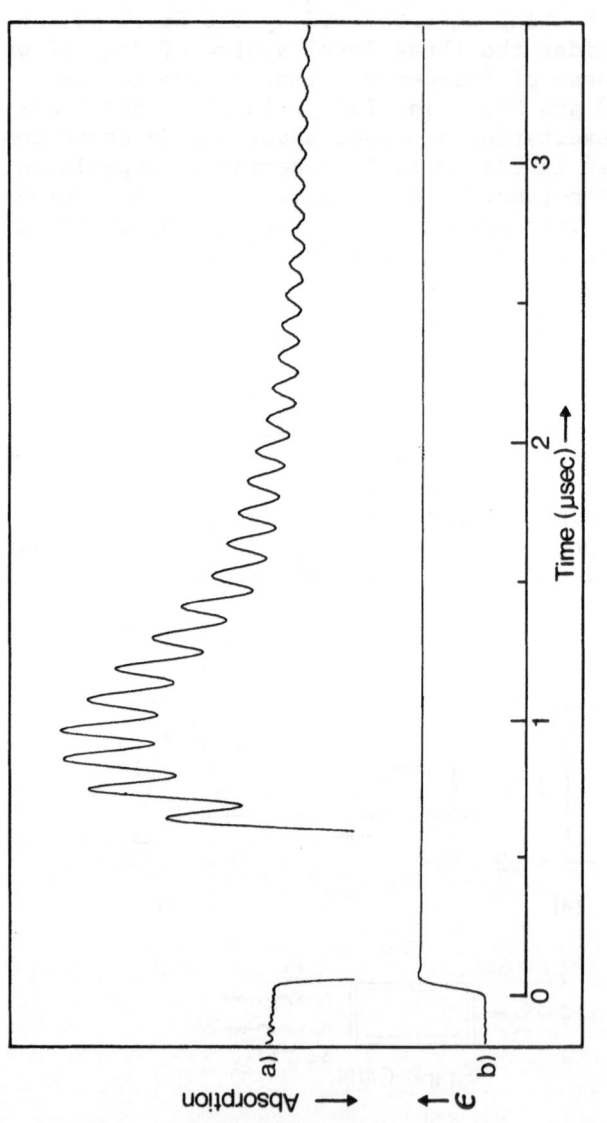

Fig. 11. Coherent Raman beat signal in $^{13}CH_3F$ following a step-function Stark pulse. See Fig. 10.

Figure 12 shows a related two-photon process which can occur in a cascade level configuration [3]. Here, two counter propagating laser beams of frequency Ω_1 and Ω_2 resonantly excite the $1 \rightarrow 2$ transition during the preparative phase such that $\Omega_1 + \Omega_2 = \omega_{21}$. When a step function Stark field shifts the levels abruptly to a new value ω'_{21}, the resonant excitation ends and the Ω_1 beam stimulates two-photon emission with a component at $\omega'_{21} - \Omega_1$, directed in the backward direction. A photodetector seeing this light monitors a beat signal with a beat frequency equal to $\omega'_{21} - \Omega_1 - \Omega_2 = \omega'_{21} - \omega_{21}$, which is just the Stark shift. The Ω_2 beam behaves similarly. The decay will also be independent of Doppler dephasing since the Doppler shifts again cancel. This effect which was predicted by Brewer and Hahn [3] was observed just recently [4,5]. The Raman beat and the stimulated two-photon emission process are fundamentally the same two-photon process obeying the same equations of motion with nearly identical solutions.

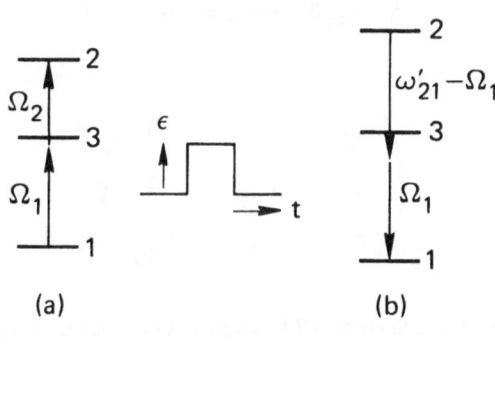

(a)   (b)

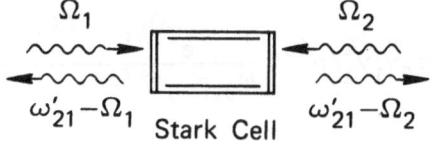

Fig. 12. Molecular energy-level diagrams illustrate (a) coherent preparation of a three-level quantum system by two oppositely directed beams (Ω_1 and Ω_2); a Stark pulse shifts the levels so that the condition $\Omega_1 + \Omega_2 = \omega_{21}$ is satisfied; and (b) transient two-photon emission (backward scattering) following the pulse.

4.1 Raman Beats

The calculation [2,3] resembles the one-photon free induction decay in that there is a steady state preparative stage and a transient regime which follows Stark switching. For this three level problem the equations of motion are of the more complicated form

$$i\hbar\dot{\rho} = [H,\rho] + i\hbar\lambda + \text{damping terms}$$

$$\rho = \begin{pmatrix} \rho_{11} & \rho_{12} & \rho_{13} \\ \rho_{21} & \rho_{22} & \rho_{23} \\ \rho_{31} & \rho_{32} & \rho_{33} \end{pmatrix}$$

$$H = \begin{pmatrix} H_{11} & 0 & -\mu_{13}E \\ 0 & H_{22} & -\mu_{23}E \\ -\mu_{31}E & -\mu_{32}E & H_{33} \end{pmatrix}$$

$$\lambda = \begin{pmatrix} \lambda_{11} & 0 & 0 \\ 0 & \lambda_{22} & 0 \\ 0 & 0 & \lambda_{33} \end{pmatrix}. \tag{4.1}$$

A perturbation treatment [2] shows that the Raman beat signal should behave as

$$\langle\tilde{\rho}_{13}(t)\rangle = 2\pi\tau_2\chi_1\chi_2^2\Delta\rho^{(0)}\frac{e^{i\omega_{21}t-t/\tau_2}}{\omega_{32}-\omega_0+i(2/T_2-1/\tau_2)}. \tag{4.2}$$

The bracket again indicates a Doppler average; the damping rate associated with ρ_{12} is $\Gamma_{12}=1/\tau_2$ and with ρ_{13} is $\Gamma_{13}=\Gamma_{23}=1/T_2$; the Rabi frequencies for the $1\to3$ and $2\to3$ transitions are χ_1 and χ_2; and we have assumed in zero order that the population difference $\Delta\rho^{(0)}=\rho_{33}^{(0)}-\rho_{22}^{(0)}=\rho_{33}^{(0)}-\rho_{11}^{(0)}$. Equation (4.2) states that a signal of beat frequency ω_{21} having a decay time τ_2 should be observed in the forward direction independent of Doppler dephasing and elastic collisions and with a $\chi_1^2 \chi_2^2$ intensity dependence.

4.2 Pulse Solution

If one considers only resonant excitation with no damping and feeding terms in (4.1), exact pulse preparation solutions [3] can be obtained. Instead of solving the eight density matrix equations (4.1), it is simpler to solve the wave equations [6]

$$\dot{\tilde{c}}_1 = \frac{i\chi_1}{2} \tilde{c}_3$$

$$\dot{\tilde{c}}_2 = \frac{i\chi_2}{2} \tilde{c}_3$$

$$\dot{\tilde{c}}_3 = -i\Delta\tilde{c}_3 + \frac{i\chi_1}{2} \tilde{c}_1 + \frac{i\chi_2}{2} \tilde{c}_2 \qquad (4.3)$$

where $\tilde{c}_1 = c_1 e^{iE_1/\hbar t}$, $\tilde{c}_2 = c_2 e^{iE_2/\hbar t}$, $\tilde{c}_3 = c_3 e^{i(\Omega + E_1/\hbar)t}$, Ω is the optical frequency, levels 1 and 2 are degenerate so that $\omega_{12} = 0$, and $\Delta = \omega_{31} - \Omega$.

The resulting solutions written in terms of the Bloch-like variables, where $\rho_{ij} = c_i c_j^*$, is

$$u_{12} = \tilde{\rho}_{12} + \tilde{\rho}_{21}$$

$$iv_{12} = \tilde{\rho}_{12} - \tilde{\rho}_{21}$$

$$w_{12} = \rho_{11} - \rho_{22} \qquad (4.4)$$

are

$$u_{12} = \frac{8\chi_1\chi_2 K}{\varepsilon} (\cos\gamma t - 1)$$

$$- \left(\frac{\chi_1\chi_2}{\delta}\right)\left(\frac{\chi_1^2 - \chi_2^2}{\chi_1^2 + \chi_2^2}\right) w_{12}(0) \left[\frac{\cos(\delta - \Delta/2)t - 1}{\delta - \Delta/2} + \frac{\cos(\delta + \Delta/2)t - 1}{\delta + \Delta/2}\right]$$

$$v_{12} = \frac{\chi_1 \chi_2 w_{12}(0)}{\delta} \left[\frac{\sin(\delta - \Delta/2)t}{\delta - \Delta/2} - \frac{\sin(\delta + \Delta/2)t}{\delta + \Delta/2} \right]$$

$$w_{12}(t) = w_{12}(0) + \frac{2\chi_1^2 \chi_2^2}{\delta(\chi_1^2 + \chi_2^2)} w_{12}(0) \left[\frac{\cos(\delta - \Delta/2)t - 1}{\delta - \Delta/2} + \frac{\cos(\delta + \Delta/2)t - 1}{\delta + \Delta/2} \right]$$

$$+ \frac{K\epsilon(\chi_1^2 - \chi_2^2)}{\chi_1^2 + \chi_2^2} (\cos\gamma t - 1) . \qquad (4.5)$$

Here, we have defined $\delta = (\chi_1^2 + \chi_2^2 + \Delta^2/4)$, $\gamma = \sqrt{4(\chi_1^2 + \chi_2^2) + \Delta^2}$, $\epsilon = 2\sqrt{\chi_1^2 + \chi_2^2}$ and $K = \epsilon w(0)/(\Delta^2 + \epsilon^2)$.

These solutions which are of interest for pulse preparation apply to the Raman beat case when the resonance condition

$$\omega_{12} = 0 \qquad (4.6)$$

is satisfied and to the stimulated two-photon emission case when the resonance condition

$$\omega_{21} - \Omega_1 - \Omega_2 = 0 \qquad (4.7)$$

is satisfied.

Equations (4.5) contain terms oscillating at low frequency $(\delta - \Delta/2)$ but having a large amplitude and terms at high frequency $(\delta + \Delta/2)$ with a small amplitude. Physically, these are the Rabi side bands which have now been observed in recent two-photon NMR measurements [7] and provide a quantitative test of the correctness of Eq. (4.5).

4.3 Three Level Bloch Equations

It can be shown that when the intermediate level(s) in a two-photon transition is (are) far off resonance so that

$$\delta \sim \frac{|\Delta|}{2} + \frac{\chi_1^2 + \chi_2^2}{|\Delta|} , \qquad (4.8)$$

then the general solution Eqs. (4.5) reduce to Bloch-like form

$$u_{12}(t) = -w(0)\sin\theta\cos\theta(\cos\Gamma t-1)$$

$$v_{12}(t) = -w(0)\sin\theta\sin\Gamma t$$

$$w_{12}(t) = w(0)[1+\sin^2\theta(\cos\Gamma t-1)] \qquad (4.9)$$

where

$$\sin\theta = 2\chi_1\chi_2/(\chi_1^2+\chi_2^2)$$

$$\cos\theta = (\chi_1^2-\chi_2^2)/(\chi_1^2+\chi_2^2).$$

$$\Gamma = |\Delta|+(\chi_1^2+\chi_2^2)/|\Delta|$$

If we compare the three level case (4.9) with the two-level problem (2.1), we see that they exhibit exactly the same functional behavior.

Adiabatic solutions by Grischkowsky et al. [8] predict behavior similar to (4.9) for a two-photon process involving an arbitrary number of intermediate levels. A comparison of this calculation with the above has been given recently [9].

4.4 Experiments

Two independent groups [4,5] have verified the prediction, contained in the solution Eq. (4.5), that a simulated two-photon emission transient should be observable. The experimental configuration was that of Fig. 12 where Stark switching was employed. In the one case, measurements were made in sodium vapor using two cw visible dye lasers. The other observation was carried out in the vibrational $\nu_2=0$ to 2 transition of NH_3 using 10 micron infrared radiation from two CO_2 lasers; the pressure dependent decay time allowed the author to conclude that relaxation is dominated by the ground vibrational state rather and not by the second excited vibrational state. In addition, various NMR two-photon measurements have been performed just recently [7,10].

References to Lecture IV

1. R. L. Shoemaker and R. G. Brewer, Phys. Rev. Lett. 28, 1430 (1972).
2. R. G. Brewer and E. L. Hahn, Phys. Rev. A8, 464 (1973).
3. R. G. Brewer and E. L. Hahn, Phys. Rev. A11, 1641 (1975).
4. P. F. Liao, J. E. Bjorkholm, and J. P. Gordon, Phys. Rev. Lett. 39, 15 (1977).

5. M. M. T. Loy, Phys. Rev. Lett. $\underline{39}$, 187 (1977).
6. M. Sargent III and P. Horwitz, Phys. Rev. $\underline{A13}$, 1962 (1976).
7. D. G. Gold and E. L. Hahn, Phys. Rev. $\underline{A16}$, 324, 1977.
8. D. Grischkowsky, M. M. T. Loy and P. F. Liao, Phys. Rev. $\underline{A12}$, 2514 (1975).
9. D. Grischkowsky and R. G. Brewer, Phys. Rev. $\underline{A15}$, 1789 (1977).
10. H. Hatanaka and T. Hashi, J. Phys. Soc. Japan $\underline{39}$, 1139 (1975); H. Hatanaka, T. Ozawa and T. Hashi, J. Phys. Soc. Japan, $\underline{42}$, 2069 (1977).

LECTURE V: The Laser Frequency Switching Technique

Just one year ago, A. Z. Genack and I reported a very general
technique for observing coherent optical transients in the
visible-ultraviolet region [1]. The concept is the following.
A sample is resonantly excited under steady state conditions by
a continuous laser beam and therefore is coherently prepared.
The laser frequency is then abruptly switched to some new value
outside the sample's homogeneous linewidth. The frequency packet
within the inhomogeneous lineshape that was initially prepared
is now out of resonance with laser light and therefore, freely
radiates an intense coherent beam of light, the FID.
Simultaneously, a second packet is excited and exhibits the
nutation effect. If the laser frequency is switched twice in
two successive pulses, the photon echo will appear. In this way,
the entire class of coherent optical transients can be monitored.

The method obviously resembles Stark switching. From the
standpoint of the time-dependent interaction of the sample with
the optical wave, the two techniques are equivalent. But now
the method is not restricted to Stark tunable systems.
Furthermore, with a frequency-switched cw dye laser, the broad
tuning range permits general studies in atoms, molecules, and
solids - in a manner resembling the versatile pulsed NMR
techniques.

5.1   Qualitative Discussion

Laser amplitude and frequency modulating techniques have been
studied extensively since the development of the laser. A large
body of literature exists on the subject [2]. It is interesting,
however, that rapid electro-optic switching by means of square
wave pulses has received little attention [3], presumably because
applications involving coherent transients or other phenomena
had not been considered.

Perhaps the simplest way of shifting the laser frequency is
to move the end mirror of the optical cavity. This of course is
a steady-state argument. A dynamic argument would require that
the mirror be set into uniform motion. Then the laser beam
striking the moving mirror experiences a Doppler shift, which
incidentally occurs instantaneously.

If we replace the moving end mirror idea by a phase modulator
as in Fig. 13, the same Doppler shift can result. Imagine that
the light wave passing through the phase modulator sees a time
varying phase which is a ramp function (Fig. 13). At the onset

Laser Cavity

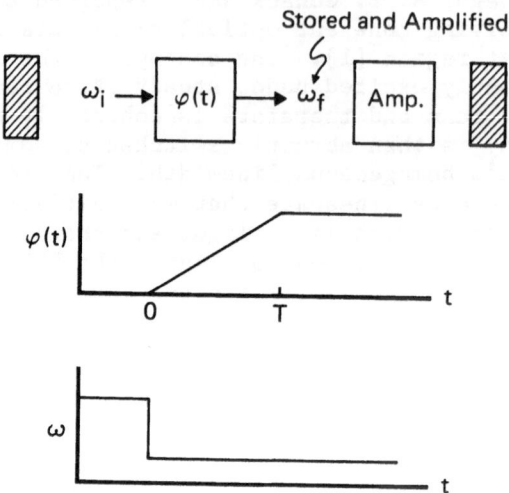

Fig. 13. Laser frequency switching technique. An electro-optic
modulator shifts the optical frequency instantaneously to a new
value when the phase-time behavior of the modulator is a simple
ramp function.

of the ramp, the light will be immediately frequency-shifted
since the instantaneous shift will be given by the time derivative
of the phase or $\Delta\omega=\dot{\phi}$. Assuming that the modulator is at one end
of the cavity and that the ramp time T is the time required for
light to travel from one end of the cavity to the other, all of
the light will be converted to the new frequency at time T.
Thereafter for t>T, since $\dot{\phi}=0$ no additional frequency shift occurs
due to multiple passage through the phase modulator. The shifted
light is now a cavity mode and will be stored and amplified in
the same way that the unshifted light was prior to switching.

5.2 Switching Theory

To elucidate these ideas further, consider a ring laser [4]
having a phase modulator, and assume (1) unit gain, (2) a cavity
round trip time T, and (3) a phase ramp which reaches its final
and maximum value ϕ_m in time T.

We first treat the steady state. The cavity frequency is given by

$$\omega_0 = \frac{\phi}{T} = n \frac{2\pi}{T} \qquad (5.1)$$

where n is an integer, $2\pi/T$ the axial mode spacing, and ϕ is the phase the light wave accumulates in time T. Thus, a phase change $\Delta\phi$ implies a frequency change

$$\Delta\omega = \Delta\phi/T . \qquad (5.2)$$

For the ramp phase function

$$\Delta\omega = \phi_m/T . \qquad (5.3)$$

For the dynamic case, the basic cavity equation

$$E(t) = E(t-T)e^{i\phi(t)} \qquad (5.4)$$

applies where the phase modulation is $\phi(t)$.

Now assume that the laser field is of the form

$$E(t) = Ae^{i(\omega_0 t+\theta(t))} , \qquad (5.5)$$

$\theta(t)$ being the resultant phase. By equating (5.4) with (5.5) and introducing (5.1), we obtain the basic phase relation

$$\theta(t) = \theta(t-T) + \phi(t) . \qquad (5.6)$$

Equation (5.6) states that the resultant phase $\phi(t)$ at time t is given by the phase at the earlier time (t-T) plus any additional phase modulation $\phi(t)$.

With the initial condition $\theta(0)=0$ and assuming a phase ramp function $\phi(t)$ where the ramp time is again T, it follows that

$$\theta(t) = \frac{\phi_m}{T} t ,$$

$$E(t) = Ae^{i(\omega_0+\phi_m/T)t} . \qquad (5.7)$$

Hence, the laser frequency is shifted instantaneously at t=0 by

$$\Delta\omega = \phi_m/T , \qquad (5.8)$$

in agreement with the steady-state result (5.3).

The conclusion that the frequency shift is instantaneous
deserves further comment. According to (5.8), the shift is not
limited by the rise time of the phase ramp T but rather by its
slope. It also appears that the cavity ringing time does not
limit the response time. On the other hand, the finite length
of the phase modulator element certainly will impose a limit.
In a typical modulator crystal, the transit time of light is
about 100 picoseconds, which is perhaps a reasonable estimate of
the switching time.

Suppose now that the rise time of the phase ramp is longer
than the cavity round trip time T. For simplicity, let the ramp
time be qT where q is an integer and let ϕ_m be the ramp's phase
excursion in time T. Successive applications of (5.6) reveal
that the total shift following the qth traversal through the
modulator will be

$$\Delta\omega = q\phi_m/T .   \qquad\qquad (5.9)$$

Hence, each round trip produces one frequency kick ϕ_m/T.

Next consider the situation where q is not an integer. For
example, for q=0.5, one half of the light will be frequency
shifted and the other half will be unshifted (the initial
frequency). The one wave chases the other as they circulate
through the cavity. These waves remain in phase with another
according to (5.6) and therefore satisfy the cavity equation
(5.4). Therefore, one concludes that this case is a stable cavity
solution.

5.3 Apparatus

The laser frequency switching apparatus shown schematically
in Fig. 14 consists of a cw dye laser, Spectra-Physics 580A or
Coherent Radiation 599. The dye is Rhodamine 6G. The laser is
phase-modulated by an electro-optic intracavity crystal, an x-cut
ammonium dideutrohydrogen phosphate (AD*P) crystal which is driven
repetitively by low voltage pulses. The frequency shift is 0.6
MHz/V. The output beam is single mode, linearly polarized and
has a power up to 100 mW, in a beam diameter of 0.5 mm. A p-i-n
photodiode monitors the forward beam which emerges from the sample
and transients are observed with a Tektronix 7904 sampling
oscilloscope. The experiment is also hard wired to a small
computer, IBM System 7, which stores the real time data in digital
form. Transfer of this data to the IBM 360 Model 195
computer permits Fourier transform and other data handling
operations.

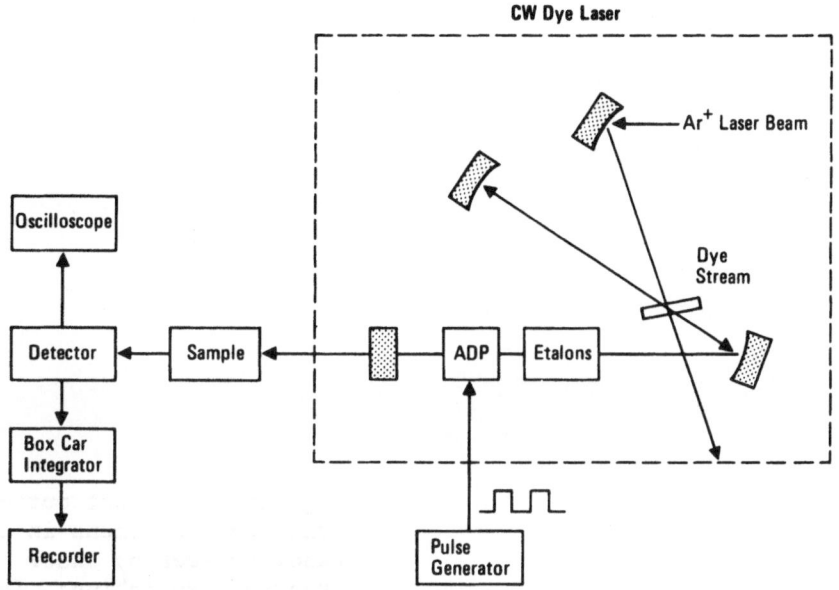

Fig. 14. Schematic of the apparatus for observing coherent
optical transients using a frequency-switched cw dye laser.

Since the experiment is controlled electronically, all the
advantages inherent in the Stark switching technique are preserved
here as well. We find, therefore, the following: (1) the only
transient observed is the desired coherent transient itself and
without undesirable background signals as occur with pulse laser
sources. (2) Heterodyne detection occurs automatically and
facilitates measuring the decay of emission signals. (3) Signal
averaging is possible because the pulse sequence is repetitive.
(4) The entire class of coherent optical transients can be
monitored since the electronic pulse sequence can be tailored to
the particular experiment of interest. Moreover, when these
features are combined with the broad tuning range available in
a dye laser, it is apparent that coherent transient phenomena
can be observed with ease in a large number of optical transitions
in various atomic, molecular and solid state systems.

A direct measurement [5] of the frequency shift has been
observed in our laboratory by monitoring the heterodyne beat
signal of the unshifted light, which is delayed compared to the
frequency switched light by means of an optical delay line. The
beat frequency was ~300 MHz.

The FID, nutation effect and photon echo are shown in Fig.
15 for a visible electronic transition of I_2, corresponding to

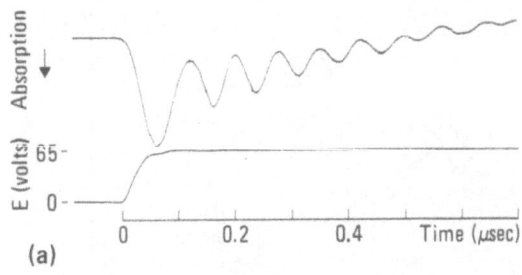

(a)

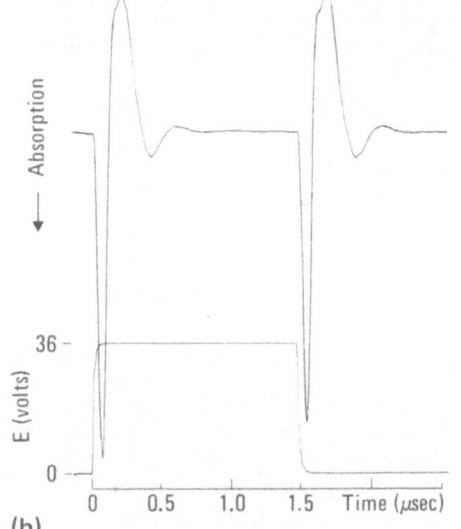

(b)

Fig. 15. Coherent optical
transient phenomena in I_2
vapor as seen by laser
frequency switching: (a)
FID, (b) two optical nutation patterns, and (c)
photon echoes observed for
three different pulse delay
times. Transition frequency
is 16956.43 cm^{-1}.

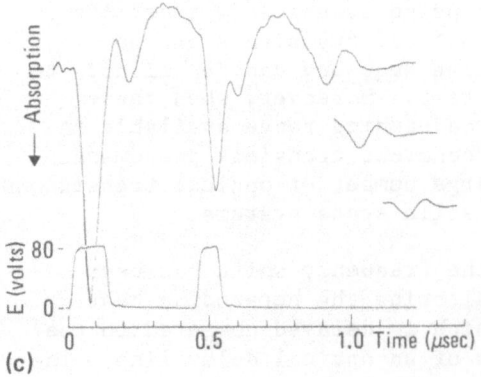

(c)

three different laser frequency switching pulse sequences [1].

Figure 16 shows a computer plot of the iodine FID and nutation signals under conditions where the two effects appear simultaneously. Here, the beat frequency is 28.11 MHz and the laser power is 7 mW. While the nutation and FID signals overlap in time, it is possible to separate them in frequency using a digital computer and the Fourier transform program mentioned in Lecture III. The spectral lineshape of the FID signal presented in Fig. 17, is a perfect Lorenztian as it clearly matches the Fourier transform of a damped cosine (solid line) [5]. The least squares fit of the linewidth determined by this method has an uncertainty of about one part in 300. Hence, these methods can be made highly quantitative.

As indicated in Eq. (2.7b) the FID decay rate is power-dependent. However, since the sensitivity of the laser frequency switching method is quite high, measurements can be performed at laser powers as low as ~50 μW where the power broadening is negligible. In that case, the FID decay rate is simply $2\Gamma=2/T_2$ where the dipole dephasing rate is

$$\Gamma = \frac{1}{2} (\gamma_1 + \gamma_2) + \gamma_\phi \ , \tag{5.10}$$

and γ_ϕ is the rate of phase-interrupting collisions.

Our studies of I_2 in the visible region support (5.10) and provide the first direct coherence measurements of elastic collisions due to phase interruptions resulting from perturber-induced frequency shifts. We note that this result contrasts with the effect of elastic velocity-changing collisions in the infrared (Lecture III). Apparently, excited electronic states are more sensitive to phase interruptions than excited vibrational states are.

Laser frequency switching has been applied in other laboratories [6,7] and to other systems, such as Na vapor [8] where the decay time is a few nanoseconds and also to impurity ion crystals [9] such as Pr^{3+} ions in a LaF_3 host at liquid helium temperature ($T_2=0.76$ μsec). In the future, it is felt that these studies will embrace decay time measurements from the 50 picosecond to the millisecond time scale.

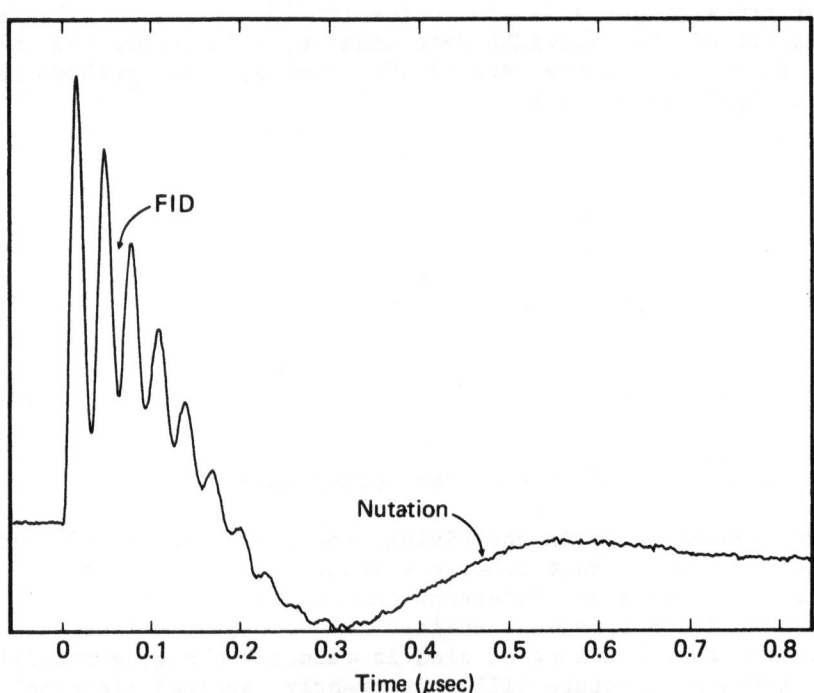

Fig. 16. Free induction decay and nutation in I_2 vapor observed under conditions where both appear simultaneously. Dye laser power is 7mW and the best frequency is 28.11 MHz. Other conditions are the same as Fig. 15.

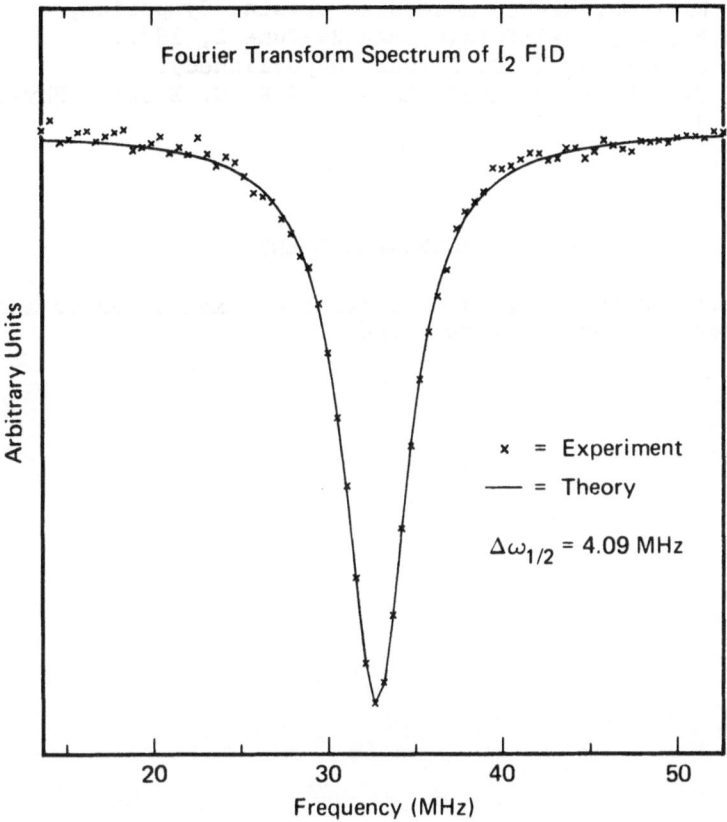

Fig. 17. Fourier transform spectrum of FID-nutation signal similar to Fig. 16 for I_2 vapor. Here, the FID portion of the spectrum is shown.

References to Lecture V

1. R. G. Brewer and A. Z. Genack, Phys. Rev. Lett. <u>36</u>, 959 (1976).
2. I. P. Kaminow, <u>An Introduction to Electrooptic Devices</u>, Academic Press, New York (1974).
3. An exception is the paper by A. Yariv, Proc. of the IEEE <u>52</u>, 719 (1964).
4. I wish to thank A. E. Siegman for bringing this model to my attention and for interesting discussions which led to several of the arguments presented in this section.
5. A. Z. Genack and R. G. Brewer (to be published).
6. A. H. Zewail, T. E. Orlowski, D. R. Dawson, Chem. Phys. Lett. <u>44</u>, 379 (1976).

7. P. A. Wiersma, Eighth Molecular Crystal Symposium,
    Santa Barbara, California, May 29-June 2, 1977.
8. A. Z. Genack and R. G. Brewer (unpublished).
9. A. Z. Genack, R. M. Macfarlane, and R. G. Brewer, Phys. Rev.
    Lett. _37_, 1078 (1976).

ACKNOWLEDGEMENT

The careful reading of this manuscript by Axel Schenzle and
Ralph DeVoe is greatly appreciated.

MEAN FIELD THEORY OF OPTICAL BISTABILITY AND RESONANCE FLUORESCENCE

L.A. Lugiato and R. Bonifacio

Istituto di Fisica dell'Università

Via Celoria,16 – Milano – Italy

INTRODUCTION

Most of the material of this notes is taken from a few papers which are enlisted in Ref. 1. Here we concentrate on a physical discussion, omitting the mathematical details. The aim of these lectures is to present and analyze a mean field model, which describes the dynamics of a system of two – level atoms driven by a coherent resonant field. On the basis of this model, we shall see that above a suitable critical density of atoms the system exhibits a bistable behaviour which concerns both the stationary situation and the transient, both the light transmitted in the forward direction and the fluorescent light. More specifically, we shall analyze several features of Optical Bistability and Resonance Fluorescence.

1) The phenomenon of Optical Bistability (OB) has been predicted mainly by McCall[2] . This prediction was based on a very complete but partially numerical analysis of the Maxwell-Bloch equations. Hence our first aim will be to give a completely analytical description of OB in its stationary aspects. In particular, this will allow us to obtain explicit analytical conditions for the observation of this phenomenon. In the framework of this description we shall analyze the role of atomic cooperation in OB, which is an absorption phenomenon. This feature will provide a direct link with Superfluorescence ,

which arises from atomic cooperation in emission
(3). Finally the theory I shall illustrate allows
making new predictions concerning the transient
behaviour of the system and in particular of the
transmitted light.

2) The usual theories of Resonance Fluorescence (RF)
(4) analyze this phenomenon as the interaction
of the incident field with a <u>single</u> atom. In doing
that, one completely neglects all cooperative effects. On the contrary our model shows that above
the critical density of atoms the cooperative effects are very important and change completely
the usual picture of RF. In fact both the total
fluorescent intensity and the spectrum of the fluorescent light show bistability, with discontinuous changes and a hysteresis cycle.

WHAT IS OPTICAL BISTABILITY

This phenomenon is clearly going to receive more and
more attention from the physicist community, especially
after it has been beautifully observed by Gibbs, McCall
and Venkatesan (5). Fig.1 plots the light E_T transmitted
by a Fabry-Perot versus the incident field amplitude E_I.
When the Fabry-Perot is empty, assuming that the resonant cavity is perfectly tuned to the frequency of the
incident field, the plot is simply the straight line at
45° (i.e. $E_T = E_I$). On the contrary when the Fabry-Perot
is filled with atomic material in such a way that the sample is optically thick ($\alpha L > 1$, α = absorption coefficient, L = length of the cavity) the transmitted light
varies discontinuously with the incident light. In fact,
for small incident field the transmission is very low,so
that almost all the incident light is reflected. E_T increases very slowly with E_I, until at a certain value
$E_I^{(+)}$ there is a sudden change of E_T, such that a large fraction of E_I is transmitted. Increasing E_I further, E_T
increases continuously approaching the empty cavity behaviour. If one conversely decreases E_I starting from values $E_I > E_I^{(+)}$, nothing spectacular happens when E_I crosses the value $E_I^{(+)}$. The transmitted light decreases continuously until E_I crosses a second discontinuity point
$E_I^{(-)} < E_I^{(+)}$, where E_T suddenly jumps downwards.Thus
the plot of E_T versus E_I exhibits a hysteresis cycle.This
behaviour is due to the fact that the two branches in
Fig. 1 - the low transmission branch and the high transmission branch - correspond to two different stable stationary states of the system, which is precisely <u>bista-
bility</u>.

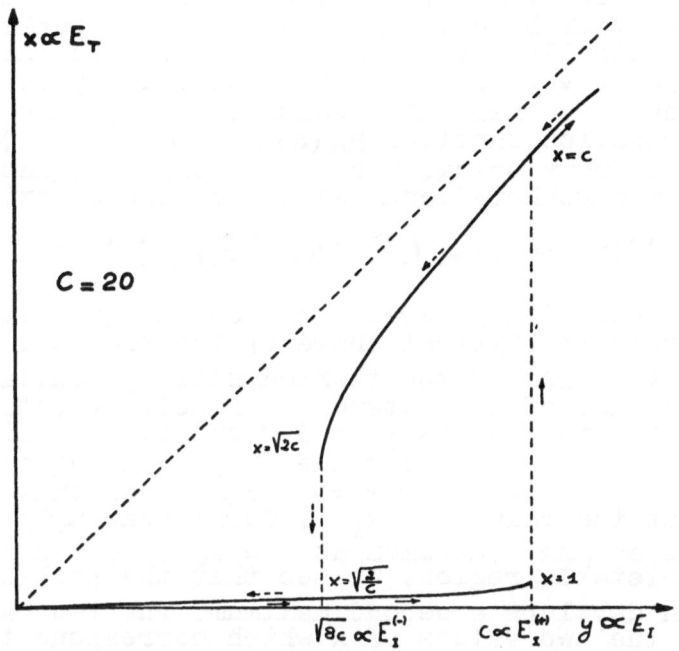

Fig. 1 – Hysteresis cycle of the transmitted light.
Full (dotted) line arrows indicate the variation obtained
increasing (decreasing) the incident field.

BISTABILITY AND FIRST-ORDER PHASE TRANSITIONS

As it is well known, hysteresis is one of the main
characteristic features of first order phase transitions.
Hence OB provides a remarkable example of first order phase
transition in an open system far from thermal equilibrium. In this connection, let us now illustrate a little
bit the general link between bistability and first order
phase transitions. I stress that the following discussion
is not restricted to OB, but refers to a general bistable
system.

Let x be the observed quantity; typically x is the
order parameter of the problem or a quantity related to
it. On the other hand let y be an external parameter,
which is controlled by the observer. Fig. 2a shows the
variation of x versus y with a hysteresis cycle. For any
given value of y there is a suitable stationary probability distribution function $P_{st}(x)$. Typically, $P_{st}(x)$
is the stationary solution of a stochastic equation as
a master or a Fokker-Planck equation. The quantity

$$V(x) = -\ln \left(P_{st}(x) / P_{st}(0) \right) \tag{1}$$

plays the role of a generalized free energy. Fig.3 shows
qualitatively the typical shape of the probability distribution $P_{st}(x)$ and of the thermodynamic potential V(x)
for some values of the external parameter y (cfr.Fig.2a).
Note that for the values y_1 and y_4 one finds a single value of x, whereas for the values y_2, y_c and y_3 one finds
a bistable situation. For $y = y_1$ $P_{st}(x)$ has only one peak
centered at the value x = x_1. Correspondingly the free
energy has a single minimum at x = x_1. Let us now enter
into the bistable region. We see that the probability distribution develops a second maximum. The two maxima coincide with the two values of x which correspond to the given value of y. For y = y_2 the second maximum is dominated by the first one. However, increasing the external
parameter the second maximum grows. For $y = y_c$ the two maxima have the same height, whereas for y = y_3 the second
maximum dominates the first one. Finally for y = y_4 we
are out of the bistable region, the first maximum has di-
sappeared and we are left again with a one-peaked distribution function.

In the bistable situation the absolutely stable state corresponds to the absolute minimum of the free energy,
or equivalently to the absolute maximum of the probability distribution. Therefore if we plot the absolutely
stable value x_{st} of x versus the external parameter y we
find the behaviour of Fig. 2b, which shows the first order phase transition. The critical value y_c, at which
the transition occurs, is defined as that value of y in
correspondence of which the free energy has two minima
of equal height (cfr. Fig.3). Clearly Fig.2b is very different from Fig. 2a. However, Fig.2a still retains a well
defined physical meaning. In fact the relative minimum
of the free energy corresponds to a metastable state.
Therefore the part of Fig.2a which does not overlap Fig.2b
is formed by metastable states.

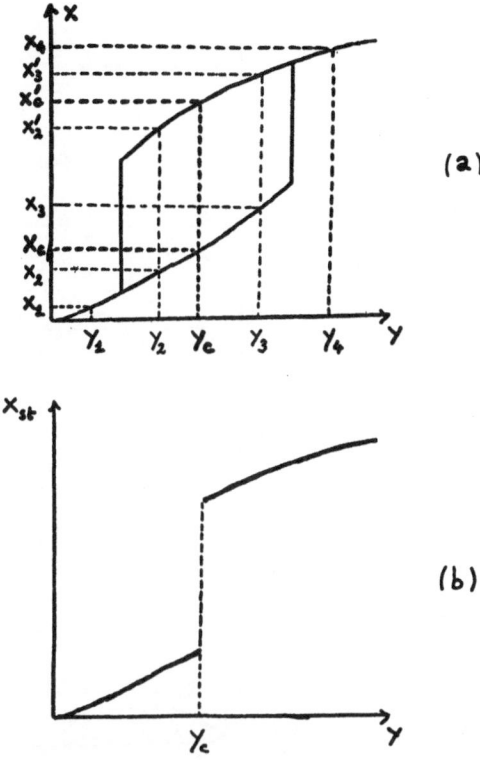

Fig. 2 - (a) Hysteresis cycle of the observed quantity x versus the external parameter y. (b) Absolutely
stable value of x versus y.

Let us briefly illustrate why the relative minimum
corresponds to a metastable state. The key point is that
the problem is similar to that of a brownian particle moving in the potential profile $V(x)$. In the bistable situation, this profile has two wells separated by a barrier. Now let us assume that the particle is in the relative minimum. This situation is clearly stable against
small fluctuations. However some unusually large fluctuation can make the particle overcome the barrier jumping
to the absolute minimum, or - using the language of quantum mechanics - a large fluctuation can make the particle tunnel the barrier. Therefore the relative minimum

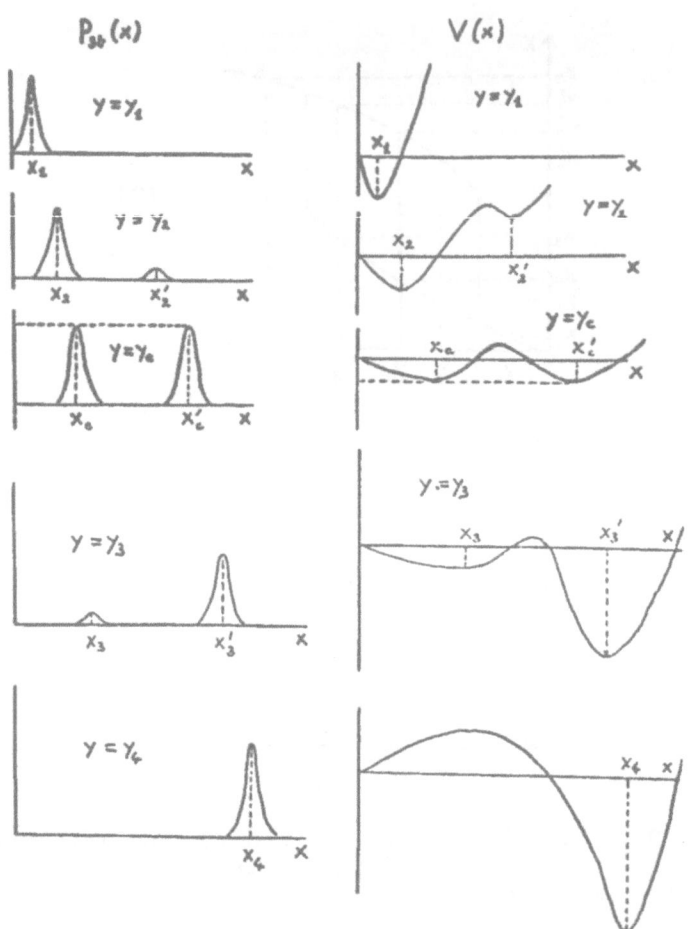

Fig. 3 – Qualitative shape of the stationary probability distribution $P_{st}(x)$ and of the generalized free
energy $V(x)$ for the values of y indicated in Fig. 2a.

does not correspond to a state with an infinite lifetime,
but to a long-lived state, i.e. to a metastable state.
These states can be experimentally observed by using a
suitable preparation. E.g. to observe the metastable state x_3 one can start with the system in the stable state
x_1 corresponding to y_1 and then one must change y rapidly to the value y_3 . In this way the system remains
temporarily trapped in the metastable state x_3. Following this procedure one can in principle experimentally
observe the hysteresis cycle.

I stress that the following treatment of OB is semiclassical, so that we shall not deal with large fluctuations. Therefore we shall never distinguish between stable and metastable states, i.e. we shall use the term
"stable" both for absolutely stable and for metastable
states. Accordingly, we shall always deal with drawings
of the kind of Fig.2a (i.e. drawings with hysteresis cycles) and we shall never deal with drawings of the type
of Fig.2b.

THE MEAN FIELD MODEL

We consider a coherent monochromatic field E_I of
frequency ω_0 which is injected into a resonant cavity
of length L, of volume V and with mirrors of transmittivity T (see Fig. 4). The Fabry-Perot is filled with N
atoms, which are initially in the ground state. Once entered into the cavity, E_I interacts with the atoms giving rise to a selfconsistent internal field E. The net
result of this interaction is described by the field
transmitted in the forward direction and by the field
reflected backwards.

We put ourselves in the simplest situation:
i)   the atoms are described as two-level systems,
ii) we consider the case of a homogeneously broadened
     atomic system, with zero detuning between the atomic
     transition frequency and the frequency ω_0 of the in-
     cident field,
iii) we assume zero cavity mistuning, i.e. that $L = n \lambda_0$
     with λ_0 = wavelength, n = integer number.

Note that by assumption ii) in our problem only absorption will play a role, while dispersion will not play
any role. I stress in this connection that in the experiments of Ref.5 dispersion is dominant over absorption.
Also the presence of a nonzero mistuning is essential
for the interpretation of these experiments. However we
shall consider the purely absorbitive case because it illustrates in the simplest way the main features of the
phenomenon.

Our mean field model is deduced from the Maxwell-
Bloch equations with two coherently coupled directions
of propagation by using the mean field approximation and
taking into account the proper boundary conditions for
the propagating field.

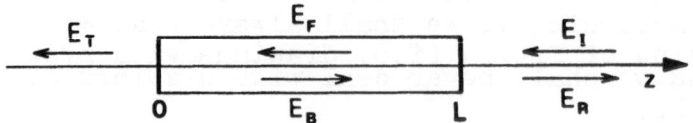

Fig. 4 – E_I is the incident field, E_T and E_R the transmitted and reflected fields respectively.

We call "mean field" the models which are entirely expressed in terms of space averages of the fields. In other words, one gives up describing the spatial variation of the fields inside the sample, but on the other hand one has the advantage of dealing with a much simpler model. Of course one cannot deduce exactly from the Maxwell-Bloch equations closed time evolution equations for the space averages. Therefore a mean field theory is necessarily an approximate one. However, in many cases the mean field approximation turns out to be good. This holds whenever the scale which characterizes the spatial variation of the fields inside the sample has the same order of magnitude of the length of the sample itself.

The deduction of the model is shown in ref.1. It consists in three coupled time evolution equations for the quantities S, Δ and E, which are respectively the space average of the polarization, of the difference between the populations of the lower and the upper level, of the total internal field:

$$\dot{S} = \frac{\mu}{\hbar} E \Delta - \gamma_\perp S \quad , \tag{2a}$$

$$\dot{\Delta} = - \frac{\mu}{\hbar} E S - \gamma_\| \left(\Delta - \frac{N}{2} \right) \quad , \tag{2b}$$

$$\dot{E} = - g S - \kappa \left(E - E_I / \sqrt{T} \right) \quad . \tag{2c}$$

μ is the modulus of the dipole moment of the a-toms; g is a suitable coupling constant in dipole approximation:

$$g = \left(4 \pi \omega_0 \mu \right) / V \tag{3}$$

κ is the width of the modes of the empty Fabry-Perot

$$\kappa = \left(c T / 2 L \right) \quad ; \tag{4}$$

$\gamma_\|$ and $\gamma_\perp$ are respectively the inverse of the atomic relaxation times T_1 , T_2 . Eqs (2) are supplemented by two relations which give the transmitted and the reflected field

$$E_T = \sqrt{T} \, E \quad , \tag{5a}$$

$$E_R = E_T - E_I \quad . \tag{5b}$$

REMARKS ON THE MODEL

1) Clearly (2) is a generalization of the well known one-mode Laser model [6] to take into account the effect of the external field. In fact the one - mode Laser model (for a ring Laser cavity) in the semiclassical approximation is immediately obtained from (2) by putting $E_I = 0$ and replacing $-N/2$ by a positive inversion. Furthermore, (2) is also a generalization of the mean field model for Superfluorescence [3] , again in the semiclassical approximation. In fact, one obtains the latter model from (2) by putting $E_I = 0$ in eq.(2c) and $T = 1$ in definition (4) (because in Superfluorescence the cavity has no mirrors).

2) For an empty Fabry-Perot, we must drop the term $g S$ in eq. (2c), so that the stationary solution is $E = E_I / \sqrt{T}$, i.e. by (5a) $E_T = E_I$. On the other hand when the Fabry-Perot is filled with resonant atomic material the total internal field E is no longer equal to the incident field, because there is also a reaction field emitted by the atoms themselves. This reaction

field is very important because it couples together the
atoms giving rise to atomic cooperation. In fact radiation reaction is the very cause of the cooperative behaviour both in Superfluorescence and in OB. The reaction field is defined as

$$E_{react} = E - E_I/\sqrt{T} .\qquad\qquad(6)$$

From (5a,b) we obtain immediately

$$E_{react} = (E_T - E_I)/\sqrt{T} = E_R/\sqrt{T} .\qquad\qquad(7)$$

On the other hand, from the field equation (2c) and
from (6) we have that in the stationary situation

$$E_{react} = - (g/\kappa) S .\qquad\qquad(8)$$

Finally from eqs.(7), (8) we obtain the relation

$$E_R = - (g\sqrt{T}/\kappa) S ,\qquad\qquad(9)$$

which links reflection with absorption.

3) Taking the square of eq.(5b) one obtains

$$E_I^2 = E_T^2 + E_R^2 - 2 E_T E_R ,\qquad\qquad(10)$$

which is the energy balance equation. In fact, E_I^2 is
proportional to the incoming energy per unit time,
$E_T^2 + E_R^2$ is likewise proportional to the energy outgoing per unit time as transmitted or reflected light.
Finally using eqs.(5a) and (9) one has from eq. (2b) that
$- 2 E_T E_R$ is proportional to $+ g_{\parallel} N_2$ (N_2 = population
of the upper level), i.e. to the energy outgoing per u-
nit time as fluorescent light.

4) Let us for a moment neglect the reaction field, i.e.
let us put $E_{react} = 0$, which is equivalent to put $E = E_I/\sqrt{T}$.
Then our atomic equations (2a,b) reduce to the closed
system of equations for S, Δ:

$$\dot{S} = \Omega_I \Delta - g_{\perp} S ,\qquad\qquad(11a)$$

$$\dot{\Delta} = - \Omega_I S - g_{\parallel} (\Delta - \frac{N}{2}) ,\qquad\qquad(11b)$$

where Ω_I is the Rabi frequency of the incident field

$$\Omega_I = (\mu E_I)/(\hbar\sqrt{T}) .\qquad\qquad(12)$$

Note that (11a,b) are <u>linear</u> equations. These are precisely the equations which are studied in the usual one-atom theories of RF. (in the case of exact resonance and for $T = 1$) [4]. On the contrary in our model we keep the reaction field, so that we take into account the cooperative effects arising from radiation reaction via the <u>nonlinear</u> terms of our atomic equations (2a,b).

5) One of the main advantages of the mean field models is that one can easily give a fully quantum mechanical formulation of these models. This is true also for the present model. This feature is very important because by this quantum mechanical version one can study in a complete and rigorous way the fluctuations of the system. As we shall see, the analysis of these fluctuations is necessary in order to obtain the spectrum of the fluorescent light. Since I shall not use this quantum mechanical formulation in the present lectures, I shall not illustrate it in detail. I only mention that it is again a generalization of the one-mode Laser model of Haken and coworkers, of course in its fully quantum mechanical version [6a].

STATIONARY BEHAVIOUR: THE STATE EQUATION

We want now to deduce from the model (2) all the possible physical consequences. The first thing that we do is of course to analyze the stationary behaviour of the system. To this aim, we put $\dot{S} = \dot{\Delta} = \dot{E} = 0$ in eqs. (2) and solve them in order to find the stationary solutions. Then our atomic equations (2 a,b) with eq.(5a) give us immediately the expressions of S , Δ in terms of E_T:

$$S = ((N\mu)/(2\hbar\delta_\perp\sqrt{T}))\left(E_T/\left(1 + \frac{\mu^2 E_T^2}{\hbar^2\delta_\perp\delta_\parallel T}\right)\right) , \quad (13a)$$

$$\Delta = \frac{N}{2}\left(1 + (\mu^2 E_T^2)/(\hbar^2\delta_\perp\delta_\parallel T)\right)^{-1} . \quad (13b)$$

On the other hand the field equation (2 c) gives

$$E_I = E_T + (g\sqrt{T}/\kappa) S . \quad (14)$$

Substituting eq. (13a) into eq. (14) we obtain the <u>state equation</u>, which for a given value of E_I fixes the stationary values of E_T:

$$E_I = E_T + \left(2C E_T/\left(1 + \frac{\mu^2 E_T^2}{\hbar^2\delta_\perp\delta_\parallel T}\right)\right) , \quad (15)$$

where

$$C = \gamma_R / 2\delta_\perp \quad , \quad \gamma_R = (\gamma \frac{N}{V} L \lambda_0^2) / (8\pi T) \quad , \quad (16)$$

γ being the natural linewidth of the atoms. C is the ratio between a cooperative and a noncooperative constant: in fact, for $T = 1$ (cavity without mirrors) γ_R is just the inverse of the characteristic time τ_R of Superfluorescence [3]. According to the mean field theory of Superfluorescence the time duration of the hyperbolic secant pulses is of the order of τ_R, so that we can term γ_R "cooperative linewidth". Taking into account the relation which links $\delta_R, \delta_\perp$ and the absorption coefficient α [7] one has the useful expression

$$C = (\alpha L) / 2T \quad . \quad (17)$$

From (16) one sees that the parameter C is controlled by the observer by varying the atomic density N/V, the length L of the cavity and the transmittivity T of the mirrors.

Let us consider the nonlinear term of our state equation (15). As we see from eqs.(14) and (8), it arises from radiation reaction. Hence this term describes the effect of atomic cooperation, as it is clear also from the fact that it is proportional to the cooperative linewidth γ_R. This nonlinear term is absent in the one-atom theories of RF, in which one neglects radiation reaction, so that the stationary solution coincides with the empty cavity solution $E_T = E_I$.

In order to simplify notations it is convenient to introduce the two adimensional parameters

$$y = (\mu E_I) / (\hbar \sqrt{\delta_\parallel \delta_\perp T}) \quad , \quad x = (\mu E_T) / (\hbar \sqrt{\delta_\parallel \delta_\perp T}) \quad , \quad (18)$$

which are respectively the saturation parameters in the incident and in the transmitted field. Hence eqs. (13) become

$$S = \frac{N}{2} \sqrt{\frac{\delta_\parallel}{\delta_\perp}} \frac{x}{1+x^2} \quad , \quad \Delta = \frac{N}{2} \frac{1}{1+x^2} \quad , \quad (19)$$

while the state equation takes the form

$$y = x + \frac{2Cx}{1+x^2} \quad . \quad (20)$$

The nonlinear term has a typical saturable absorber structure [8] which stems from the dependence of the polarization S on the saturation parameter x. In the saturation regime ($x \gg 1$) the nonlinear term is negligible and one recovers the stationary solution of the one-atom theory $x = y$ (i.e. $E_T = E_I$). On the contrary in the unsaturated regime the nonlinear term is relevant and can completely change the physical picture. E.g. if one substitutes the "one-atom" solution $x = y$ into eq. (19), one sees that in this case polarization and population difference are extensive quantities. This is no longer the case when the nonlinear term is relevant. In fact, since $C \propto N$ (cfr. eq.(16)) the solution x of eq.(20) depends on N, so that by eq. (19) S and Δ are no longer simply proportional to N.

From eq. (15) we see that the state equation gives E_I as a function of E_T. This is just the inverse of the function that we want to find. Note that to invert the function $y(x)$ one must solve a cubic algebraic equation, so that the inverse function can be multivalued, which means bistability. To see this point in detail let us analyze the function $y(x)$. It has a qualitatively different shape according to whether $C < 4$ or $C > 4$. In other words, there is a critical value of C ($C = 4$), or equivalently a critical value of the density (since $C \propto N/V$), which separates two physically distinct situations. In fact for $C < 4$ y is a monotonic function of x, so that the inverse function is singlevalued (see Fig.5). In this situation the transmitted light varies continuously with the incident light. The interesting feature is that for $x > 1$ the differential gain dE_T/dE_I is larger than 1 (i.e. $dy/dx < 1$). This means that a variation of E_I induces a larger variation of E_T. So in this conditions our device behaves as an optical transistor. We shall not dwell on this point, which has been extensively illustrated in the lectures of Gibbs and McCall [9].

The most interesting situation occurs when $C > 4$, because in this case the function $y(x)$ has a maximum x_M and a minimum $x_m > x_M$ (see Fig.5). Hence if $y_m < y < y_M$ our state equation (20) has three stationary solutions $x_1 < x_2 < x_3$. The linear stability analysis shows that solutions x_1 and x_3 are stable, whereas solution x_2 is unstable. In other words, the points of the plot $y(x)$ which lie on the part with negative slope correspond to unstable states. The origin of this instability can be understood on the basis of eqs. (19) and (20). In fact from eq.(19) we have that S first increases with x, has a maximum at $x = 1$ with the value $S_{max} = (N/4)(\delta_\parallel/\delta_\perp)^{1/2}$

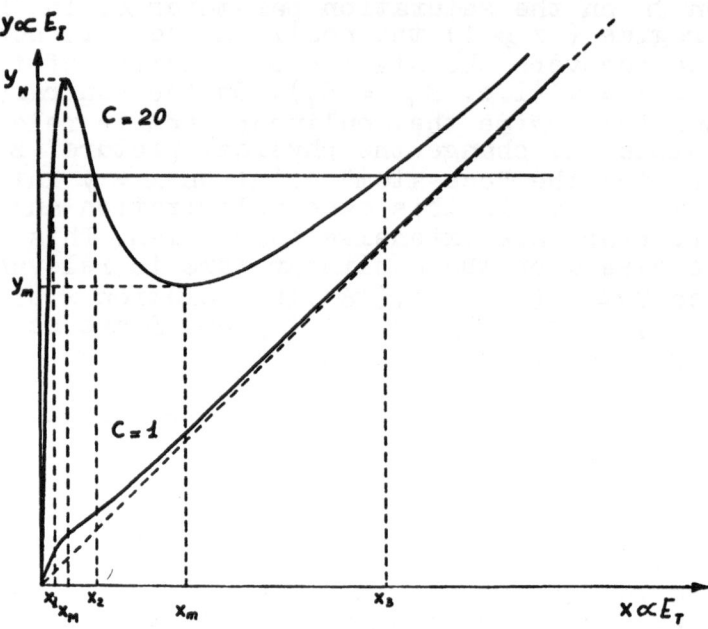

Fig.5 – Plot of the function $y = x + \dfrac{2Cx}{1+x^2}$ for $C = 1$ and $C = 20$

and finally decreases to zero. Hence for $x > 1$ an increase of x causes a decrease of S. Such a decrease causes in turn a further increase of x, as one sees rewriting eq. (20) as it follows:

$$x = y - (C/S_{max}) S .\qquad(21)$$

Clearly this runaway behaviour can originate the instability.

The stable stationary solution x_3 is near to the solution of the one-atom theory $x = y$, so that we call it "one-atom stationary state". On the other hand the

stationary solution x_1 arises from atomic cooperation,
so that we term it "cooperative stationary state". So
for $C > 4$ the system is bistable, and it is easy to see
how the hysteresis cycle arises. In fact, one must simply
exchange the axes x and y in Fig.5, in order to have a
plot of transmitted light versus incident light (remember that $x \propto E_T$, $y \propto E_I$). In such a way one obtains
a graph which is strongly similar to the numerical plot
given in Ref.2. This graph displays a negative resistan-
ce part corresponding to the unstable states x_2. For
small E_I the system is in the cooperative stationary
state x_1. Increasing E_I, E_T increases continuously
following x_1 until at $y = y_M$ the system is forced to a
discontinuous transition to the one-atom stationary state.
If we conversely start from values $y > y_M$, the system
is in the one-atom stationary state x_3. Decreasing E_I,
E_T decreases continuously following x_3 until at $y = y_m$
the system is forced to discontinuously jump to the one-
atom stationary state. In this way one gets Fig.1.
In the following, we shall always consider the case
$C \gg 1$, in which one finds simple analytical expressions
for all the quantities, as the following examples show.

1) One finds easily from eq. (20) that for $C \gg 1$

$$X_M \simeq 1 \ , \ Y_M \simeq C \ , \ X_m \simeq \sqrt{2C} \ , \ Y_m \simeq \sqrt{8C} \qquad (22)$$

In particular, the discontinuity points $E_I = E_I^{(+)}$
and $E_I = E_I^{(-)}$ correspond to $y = C$ and $y = \sqrt{8C}$ respectively. Hence using the relation

$$\Omega_I = \sqrt{\delta_\perp \delta_\parallel} \ y \ , \qquad (23)$$

which follows from (12) and (18), one can find the values
of the Rabi frequency of the external field which correspond to the discontinuity points of the hysteresis cycle.
Namely by means of (16) one gets that in correspondence
to the upper discontinuity point $E_I^{(+)}$

$$\Omega_I = \frac{\delta_R}{2} \sqrt{\frac{\delta_\parallel}{\delta_\perp}} \ , \qquad (24a)$$

while in correspondence to the lower discontinuity point
$E_I^{(-)}$

$$\Omega_I = 2 \sqrt{\delta_R \delta_\perp} \ . \qquad (24b)$$

2) For $C \gg 1$ when the system is in the cooperative
stationary state the transmitted light amplitude is inversely proportional to the number of atoms. On the other
hand when the system is in the one-atom stationary state

for $E \gtrsim E_I^{(+)}$ (i.e. $y \gtrsim C$) one has that $E_T \simeq E_I$ independently of the number of atoms.

3) As one sees from Fig.1, for $C \gg 1$ the jump in the transmitted amplitude is of a factor C at both discontinuity points.

BISTABILITY IN THE TOTAL FLUORESCENT INTENSITY

Clearly when the system undergoes a discontinuous jump all the physical quantities, which are function of the saturation parameter x, change discontinuously. E.g. the total fluorescent intensity I_F is proportional to the population of the upper level N_2 :

$$I_F \propto N_2 = N/2 - \Delta = N/2 \; (x^2/(1+x^2)) \; , \qquad (25)$$

where we have used eq.(19). Hence also I_F exhibits a hysteresis cycle, as it is shown in Fig.6 . When the system is in the one-atom stationary state I_F is proportional to N . On the contrary when the system is in the cooperative stationary state I_F is no longer an extensive quantity, because it turns out that it is inversely proportional to the number of atoms. This cooperative effect is in a sense the opposite of what occurs in Superfluorescence, in which atomic cooperation gives rise to a radiation emission proportional to N^2 .

Fig. 7 shows the hysteresis cycle of the polarization (in S_{max}/C units) versus the incident field E_I . When the system is in the one-atom stationary state S is proportional to the number of atoms. On the other hand when the system is in the cooperative stationary state the polarization is independent of N.

PHYSICAL INTERPRETATION

We have seen that when the system is in the cooperative stationary state $E_T \propto N^{-1}$, $I_F \propto N^{-1}$ and $S \propto N^0$. The physical interpretation of this cooperative behaviour in the state x_1 is the following one. In the bistable situation, the system has two possibilities. The first possibility is that the incident field interacts with the single atoms separately, as one has in the one-atom stationary state. In other words, the interaction of E_I with the single atoms is unaffected by the presence of the other atoms. In this situation, since the saturation parameter y of the incident field is $\gg 1$, the atoms are saturated so that the medium is practically transparent. The second

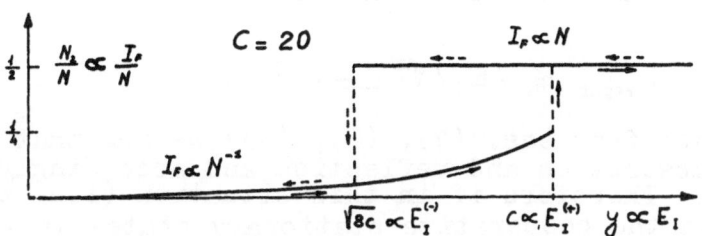

Fig. 6 – Hysteresis cycle of the fluorescence intensity per atom. Arrow convention as in Fig.1

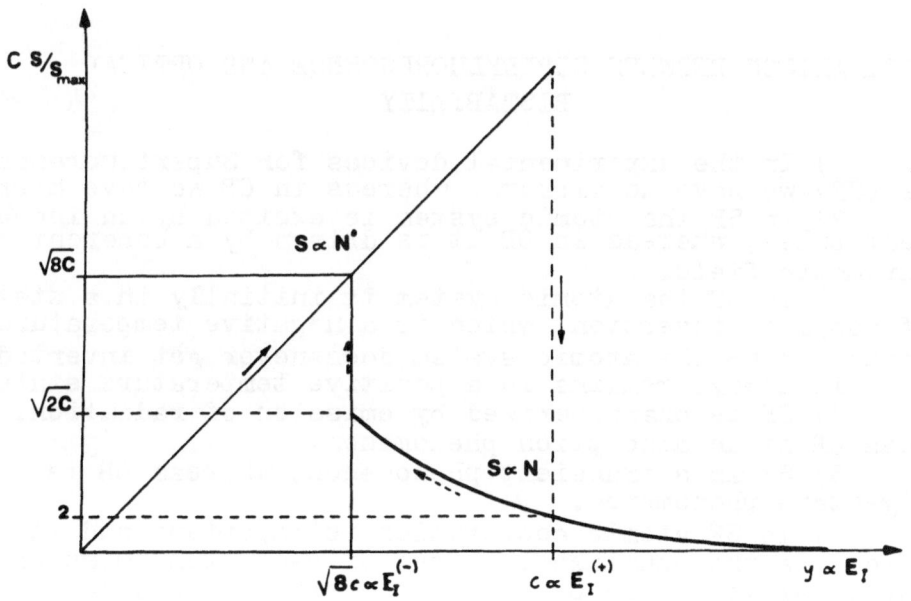

Fig. 7 – Hysteresis cycle of the polarization

possibility is that the incident field interacts with the atomic system as a whole, as it occurs in the cooperative stationary state. In this case the atoms cooperatively create a reaction field which counteracts the incident field:

$$E_{react} \simeq - E_I / \sqrt{T} \Longrightarrow E \simeq 0 \qquad (26)$$

Hence from eqs. (7), (8), (5a) we see that there is strong absorption and reflection and accordingly low transmission. Therefore if in this situation (i.e. with the system in the cooperative stationary state) we increase N keeping E_I fixed we have that the incident field is more and more completely absorbed (reflected). This explains why for $C \gg 1$, which corresponds to the limit of large number of atoms, S is independent of N. In fact in this conditions the system is in a regime of practically complete absorption, so that the polarization, which measures the absorption, does not depend on the number of atoms. Simultaneously the transmitted and fluorescent light tend to vanish, and in fact we have seen that they vanish as N^{-1} .

COMPARISON BETWEEN SUPERFLUORESCENCE AND OPTICAL BISTABILITY

1) In the experimental devices for Superfluorescence (SF) we have no mirrors, whereas in OB we have mirrors.
2) In SF the atomic system is excited by an incoherent pulse, whereas in OB it is driven by a coherent monochromatic field.
3) In SF the atomic system is initially in a state of complete inversion, which is a negative temperature state. In OB the atomic system does never get inverted, so that it always remains in a positive temperature state.
4) SF is characterized by emission of radiation, whereas OB is an absorption phenomenon.
5) SF is a transient phenomenon, whereas OB is a steady-state phenomenon.
6) In SF atomic cooperation is important and its main effect is the spontaneous onset of radiation emission proportional to N^2. In OB atomic cooperation is also relevant and its main effect is now the very appearance of the bistable behaviour.
7) A necessary (but not sufficient) condition for the observation of SF is that [3]

$$\delta_R \gg \delta_\perp \qquad (27)$$

On the other hand, we have seen that the condition for
the observation of OB is that $C > 4$, i.e. by (16)

$$\gamma_R > 8 \gamma_\perp \qquad\qquad (28)$$

Clearly conditions (27) and (28) coincide. Since $\gamma_R \propto N/V$,
condition (28) fixes a threshold value for the atomic
density which is necessary to reach in order to observe
both phenomena . The physical interpretation of this threshold is that the number of atoms must be high enough so
that atomic cooperation can overcome the incoherent one-
atom processes.

TRANSIENT BEHAVIOUR

Let us assume that initially the state of the system differs only slightly from a suitable stable stationary state. Then the approach to the stationary situation is described by the system of equations obtained
by linearizing ourmodel equations (2) around the stationary state. A situation of this kind can be experimentally created as it follows. Let us assume that the system
is initially in a stable stationary state corresponding
to some value E_I of the incident field. If we suddenly
change E_I into $E_I + \delta E_I$, with $|\delta E_I| \ll E_I$, the sys-
tem will approach the new slightly different stationary
state corresponding to the value $E_I + \delta E_I$ of the external
field. This approach is described by a solution of our linearized equations and can be experimentally observed by
looking at the transient behaviour of the transmitted light.

The solutions of the linearized equations have an exponential behaviour. In fact, since eqs. (2) are three equations, the solutions of the linearized equations are linear combinations of three exponentials $exp(-\lambda_i t)$,
i = 1,2,3. Since the decay constants λ_i are well separated,
the approach to the stationary situation will be mainly
characterized by that of the three exponentials which decays the most slowly. We call $exp(-\lambda t)$ this exponential.

Now let us summarize the main results of our analysis
of the linearized equations. As usually, these results are
obtained for $C \gg 1$.

1) The approach to the cooperative stationary state
is monotonic and is ruled by the real damping constant

$$\lambda \propto \sqrt{C^2 - y^2} \qquad (29)$$

Here the interesting feature is that $\lambda \longrightarrow 0$ in correspondence to the upper discontinuity point of the hysteresis cycle (remember that for $C \gg 1$ this point is just $y = C$). This means that in the neighbourhood of this point the approach to the cooperative stationary state becomes slower and slower, i.e. there is a critical slowing down [10].

2) Let us now cross the discontinuity point $y = C$ and jump to the one-atom stationary state. The damping constant λ changes discontinuously. This change is different according to whether $\kappa \gg \gamma_\perp, \gamma_\parallel$ or $\kappa \ll \gamma_\perp, \gamma_\parallel$.

a) When $\kappa \gg \gamma_\perp, \gamma_\parallel$ the approach to the one-atom stationary state for $y \gtrsim C$ is oscillatory and is ruled by the complex damping constants

$$\lambda = (\gamma_\perp + \gamma_\parallel)/2 \pm i\,\Omega_I \qquad (30)$$

whose real part is the mean value of the atomic relaxation rates $\gamma_\perp, \gamma_\parallel$ while the imaginary part is the Rabi frequency of the external field.

b) When $\kappa \ll \gamma_\perp, \gamma_\parallel$ the approach to the one-atom stationary state for $y \gtrsim C$ is monotonic and is ruled by a damping constant which is equal to the cavity damping constant:

$$\lambda = \kappa. \qquad (31)$$

3) Let us now decrease the external field to reach the lower discontinuity point of the hysteresis cycle $y = \sqrt{8C}$. The approach to the one-atom stationary state in the neighbourhood of this discontinuity point is always monotonic and is ruled by the real damping constant

$$\lambda \propto \sqrt{y^2 - 8C}. \qquad (32)$$

Hence we see that there is a critical slowing down also in correspondence of the lower discontinuity point.

These results on the transient behaviour are graphically illustrated in Figs. 8,9, which show qualitatively the hysteresis cycle of the damping constant λ. In these plots we consider the case $\gamma_\perp = \gamma_\parallel$ (we call $\bar{\gamma}$ the common value of $\gamma_\perp, \gamma_\parallel$) and we assume that $\kappa \gg \bar{\gamma}$.

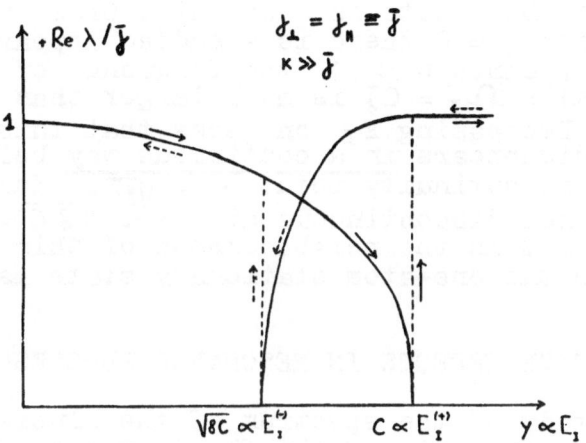

Fig.8 – Qualitative behaviour of the real part of the damping constant λ. Arrow convention as in Fig.1.

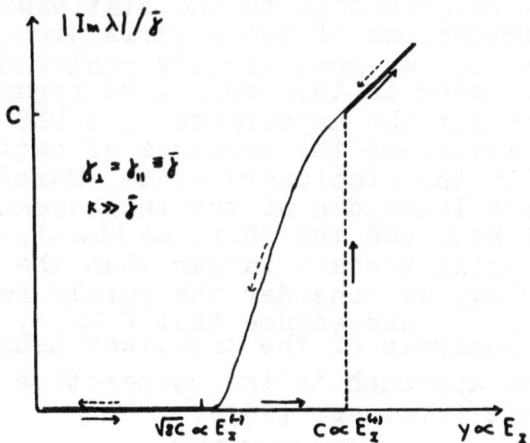

Fig.9 – Qualitative behaviour of the real part of the damping constant λ.

The real part of λ exhibits a "crossed" hysteresis cycle with the critical slowing down at the discontinuity points. The imaginary part of λ is zero as long as the system is in the cooperative stationary state because the approach to this state is monotonic. Crossing the discontinuity point $y = C$ there is a sudden appearence of oscillations. Since $C \gg 1$ the frequency of these oscillations $|\mathrm{Im}\,\lambda| = \Omega_I = C\bar{J}$ is much larger than their damping $\mathrm{Re}\,\lambda = \bar{J}$. Decreasing E_I one sees that this oscillatory character disappears in a continuous way before we reach the lower discontinuity point $y = \sqrt{8C}$. In other words $\mathrm{Im}\,\lambda$ is not discontinuous at $y = \sqrt{8C}$. In fact we have seen that in the neighbourhood of this point the approach to the one-atom stationary state is monotonic.

COOPERATIVE EFFECTS IN RESONANCE FLUORESCENCE

The study of the spectrum of the fluorescent light requires the analysis of the fluctuations of the polarization in the stationary states. This analysis can be rigorously made only in the framework of a fully quantum mechanical theory. However, by means of the regression hypothesis we can still use our semiclassical treatment to get some informations concerning the spectrum. In fact the regression hypothesis assumes that the transient approach of the polarization to the stationary state reproduces the fluctuations of the polarization in the stationary state. But we have already analyzed the transient approach [44]. Hence in this spirit we regard as a necessary condition for the appearance of a DSS (i.e. of a three-peaked spectrum) the presence of oscillations in the approach to the stationary state. More specifically, we evaluate the linewidth of the incoherent part of the spectrum as $2\,\mathrm{Re}\,\lambda$ and the shift as $|\mathrm{Im}\,\lambda|$. The DSS appears when the shift becomes larger than the linewidth. In the following, we consider the purely radiative case $\gamma_{\parallel} = 2\gamma_{\perp} = \gamma$ and assume that $C \gg 1$, $\kappa \gg \gamma$. In our previous analysis of the transient behaviour we have found that the approach to the cooperative stationary state is monotonic (cfr. eq. (29)):

$$\lambda \propto \sqrt{\gamma_R^2 - 2\,\Omega_I^2} \quad . \tag{33}$$

Therefore when the system is in the cooperative stationary state the spectrum is one-peaked. Moreover $\lambda \to 0$ in correspondence to the upper discontinuity point of the hysteresis cycle $\Omega_I = \gamma_R/\sqrt{2}$ (cfr. eq. (24a)). This means that approaching this point the spectrum becomes narrower and narrower. This cooperative line narrowing is clearly

a consequence of the critical slowing down we found previously. Let us now jump to the one-atom stationary state. We have seen that the approach to this state is o-
scillatory (cfr.eq.(30)). Hence when the system is in the
one-atom stationary state for $y \gtrsim C$ there is a DSS,which
coincides with that predicted by the one-atom theory.No-
te from eq. (30) that in the purely radiative case the
real part of λ is $3/4$, reproducing exactly the width
of the sidebands predicted by the one-atom theory. The
situation is summarized in Fig.10, which compares the physical picture in the one-atom case ($\gamma_R < \gamma$) with the picture obtained when atomic cooperation is dominant $(\gamma_R \gg \gamma)$.
In the first case one recovers the usual picture of RF:
the two sidebands emerge continuously from the central
line when $\Omega_I \sim \gamma$ in a kind of second order phase
transition. On the other hand in the cooperative case one
finds two major differences: 1) the DSS appears only when
$\Omega_I \sim \gamma_R$ and 2) it appears discontinuously in a kind
of first order phase transition. In fact the spectrum remains one-peaked as long as the system remains in the cooperative stationary state and there is a line narrowing
approaching the upper discontinuity point $\Omega_I = \gamma_R / \sqrt{2}$.
Only crossing this point there is a sudden appearence of
the DSS with two largely separated sidebands (for $C \gg 1$
one has $\Omega_I \gg \gamma$).

    One can ask what happens to the spectrum when the
system is in the one-atom stationary state and one decreases E_I to approach the lower discontinuity point. What
happens is that the DSS disappears continuously, so that
in the neighbourhood of $E_I = E_I^{(-)}$ one has a single narrow
line. In fact by eq.(32) in this region the approach to
the one-atom stationary state is monotonic and there is a
critical slowing down in correspondence of the point
$E_I = E_I^{(-)}$. This concludes the description of the spectral hysteresis cycle of the fluorescent light.

    All these results concerning the behaviour of the
spectrum have been recently substantiated by a quantum
mechanical analysis of Agarwal, Narducci, Feng and Gilmore
(42)

PHYSICAL INTERPRETATION

    The discontinuous behaviour of the spectrum of the
fluorescent light can be intuitively understood in the
following way. The atomic system is not actually driven
by the incident field E_I , but rather by the total internal field E. Therefore one can guess that the behaviour of the spectrum is still that predicted by the one-

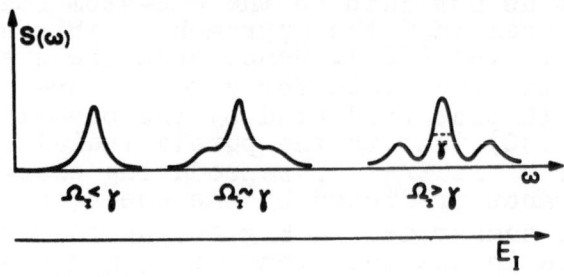

Fig.10 – Behaviour of the spectrum of the fluorescence light for increasing incident field when $\gamma_{\parallel} \simeq \gamma_{\perp} \simeq \gamma$, $\kappa \gg \gamma$.

atom theory, provided one refers to the Rabi frequency $\Omega_T = \mu E / \hbar$ of the internal field instead of the Rabi frequency Ω_I of the incident field. Now in the cooperative stationary state one has that $\Omega_T \leq \gamma / \sqrt{2}$. This explains why in the cooperative stationary state there is no DSS. On the other hand when the system jumps to the one-atom stationary state Ω_T suddenly increases to a value $\Omega_T \simeq \Omega_I \gg \gamma$ and this explains why there is an abrupt appearance of the DSS with largely separated sidebands. I stress however that this physical interpretation is limited, because it does not predict the behaviour of the spectrum in the neighbourhood of the discontinuity points. In particular, it fails to predict the cooperative

line narrowing, which is a typical phase transition effect.

NOTES AND REFERENCES

1) A brief account of this theory is given in R.Bonifacio
   and L.A. Lugiato, Opt. Comm.19, 172 (1976). An intermediate version (by the same authors) will appear in the
   Proceedings of the Fourth Rochester Conference on Coherence and Quantum Optics, edited by L.Mandel and E.
   Wolf. Finally a complete treatment will appear in Phys.
   Rev.A.
2) S.L. McCall, Phys.Rev. A 9, 1515 (1974)
3) R.Bonifacio and L.A. Lugiato, Phys.Rev. A 11, 1507
   (1975), 12, 587 (1975)
4) B.R.Mollow, Phys.Rev. 188, 1969 (1969)
   H.S. Carmichael and D.F. Walls, J.Phys. B 9, 1199 (1976)
   and references quoted therein.
5) Observation of OB in Na is reported in H.M. Gibbs, S.L.
   McCall and T.N.C. Vankatesan, Phys.Rev.Lett. 36, 1135
   (1976). More recent observations in room temperature
   ruby have been performed by T.N.C. Venkatesan and S.L.
   McCall, Appl.Phys. Lett. 30, 282 (1977)
6) (a) H.Haken, Handbuch der Physik vol. XXV/2c, Springer-
   Verlag, Berlin 1970
   (b) M.Sargent III, M.O.Scully and W.E. Lamb Jr. "Laser
   Physics", Addison-Wesley, Mass. 1974
7) R.Friedberg and S.R. Hartmann, Phys.Lett. 37A, 285
   (1971)
8) Other optical systems in which nonlinear absorption
   gives rise to bistability are
   i) the laser with saturable absorber, see R.Salomaa
   and S.Stenholm, Phys.Rev.A 8, 2695 (1973); L.A. Lugiato, P.Mandel, S.T.Dembinski and A.Kossakowski, to appear Phys.Rev. A and references quoted therein.
   ii) the dye laser, see A.Baczynski, A.Kossakowski and
   T.Marszalek, Z.Physik B 23, 205 (1976); R.B.Schaefer
   and C.R. Willis, Phys.Rev. A 13, 1874 (1976)
9) H.M. Gibbs, S.L. McCall and T.N.C. Venkatesan, Lecture
   Notes in these Proceedings.
10) This critical slowing down is similar to that one finds
   in tunnel diodes, see R.Landauer and J.W.F. Woo, in
   "Synergetics", edited by H.Haken, Teubner, Stuttgart
   1973. I stress that the existence of this critical
   slowing down is independent of the assumption C≫1.
11) By means of the linearized equations one can study
   only the "small"fluctuations, in the sense specified
   in the section entitled "Bistability and First-Order
   Phase Transitions".

12) G.S. Agarwal, L.M. Narducci, Da Hsuan Feng and R.
    Gilmore, Proceedings of the Fourth Conference on
    Coherence and Quantum Optics, edited by L.Mandel and
    E.Wolf.

OPTICAL BISTABILITY AND DIFFERENTIAL GAIN

H. M. Gibbs, S. L. McCall, and T. N. C. Venkatesan

Bell Laboratories

Murray Hill, New Jersey 07974

Abstract

A Fabry-Perot interferometer containing a nonlinear medium exhibits many fascinating operating characteristics including hysteresis and AC gain. The basic concept and physical picture are given. The experiment demonstrating optical bistability in Na is described. The dominant role often played by nonlinear phase shifts was first discovered in analyzing that experiment. A simple analytic model of such dispersive bistability is derived giving simple relations between initial plate detuning, plate reflectivity, and magnitude of the nonlinearity for bistability or gain.

Historical Introduction

The possibility of achieving optical bistability by use of a Fabry-Perot interferometer containing a saturable absorber seems to have been recognized independently by three different groups [1-3]. Hints of bistable action were seen in some of the early searches [2,4,5], but recently the effect was demonstrated clearly and studied in detail in Na [6]. However, in Na the nonlinear index of refraction rather than nonlinear absorption was found to play the dominant role. Similarly dispersive effects dominate in the bistability recently observed in room temperature ruby [7].

These demonstrations of optical bistability, AC gain, clipper and limiter action, etc. have led to considerable recent theoretical activity. Spherical mirrors and self focusing should reduce the powers required [8]. Distributed Bragg reflectors may be used in integrated optical circuits [9]. Fluctuations in the fluorescence which undergoes a phase transition may be of interest [10-12]. Experimental activity is also growing. An intracavity phase shifter driven by a signal proportional to the transmitted intensity exhibits all of the device's characteristics and is wavelength insensitive [13]. Non-Fabry-Perot versions of the optical transistor are also being tried [14].

Physical Principles

Simple physical pictures of optical bistability are useful. Consider first a Fabry-Perot with the proper plate separation for maximum transmission at the laser frequency but containing a saturable absorber. Observe the transmitted intensity I_T as a function of the incident intensity I_I. For low values of I_I, I_T is very small; most of the incident energy is reflected from the input mirror or absorbed by the intracavity medium. At sufficiently high intensity the light transmitted by the first mirror, $(1-R)I_I$ where R is the mirror reflectivity, becomes intense enough to begin saturating the absorber. As this happens, some light reaches the output mirror and is reflected back into the absorber to assist the saturation process in a runaway fashion. The device switches on and transmits almost completely for higher I_I. If I_I is reduced the large intracavity field, roughly the product of the finesse $\mathcal{N} = \pi\sqrt{R}/(1-R)$ and I_I, is able to keep the medium bleached to a much lower intensity than required to turn the device on. But when I_I is made so small that the intracavity power is insufficient to excite each absorber once each relaxation time, the device turns off abruptly.

A similar picture applies to a Fabry-Perot initially detuned by at least an instrument width and containing an intensity-dependent refractive index. However, in this case the nonlinearity need not be saturated; the index needs to be changed only enough to compensate for the initial detuning. Conversely, for a given maximum I_I, the initial detuning and nonlinear index can be adjusted to cause the device to switch on for I_I just below its maximum value. The magnitude of the nonlinear index change can be adjusted via the density or the detuning from resonance.

Experimental Verification of Optical Bistability in Na

The experimental apparatus is shown in Fig. 1. A typical bistable curve in Na is shown in Fig. 2. The low intensity absorption coefficient was $\alpha L \approx 2.75$ with the laser frequency about 150 MHz above the Na $^2S_{1/2}$, F = 2 to $^2P_{3/2}$ transitions. The Fabry-Perot plates were separated by 10.5 cm and coated to R = 0.9. The asymmetry with Fabry-Perot detuning shown in Fig. 3 indicated that phase rather than absorptive changes were dominant. Under other operating conditions, i.e., densities and detunings, other device characteristic curves are obtained such as transistor (Fig. 4) or clipper (Fig. 5) action. Note that these characteristics are obtained with only optical inputs, i.e., these are purely optical and yet passive devices.

In Na the nonlinear mechanism was optical pumping from one ground state hyperfine state to the other. The switch-on time was then determined by the optical pumping time, i.e., it depended upon the intensity of a step function input. The turn-off time was the beam transit time for replacement of pumped atoms by thermally distributed atoms. Both switching times were a few microseconds. By using a foreign gas to homogeneously broaden the Na transition and a spherical cavity to increase the intensity, it should be possible to achieve switching times close to the 16 ns lifetime. Clearly it

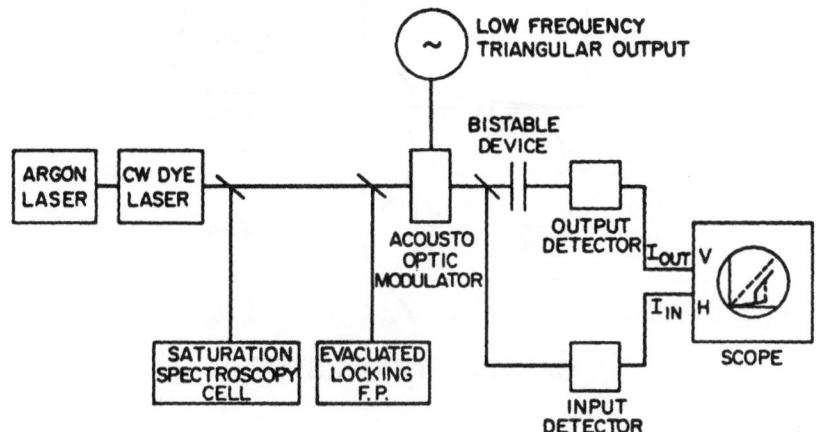

Fig. 1 Experimental apparatus for observing optical bistability in sodium.

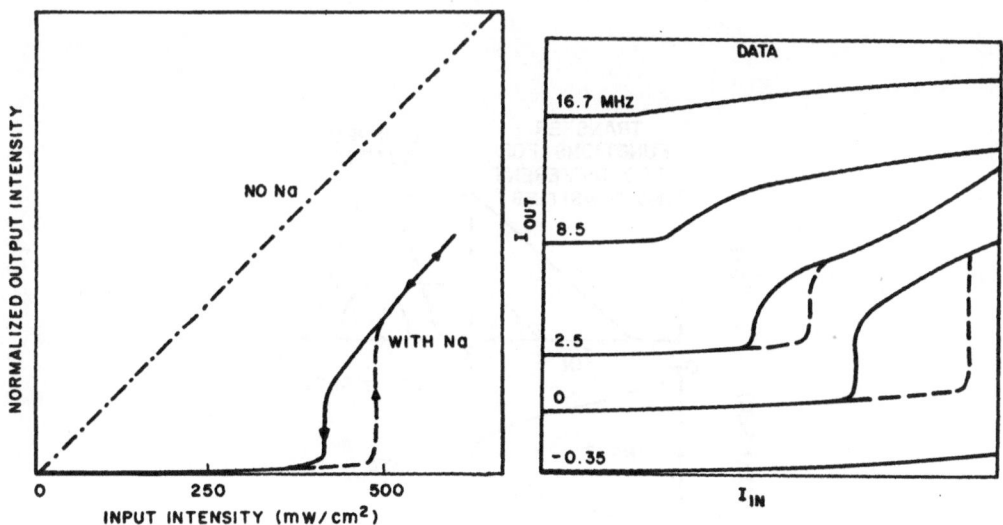

Fig. 2 Optical bistability in Na vapor. From Ref. 6.

Fig. 3 Characteristic curve dependence on Fabry-Perot plate detuning (in MHz).

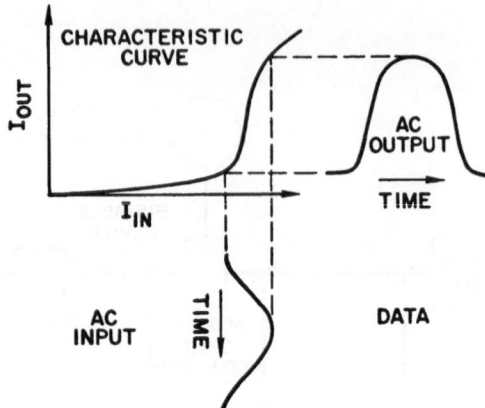

OPTICAL TRANSISTOR

Fig. 4 Differential gain in Na. AC gains at 1 kHz exceeding 2 were observed (with no Fabry-Perot losses, the differential gain would have been nearly 5). From Ref. 6.

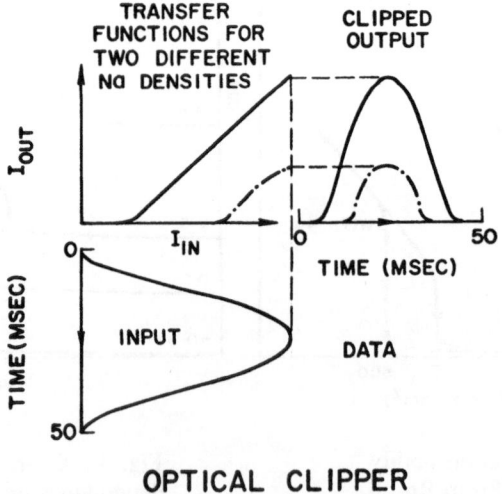

OPTICAL CLIPPER

Fig. 5 Optical clipper action in Na.

would be desirable to have not only fast (approx ns) devices but also room-temperature, solid-state, convenient-wavelength devices. The room-temperature device in ruby [7] meets some of these goals. Perhaps research on semiconductors will succeed in satifying all of them.

Differential Gain Without Population Inversion

The sign reversal of the absorption coefficient of a two-level system subjected to powerful monochromatic radiation was predicted to occur without the necessity of population inversion [15]. This AC gain was demonstrated beautifully on an optically pumped rf transition in Cd [16]. The effect has been shown in the optical region, first implicitly in the Na optical transistor experiment [6] and then directly in the recent Na gain measurements using a weak probe beam at a different frequency [17].

It is easy to show that AC gain can occur without population inversion. The coupled Maxwell-Bloch equations (after slowly varying envelope, rotating wave, and forward-only approximations) for an electric field with slowly varying amplitude ξ and phase ϕ,

$$E(z,t) = \xi(z,t)e^{i(\omega t - kz - \phi(z))} + \text{c.c.} \tag{1}$$

and polarization with in-phase dispersive component u and out-of-phase absorptive component v,

$$P(z,t) = [u(\Delta\omega,z,t) - iv(\Delta\omega,z,t)]e^{i[\omega t - kz - \phi(z)]} \tag{2}$$

consist of

$$\frac{\partial\xi}{\partial z} + \frac{\eta_0}{c}\frac{\partial\xi}{\partial t} = -\frac{2\pi\omega}{\eta_0 c}\,Npv\,, \tag{3}$$

$$\dot{u} = -(\omega_0 - \omega)v - u/T_2'\,, \tag{4}$$

$$\dot{v} = (\omega_0 - \omega)u - v/T_2' - w\kappa\xi\,, \tag{5}$$

$$\dot{w} = -(w+1)/T_1 + v\xi\kappa\,. \tag{6}$$

In steady state and on resonance, the equilibrium values are

$$u_0 = 0 \tag{7}$$

$$v_0 = \frac{T_2'\kappa\xi_0}{1+T_1 T_2'\kappa^2\xi_0^2} = \left(\frac{F}{1+F^2}\right)\left(\frac{T_2'}{T_1}\right)^{1/2} \tag{8}$$

$$w_0 = \frac{-1}{1+T_1 T_2'\kappa^2\xi_0^2} = \frac{-1}{1+F^2} \tag{9}$$

where

$$F = (T_1 T_2')^{1/2}\kappa\xi_0\,.$$

These solutions are shown in Fig. 6. Note that for all input field strengths, w_0 is always

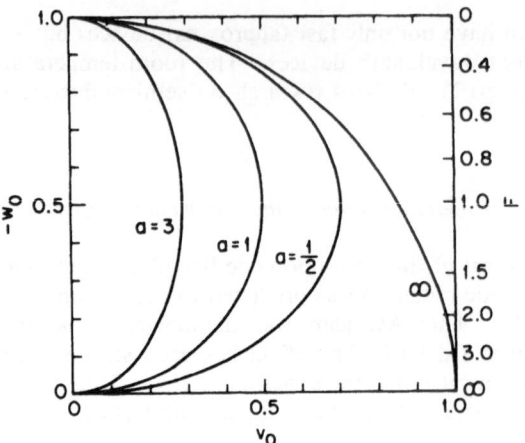

Fig. 6 $w < 0$, $v > 0$ quadrant of the Bloch circle $w^2 + v^2 = 1$. The circle represents the motion of an undamped Bloch vector subject to an electromagnetic field which is pulsed with temporal width $\ll T_2'$. The ellipses represent the steady state solutions given by Eqs. (7)-(9). Note that a differential increase in F, for $F > 1$, leads to a decrease in v_0, and consequently to instability and differential gain. $a = T_1/T_2$. From Ref. 3.

negative, i.e., population inversion never occurs. Also v_0 is always positive so absorption from the incident field always occurs (Eq. 3). However, a small modulation on ξ can experience gain:

$$\xi = \xi_0' + \Delta\xi \tag{10}$$

where $\Delta\xi$ varies slowly compared with $\kappa\xi_0$, $1/T_1$, and $1/T_2'$. For then

$$\frac{\partial\xi_0}{\partial z} + \frac{\eta_0}{c} \frac{\partial\xi_0}{\partial t} = - \frac{\alpha\xi_0}{2} \frac{1}{1+F^2} \, , \tag{11}$$

so ξ_0 experiences absorption for all values of F, but

$$\frac{\partial(\Delta\xi)}{\partial z} + \frac{\eta_0}{c} \frac{\partial(\Delta\xi)}{\partial t} = - \frac{\alpha\Delta\xi}{2} \frac{1-F^2}{(1+F^2)^2} \, , \tag{12}$$

so for $F > 1$ there is differential gain. The fact that there is AC field gain is no assurance that there is AC intensity gain. Placement of the nonlinear atoms within a Fabry-Perot makes it possible to assure that.

Cavity Boundary Conditions

For simplicity consider the case of a Fabry-Perot cavity containing a medium exhibiting negligible absorption but a round-trip intensity dependent phase shift ϕ at the frequency of the incident field E_I; see Fig. 7. Consider the boundary conditions at the mirrors having reflectivity R and transmission T. At the exit mirror $(z = L)$ the transmitted field E_T is simply related to the intracavity forward field E_F:

$$E_T = \sqrt{T} \, E_F(L) \, . \tag{13}$$

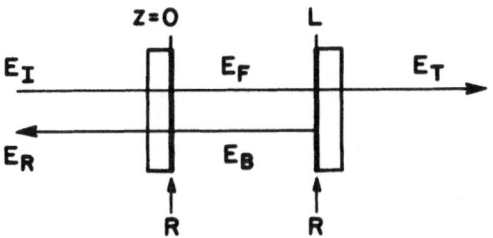

Fig. 7 Nonlinear optical device electric fields.

The intracavity backward field at the exit mirror is

$$E_B(L) = -\sqrt{R}\, E_F(L) \ .\tag{14}$$

At the entrance mirror

$$E_F(0) = \sqrt{T}\, E_I - \sqrt{R}\, e^{i\phi/2}E_B(L).$$

which becomes at the exit mirror

$$E_F(L) = \sqrt{T}\, E_i e^{i\phi/2} - \sqrt{R}e^{i\phi}E_B(L) \ .\tag{15}$$

Combining (13)-(15)

$$E_T = Te^{i\phi/2}E_I + Re^{i\phi}E_T$$

$$E_I = (1-Re^{i\phi})E_T/(Te^{i\phi/2})$$

$$I_I = |1-Re^{i\phi}|^2 I_T/T^2$$

$$\boxed{I_I \underset{\phi^2 \ll 12}{\longrightarrow} (1 + R\phi^2/T^2)I_T}\tag{16}$$

Equation (16) relates the transmitted and incident intensities as a function of the plate reflectivity $R = 1-T$ and the round-trip phase ϕ which may be intensity dependent.

Simple Model of Dispersive Bistability

Suppose the phase shift is linearly dependent upon the intracavity intensity which is proportional to I_T, i.e.,

$$\phi = \beta I_T - \phi_o\tag{17}$$

where ϕ_o contains all intensity independent phase shifts such as background index or plate detuning contributions. Phase shifts of the form of Eq. (17) arise when there is a nonlinear refractive index $\eta = \eta_o + \eta_1\xi^2$ or for Bloch equations far enough off resonance.

For maximum transmission, ϕ must vanish, i.e., $\beta I_T = \phi_o$. In order for the transmission to be low at low intensities, the initial detuning must be larger than the cavity instrument width, i.e., $\phi_o \geqslant 2\pi/N = 2T/\sqrt{R}$, where N is the finesse or free spec-

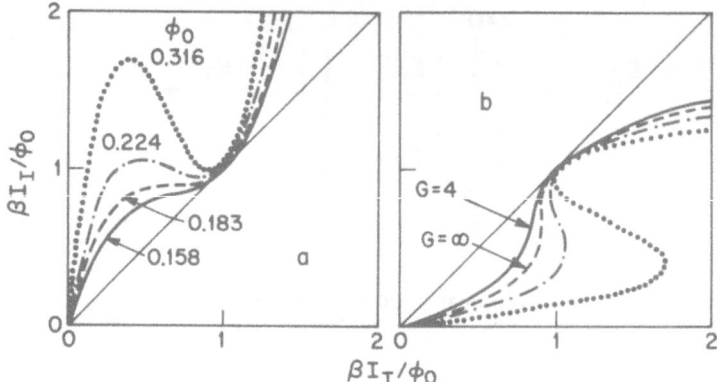

Fig. 8 Characteristic transfer functions calculated from the simple dispersive model of Eqs. (16) and (17). The equations are most easily solved as I_I versus I_T (a), but their effect as a device is best shown as I_T versus I_I (b). $R = 0.9$ so $\mathcal{N} \approx 30$. The $\phi_0 = 0.183$ curve in (a) and $G = \infty$ curve in (b) separate the AC gain and bistable regions.

tral range divided by instrument width. Consequently, $R\phi^2/T^2$ is of order unity and is as important as the unity term in Eq. (16). For $R = 0.9$, $R\phi^2/T^2 = 1$ implies $\phi^2 \approx 10^{-2}$ which is much smaller than 12 as needed in Eq. (16).

This simple model, Eqs. (16) and (17), yields both bistable (two equilibrium values of I_T for one value of I_I) and differential gain ($dI_T/dI_I > 1$) characteristic curves as shown in Fig. 8. For fixed reflectivity R, only the initial plate detuning ϕ_0 is varied to go from conditions of no gain, to high gain, to infinite gain, and finally to bistability. The conditions on ϕ_0 are easily derived. For differential gain one needs $0 < dI_I/dI_T < 1$ and $d^2I_I/d^2I_T = 0$, implying

$$0 < |\phi_0| < \sqrt{3}\, T/\sqrt{R} = \sqrt{3}\, \pi/\mathcal{N}. \quad \text{AC GAIN} \qquad (18)$$

At the inflection point $\beta I_T/\phi_0 = 2/3$ and the gain is

$$G = \frac{dI_T}{dI_I} = (1 - R\phi_0^2/3T^2)^{-1} . \qquad (19)$$

Transfer functions for gains of 4 and ∞ are shown in Fig. 8. For bistability one needs a region of $dI_I/dI_T < 0$, i.e., the phase shift term in Eq. (16) is large enough to produce a valley in the I_I versus I_T plot. This implies

$$|\phi_0| > \sqrt{3}\, T/\sqrt{R} = \sqrt{3}\, \pi/\mathcal{N} . \qquad \text{BISTABILITY} \qquad (20)$$

Note that the phase shift required for bistability, $\sqrt{3}\, \pi/\mathcal{N}$, is close to the value $2\pi/\mathcal{N}$ argued physically above.

Clearly this simple model can be extended to include many more details such as standing wave effects, nonlinear absorption, medium complexities such as many transitions on and off resonance, inhomogeneous broadening, etc. Nonetheless the basic principles of the device are described well by this simple model which agrees with the simple physical picture.

REFERENCES

1. H. Seidel, U.S. Patent No. 3,610,731 (Oct. 5, 1971).
2. A. Szöke, V. Daneu, J. Goldhar, and N. A. Kurnit, Appl. Phys. Lett. **15,** 376 (1969).
3. S. L. McCall, Phys. Rev. A **9,** 1515 (1974).
4. J. W. Austin and L. G. DeShazer, J. Opt. Soc. Am. **61,** 650 (1971).
5. E. Spiller, J. Opt. Soc. Am. **61,** 699 (1971) and J. Appl. Phys. **43,** 1673 (1972).
6. S. L. McCall, H. M. Gibbs, G. G. Churchill, and T. N. C. Venkatesan, Bull. Am. Phys. Soc. **20,** 636 (1975) and J. Opt. Soc. Am. **65,** 1184 (1975). H. M. Gibbs, S. L. McCall, and T. N. C. Venkatesan, Phys. Rev. Lett. **36,** 1135 (1976) and U. S. Patent No. 4,012,699.
7. T. N. C. Venkatesan and S. L. McCall, Appl. Phys. Lett. **30,** 282 (1977) and T. N. C. Venkatesan, Ph.D. Dissertation, City University of New York, 1977.
8. F. S. Felber and J. H. Marburger, Appl. Phys. Lett. **28,** 731 (1976).
9. M. Okuda, M. Toyota, and K. Onaka, Opt. Commun. **19,** 138 (1976).
10. R. Bonifacio and L. A. Lugiato, Opt. Commun. **19,** 172 (1976) and in Ref. 11.
11. L. Mandel and E. Wolf, eds., Fourth Rochester Conference on Coherence and Quantum Optics, 1977.
12 L. M. Narducci and R. Gilmore in Ref. 11.
13. P. W. Smith and E. H. Turner, Appl. Phys. Lett. **30,** 280 (1977) and J. Opt. Soc. Am. **67,** 250 (1977).
14. K. Jain and G. W. Pratt, Jr., Appl. Phys. Lett. **28,** 719 (1976).
15. S. G. Rautian and I. I. Sobel'man, Zh. Eksp. Teor. Fiz. **41,** 456 (1961) [Sov. Phys. JETP **14,** 328 (1962)]. See also B. R. Mollow, Phys. Rev. A **5,** 2217 (1972); S. Haroche and F. Hartmann, Phys. Rev. **6,** 1280 (1972); Ref. 3.
16. D. W. Aleksandrov, A. M. Bonch-Bruevich, V. A. Khodovoi, and N. A. Chigir, Pis'ma Zh. Eksp. Teor. Fiz. **18,** 102 (1973) [JETP Lett. **18,** 58 (1973)].
17. F. Y. Wu, S. Ezekiel, M. Ducloy, and B. R. Mollow, Phys. Rev. Lett. **38,** 1077 (1977).

SUPERFLUORESCENCE EXPERIMENTS

H. M. Gibbs

Bell Laboratories

Murray Hill, New Jersey 07974

Abstract

Superfluorescence is the cooperative emission of an initially inverted system of atoms. Since no macroscopic polarization is excited to define a preferred direction of emission, a long cylinder of inverted atoms is generally prepared to provide such a direction.

There have been several experiments providing information on superfluorescence. Experiments in hydrogen fluoride, sodium, cesium, thallium, and methyl fluoride are discussed and compared. Particular attention is paid to the Cs experiment in which single pulse emission has been observed under conditions of negligible dephasing or relaxation losses.

Preparation of a Superfluorescence System

It is quite clear classically that a phased array of electric dipole moments possesses a macroscopic polarization which radiates in the phased direction with an intensity proportional to the square of the number N of such moments [1-3]. If the phased array extends over a sample of dimensions much larger than the wavelength then the individual emissions no longer contribute optimally to the superradiant emission. Nonetheless, Rehler and Eberly [3] have shown that many of the features of superradiance persist in such an extended sample, with the effective number of cooperating atoms reduced to μN (the shape factor μ is typically of order 10^{-5}).

Free induction decay and photon echo measurements are examples of classical N^2 emission of a macroscopic polarization prepared by coherent optical excitations [2]. However for simple interpretation of the results the absorption must be kept low so all

121

atoms are excited by the same amount. This is equivalent to requiring that only a small amount of the stored energy be emitted cooperatively or that the dephasing time T_2^* be much shorter than the superradiance or superfluorescence time $\tau_R = 8\pi\tau_0/3n\lambda^2L$. Here τ_0 is the radiative lifetime of the transition, n is the inversion density, λ is the wavelength, L is the sample length, and the Fresnel number $N \equiv A/\lambda L$ is assumed to be one or larger (A is the cross sectional area) [2]. One can show that $\alpha_l L = 2T_2^*/\tau_R$, so that low absorption implies that dephasing occurs before much energy is emitted cooperatively; such superradiance has been called limited superradiance [2].

It would also be desirable to study strong superradiance in which most of the energy is emitted cooperatively before dephasing can occur. However, if coherent optical excitation of a two-level system is attempted to produce the phased array, the coherent pulse is incident on a highly absorbing system as in self-induced transparency. If the input pulse length τ_p is longer than T_2^*, ordinary broadline self-induced transparency results with pulse reshaping and nonuniform excitation of the sample. However, for superradiance studies one should have $\tau_p < T_2^*$ and $\tau_p \lesssim \tau_R$ for which sharp-line pulse propagation applies. It may then be possible to excite the sample uniformly and Burnham-Chiao [4] ringing should result. There appears to be no experimental confirmation of such cooperative ringing of an absorber excited by a very short pulse. Also it would be extremely difficult to excite a two-level system with a pulse with area close enough to π throughout the sample to permit the study of the evolution of a system from a state of complete inversion emitting quantum noise to a state emitting classically.

The first breakthrough in experiments on cooperative emission was in HF by the MIT group [5]. They incoherently excited two levels of a three-level system. With complete initial inversion there is no initial macroscopic polarization; instead it must build up from the quantum noise of spontaneous emission. Furthermore, only spatial asymmetries in gain introduced by the excitation geometry give rise to a preferred emission direction, i.e., it is not built-in by initial phasing. Even though this cooperative emission from initial inversion, called superfluorescence (SF) by Bonifacio and Lugiato [6], was mentioned briefly by Dicke in his first paper on superradiance [1a], it is clearly a phenomenon quite different from cooperative emission of an initially phased array because of its quantum initiation. It must be closely related to the coherence brightened laser described qualitatively by Dicke in 1962 [1b].

Ringing and Initial Tipping Angle

The MIT group also introduced the concept of ringing and stressed the importance of propagation effects in SF. Theories neglecting propagation predicted symmetric emission after a delay time τ_D: $I \propto \text{sech}^2(t-\tau_D)/2\tau_R$ [2,6]. In contrast, the HF outputs were often asymmetric with long tails containing three or four modulations or rings [5]. This ringing was associated with the ringing discussed by Burnham and Chiao [4] in absorbers in which regions which have already returned to their ground states by radiating are reexcited by radiation from other regions and subsequently radiate again with opposite phase. The MIT simulations of coupled uniform plane-wave Maxwell-Bloch equations yielded ringing much stronger than observed. A linear loss term, $-\kappa\xi$, was introduced in the field equation to account for linear diffraction losses: $\kappa L = 1/2 \ (1+N^{-2})$; $\kappa L \approx 2.5(N=0.08)$ was found to suppress the ringing to the observed level.

Coupled Maxwell-Bloch equations require an initial tipping angle θ_0 or a fluctuating polarization source to initiate the semi-classical evolution. Ringing increases with decreases in θ_0 for fixed density or with increases in density for fixed θ_0. From thermal equilibrium and gain narrowing considerations MacGillivray and Feld derive $\theta_0^{MF} = (2\pi)^{1/4} N^{-1/2} (\alpha_0 L)^{-3/4}$ where α_0 is the electric field gain coefficient and N is the number of inverted atoms. A much larger value $2/\sqrt{\mu N}$ is obtained as the angle for which classical cooperative emission becomes equal to quantum random emission in all directions. An intermediate value $2/\sqrt{N}$ is the tipping expected by the first photon emitted along the cylinder. These values of θ_0 can differ by many orders of magnitude, so that definitive measurements may distinguish between them even though the delay depends on θ_0 only logarithmically.

Superfluorescence Experiments

There have been several recent SF experiments based on the MIT 3-level scheme: methyl fluoride (CH_3F) [8,9], sodium (Na) [10], thallium (Tl) and alkalis (11), and cesium (Cs) [12]. The CH_3F experiment in the far infrared is in the homogeneously broadened limit, complementing the inhomogeneously broadened HF case. An abruptly terminated excitation pulse permits one to look for swept-gain effects, which might relax the need to have the sample length less than the cooperation length [2]. No simulation of the CH_3F data by the Rehler-Eberly, Bonifacio-Lugiato, or Maxwell-Bloch approaches gives good agreement in either the disk or needle limits. The Na and Tl experiments illustrate the relative ease with which SF may be seen now on a multitude of transitions. In the Na article there is a discussion of the interesting transition at high density from superfluorescence emission with an intensity proportional to N^2 to a regime with intensity proportional to N; apparently no data are presented to illustrate this transition, however. The Cs experiment was specifically designed to study SF under well defined conditions in order to compare data with theories. In particular, an atomic beam was used to reduce the Doppler dephasing rate so that output pulse shapes could be studied free of dephasing and relaxation damping which might suppress ringing. Degeneracies were removed, and excitation diameters and inversion densities were measured.

Comparison of Experiments

A comparison of the experiments is made in Table I. The wavelengths of the SF and pump transitions are λ_{SF} and λ_p, respectively. The pump pulse has temporal width τ_p and spectral width $\Delta\nu_p$. The SF region of length L and cross sectional area A has Fresnel number $\mathcal{N} = A/\lambda L$. The times are all given in ns: the escape time is $\tau_E = L/c$, $\tau_R = 8\pi\tau_0/3n\lambda^2 L$, τ_D is the observed delay time, T_1 and T_2' are the longitudinal and transverse relaxation times, and T_2^* is the inhomogeneous dephasing time.

The Cs experiment is the only experiment on a nondegenerate transition, granting that the HF degeneracy may be of only minor consequence. It is also the only one for which the emission evolution time, τ_D, is clearly much shorter than all of the relaxation times T_1, T_2', T_2^*. It is the only one which has a superfluorescence time τ_R much

TABLE 1

COMPARISON OF EXPERIMENTS ON SUPERFLUORESCENCE

System	λ_{SF} (μm)	λ_p (μm)	τ_p (ns)	$\Delta\nu_p$ (GHz)	$\mathcal{N}_{SF}$	τ_E	τ_R	τ_D	T_1, T_2'	T_2^*
HF v=1 J=3 to 2	84	2.5	50 to 400	Broad 0.03?	0.1 to 1.0	3.3	6	$\leq$2000	5400,5400	>48 $\frac{}{}$ 330?
CH$_3$F	496	9.55	100 0.1ns fall	0.01	0.23	11.7	0.6	187	67,67	366
R(J=±1,K=1 or 2)										
Na 5S-4P 4P-4S	3.41 2.2	0.589+ 0.616	2	10	~1	0.47	~0.5	$\leq$7	long	1.7
Tl 7P-7S	1.3	3-step: 0.3791+ SRS+0.3791	5-6	9	?	0.5	~0.5	$\leq$12	long	1
Cs 7P($M_J=-\frac{3}{2}$,$M_I=-\frac{5}{2}$) to 7S($M_J=-\frac{1}{2}$,$M_I=-\frac{5}{2}$)	2.93	0.455	2	0.5	1 >1 also	0.067 (0.033–0.167)	0.5	10 $\leq$40	70,80	32 (5–80)

longer than the escape time. This is equivalent to a cooperation length L_c much longer than the sample length; since the cooperation time τ_c is just τ_R for a sample length $c\tau_R$, $\tau_c = \sqrt{\tau_E \tau_R}$ [13]. Bonifacio and Lugiato have specified the conditions for single-pulse emission to be $\tau_E < \tau_c < \tau_R < \tau_D < T_1, T_2', T_2^*; \tau_p \ll \tau_D;$ and $\mathcal{N}= 1$. These conditions would seem to optimize the study of the superfluorescence process free of damping, excitation details, and mode competition. It is not so clear that propagation effects and dynamic transverse effects are of no consequence under these conditions. Since the Cs experiment satisfied these conditions best and is the most quantitative in all the relevant parameters, it will be described in some detail (see Ref. [12]) and compared with various theories.

The Experimental Case for Single-Pulse Superfluorescence

Setting aside quantum beat modulations [12a, 12c,11], the outputs of these experiments have considerable in common. In all of the experiments single pulses were observed under some conditions; see Figs. 1 to 4. In CH_3F homogeneous relaxation prevented ringing. In simulations of the HF case, strong ringing persists using the experimental values of T_1, T_2' and T_2^*; the linear loss term discussed above is required to suppress ringing to the observed level. In Cs there is no question about the fact that dephasing is unimportant, yet single-pulse emission still occurs. It is true that the typical output pulse in the Cs case is asymmetric with a slower fall than rise time. But it is also true that single symmetric pulses are often seen and that they are *narrower than the asymmetric pulses*. Therefore, the asymmetric pulses are more likely to be averaged single pulses than averaged ringing pulses. Also uniform plane-wave simulations of the Maxwell-Bloch equations for the Cs case indicate ringing so strong that it should have been seen. It now seems likely that Burnham-Chiao ringing is absent under the Bonifacio-Lugiato single-pulse conditions, i.e., it is totally obscured by transverse effects.

This disagreement between SF data and coupled Maxwell-Bloch simulations is in sharp contrast with the close agreement between simulations and data for coherent pulse propagation in absorbers, i.e., self-induced transparency (SIT) [2]. Comparing SF and SIT emphasizes the difficulties and opportunities for study of the SF phenomenon. First, the quantum-classical transition is absent from SIT since a strong field is always present; it is not yet clear that this transition can be reduced to a few spatially independent parameters such as θ_0 or a random polarization source. Second, in both SF and SIT simulations, the uniform plane wave approximation is usually made. In SF a Fresnel number $\mathcal{N}$ of one is usually selected to avoid the mode competition of large $\mathcal{N}$ and the poor lengthwise communication of small $\mathcal{N}$. Recently Vrehen has performed experiments with large $\mathcal{N}$ [12d]. Phase and dynamical transverse variations have not yet been included in SF computations; they have been shown to dominate SIT pulse evolution under some conditions [14]. For $\mathcal{N}= 1$, nondynamical diffraction effects are quite small ($\kappa L = 0.35$ in notation of Ref. [5]). The seemingly large value $\kappa L = 2.5$ used to simulate the HF experiment may effectively dampen ringing much like dynamic transverse effects, but the importance of the latter have not yet been determined. Finally, SIT simulations consider only one direction of propagation; two-direction SF simulations [5,15] indicate little change in predicted output except that ringing may be suppressed by a factor of two.

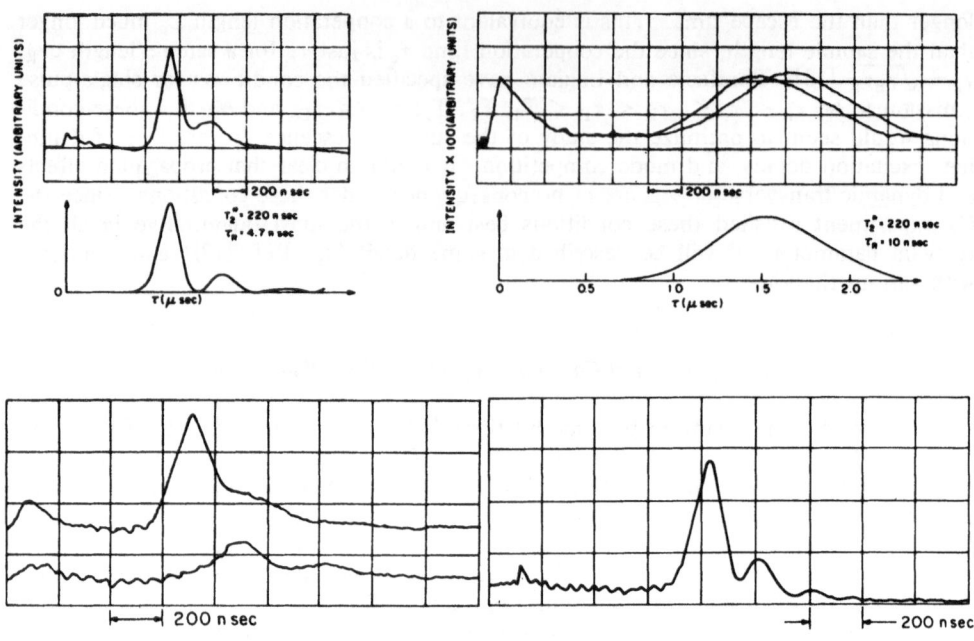

Figure 1
Oscilloscope traces of HF superradiant pulses at 84 μm (J = 3 → 2). The center curves are Maxwell-Schroedinger simulations of the upper experimental curves. From Ref. 5.

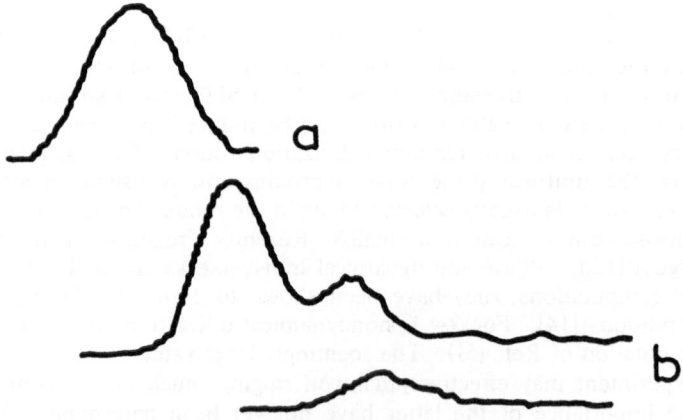

Figure 2
Na experiment pulses. (a) exciting pulse, (b) 3.41 μm superfluorescence for two different excitation intensities. From Ref. 10.

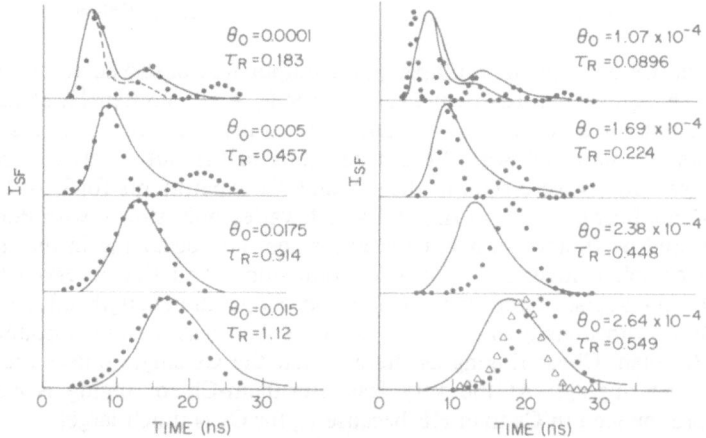

Figure 3

Comparison of normalized Cs data (solid curve) with simulations of one-way Maxwell-Bloch equations initiated by a short input pulse of area θ_0. The experimental values of $\tau_R = 8\pi\tau_0/3n\lambda^2 L$ are from the top down 0.14, 0.35, 0.71, and 0.87 ns; experimental relaxation times are used except for the triangular point curve with 10^5 ns times. In the left column θ_0 values were chosen to give approximate agreement between the simulated and observed delays with the 0.78 correction to μ for $\mathcal{N}= 1$ applied. In the right column the τ_R values are about 63% of the μ-uncorrected experimental values as required so that θ_0 could be given by $2/\sqrt{N}$ and some agreement with data obtained. For more details see Ref. 12e.

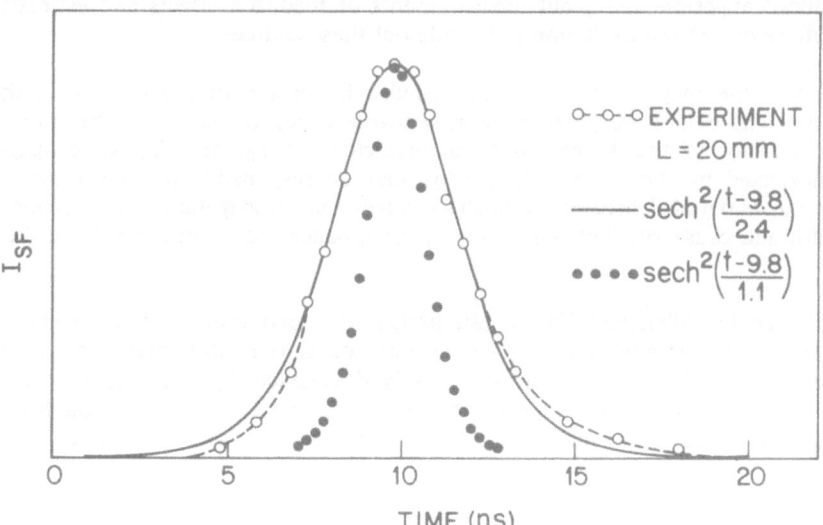

Figure 4

An example of the almost symmetrical Cs pulses frequently observed (circle points) and comparison with best-fit sech² (solid line) and mean-field theory sech² (solid points). The 25% reduction in the observed density required to make the mean field delay agree with the data is well within experimental uncertainty.

Multiple Pulse Output: Ringing or Transverse Effects?

In the Cs experiment multiple pulse output was observed to commence rather abruptly when τ_R became less than $2\tau_E$, i.e., for $L_c = L$. Vrehen [12d] has established that this multiple pulse output is associated with transverse effects; in a single shot the output is not the same in two different positions in the output cross section. Similar multiple-pulse outputs were seen in the HF and Na experiments for $L \approx L_c$; see Figs. 1 and 2. Perhaps Burnham-Chiao ringing, which varies only slowly with density, is completely obscured by transverse averaging and smoothing occurring in the initiation transition from complete inversion to classical emission. And the observed "ringing" may very well be transverse mode competition and cooperation length effects which set in rather abruptly with density at $L \approx L_c$. So there is at least as much doubt that the HF ringing is Burnham-Chiao ringing as there is that the Cs single pulses are not smeared Burnham-Chiao ringing. At the very least Burnham-Chiao ringing is expected to be much less pronounced in Cs than HF because θ_0 for Cs is much larger.

Comparison of Cs Data with Maxwell-Bloch
and Mean-Field Predictions

If, in spite of the above discussions to the contrary, one tries to fit the Cs data with uniform-plane-wave Maxwell-Bloch simulations one finds the results shown in Fig. 3. In particular, if one fits the 13 ns delay curve allowing the density to vary over uncertainties of +60% and -40%, one finds θ_0 varies from 0.05 to 0.0025. For comparison $2/\sqrt{\mu N} = 0.08$, $2/\sqrt{N} = 0.00024$, and $\theta_0^{MF} = 3 \times 10^{-6}$; thus θ_0^{MF} is inconsistent with the data, and the other values are not in good agreement either. Of course, a large θ_0 might be introduced experimentally, but measurements of feedback effects and estimates of Raman effects during the excitation pulse rule out these sources.

Alternatively, one may compare the single-pulse Cs data with the mean-field theory sech^2 [3,6]. The delays are given by $\tau_R \ln N$ within experimental error; the widths are just over a factor of two longer than the predicted $3.5 \tau_R$; see Fig. 4. Propagation effects, neglected by the mean field theory, may be responsible for this discrepancy. Certainly the Maxwell-Schroedinger analysis shows that propagation effects change the pulse width and cause ringing; the larger θ_0 the broader the output pulse and the less ringing.

As emphasized by Allen and Eberly [2], perhaps the evolution from pure inversion, through the quantum initiation, to classical emission is the most interesting aspect of superfluorescence. This problem is not yet solved satisfactorily. Its solution may well support the contention that the theory, not the experiment, is lacking, although further experiments to elucidate this quantum-classical transition should be most welcome.

REFERENCES

[1]   (a) R. H. Dicke, Phys. Rev. **93**, 99 (1954). (b) R. H. Dicke, p. 35 of P. Grivet and N. Bloembergen, eds, *Quantum Electronics Vol. 1,* (Dunod, Paris, 1963).

[2]   For a general introduction see L. Allen and J. H. Eberly, *Optical Resonance and Two-Level Atoms,* (John-Wiley, N.Y., 1975).

[3]   N. E. Rehler and J. H. Eberly, Phys. Rev. A **3**, 1735 (1971).

[4]   D. C. Burnham and R. Y. Chiao, Phys. Rev. **188**, 667 (1969) and S. L. McCall, Thesis, University of California, 1968, unpublished.

[5]   N. Skribanowitz, I. P. Herman, J. C. MacGillivray, and M. S. Feld, Phys. Rev. Lett. **30**, 309 (1973). I. P. Herman, J. C. MacGillivray, N. Skribanowitz, and M. S. Feld, in *Laser Spectroscopy,* edited by R. G. Brewer and A. Mooradian (Plenum, N.Y., 1974).

[6]   R. Bonifacio and L. A. Lugiato, Phys. Rev. A **11**, 1507 (1975) and **12**, 587 (1975) and reference therein. Also see Ref. 7.

[7]   C. M. Bowden, D. W. Howgate, and H. R. Robl, eds., *Cooperative Effects in Matter and Radiation* (Plenum, N.Y., 1977).

[8]   A. T. Rosenberger, S. J. Petuchowski, and T. A. DeTemple, in Refs. 7 and 9.

[9]   L. Mandel and E. Wolf, eds., Proceedings of the Fourth Rochester Conference on Coherence and Quantum Optics, 1977.

[10]   M. Gross, C. Fabre, P. Pillet, and S. Haroche, Phys. Rev. Lett. **36**, 1035 (1976).

[11]   A. Flusberg, T. Mossberg, and S. R. Hartmann, Phys. Lett. **58A**, 373 (1976) and Ref. 7.

[12]   (a) H. M. Gibbs in Ref. 7. (b) Q. H. F. Vrehen in Ref. 7. (c) Q. H. F. Vrehen, H. M. J. Hikspoors, and H. M. Gibbs, Phys. Rev. Lett. **38**, 764 (1977) and (d) in Ref. 9. (e) H. M. Gibbs, Q. H. F. Vrehen, and H. M. J. Hikspoors, in *Laser Spectroscopy,* J. L. Hall and J. L. Carlsten, eds., (Springer-Series in Optical Sciences, Vol. 7), Springer-Verlag, N.Y., Heidelberg, 1977 and Phys. Rev. Lett. **39**, 547 (1977).

[13]   F. T. Arecchi and E. Courtens, Phys. Rev. A **2**, 1730 (1970).

[14]   H. M. Gibbs, B. Bölger, F. P. Mattar, M. C. Newstein, G. Forster, and P. E. Toschek, Phys. Rev. Lett. **37**, 1743 (1976).

[15]   R. Saunders, S. S. Hassan, and R. K. Bullough, J. Phys. A **9**, 1725 (1976) and private communications.

COHERENT TWO-PHOTON AMPLIFICATION IN AN INVERTED MEDIUM*

L.M. Narducci, L.G. Johnson, E.J. Seibert, W.W. Eidson

Department of Physics and Atmospheric Science
Drexel University
Philadelphia, Pennsylvania 19104

Abstract: We study the propagation of an electromagnetic
pulse in an inverted medium where atomic levels of interest have
the same parity and an energy separation equal to twice the energy
of the incident photons. We analyze the coupled atom-field evolution equations in the limit when incoherent atomic relaxation
processes can be neglected, and derive the threshold conditions
for amplification and a few exact conservation relations. The
detailed evolution of the pulse through the inverted medium is
investigated with the help of a hybrid computer.

I. INTRODUCTION

The subject matter of these lectures concerns the interaction
of coherent electromagnetic radiation and matter. This problem
of course, has played a central role in Quantum Optics and continues to be of great current interest as indicated by the numerous recent advances discussed in this volume. In the traditional
view of radiation-matter interaction, the atoms are pictured as
two-level systems undergoing transitions between dipole-coupled
energy levels[1]. Our setting, instead, will be substantially
different.

The two active levels in our medium are assumed to have the
same parity. All the other energy levels play the role of
virtual states in the atomic dynamics. Because no direct
transitions are allowed between the two levels of interest,

*Work partially supported by the Office of Naval Research
 contract number N00014-76-C-1082.

spontaneous or stimulated emission of radiation must be provided
by other processes such as, for example, inelastic atomic
collisions, spontaneous decay through a ladder of lower lying
states, or higher-order non-linear effects induced by an
external field.

The latter mechanism and, in particular, the two-photon
emission process will be explored in these lectures as a means
for producing amplification of an incident pulse of radiation.

Two-photon transitions have been the subject of considerable
activity in the last few years[2], yielding for example, such
spectacular pay-offs as Doppler-free spectroscopy[3]. Early in
the development of the theory of two-photon processes considerable interest was directed to the study of coherent atomic
evolution[4]. As a result of these efforts, many well known one-
photon coherent atomic effects were shown to have two-photon
analogues[5].

Much of the recent work has focused on the evolution of
atoms from their ground state. More limited attention has been
paid to the propagation of an incident pulse in an inverted
medium[6]. We address ourselves to this problem with the added
restriction that the carrier frequency of the incident pulse
be approximately one half the atomic transition frequency. The
more general situation involving two incident pulses whose
carrier frequencies add up to the atomic transition frequency
has also been considered in the literature[7]. This generalization, however, will not be discussed in these lectures.

Our analysis evolves along the lines mapped out by Estes et
al. in the context of the so-called self-consistent field
approximation[8]. Schematically, one can picture the process as
follows: the incident field drives the excited atoms and
generates a non-linear polarization which oscillates with a
carrier frequency equal to that of the incident field. The
radiated and incident fields add coherently and drive another
group of atoms further down into the active medium. This
picture is oversimplified because additional non-linear
polarization components also come into play and competition
effects may become important. Additional comments on this
problem will be made in subsequent sections of this paper.

Mathematically, the atomic evolution is described by the
Schrödinger equation; the field propagation, instead, is
governed by Maxwell's equations. The macroscopic non-linear
polarization is calculated self-consistently in terms of the
quantum mechanical atomic amplitudes[9].

The resulting atomic equations can be cast into a form that
is formally identical to that of the familiar Bloch equations
describing the one-photon evolution of two-level systems. Two
main differences will be noted: the driving field amplitude
enters quadratically, rather than linearly, in the atomic
equations of motion, and intensity-dependent dispersion effects
are predicted which will be interpreted as dynamic Stark shifts.

II. ATOMIC EVOLUTION

In order to fix our attention on a concrete example,
consider the lowest lying levels of the singlet spectrum of
Calcium (Fig. 1)

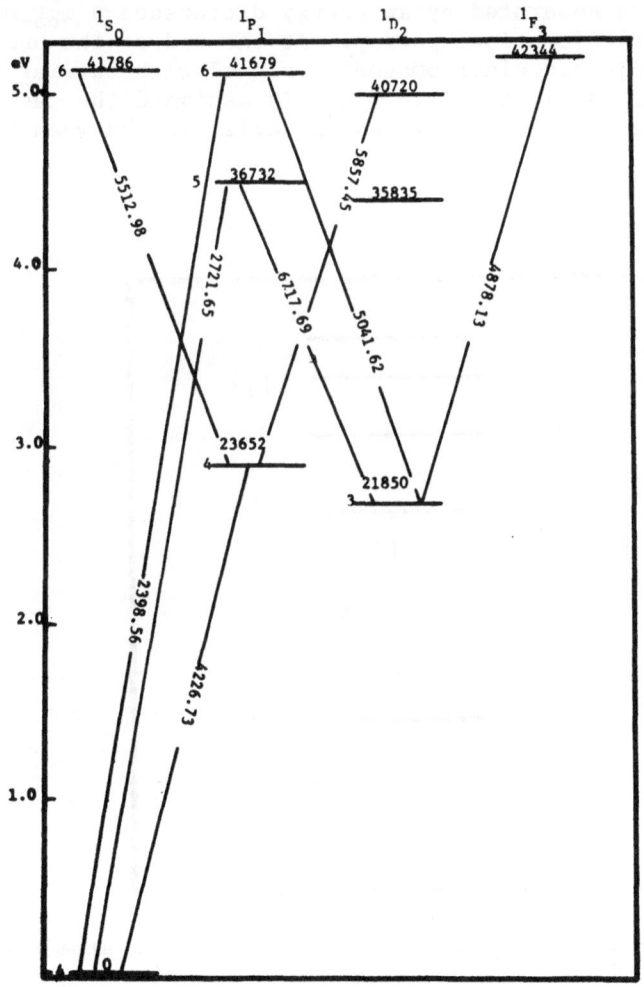

Figure 1. Lower lying singlet levels of Calcium

We assume the atoms to be initially prepared in a state of inversion between the $^1D_2$ level at 21850 cm^{-1} and the ground state. The $^1D_2$ level is the lowest excited state of the singlet manifold. It is weakly coupled to the ground state by a quadrupole transition, and to lower lying triplet states by collisional mixing. The absence of lower lying P states is an important practical advantage of Calcium atoms, if one wants to maintain a sufficient inversion prior to the arrival of the signal to be amplified.

The schematic energy level diagram to be kept in mind for the following development is illustrated in Fig. 2. The active levels $|a>$ (e.g. the ground $^1S_0$ state) and $|b>$ (e.g. the excited $^1D_2$ state) are separated by an energy difference $\hbar\,\omega_{ba}$ and are assumed to have identical parity. At the end of the pumping process an injected electromagnetic signal with carrier frequency approximately equal to $\frac{1}{2}\,\omega_{ba}$ is assigned the task of producing the required non-linear polarization between the active levels.

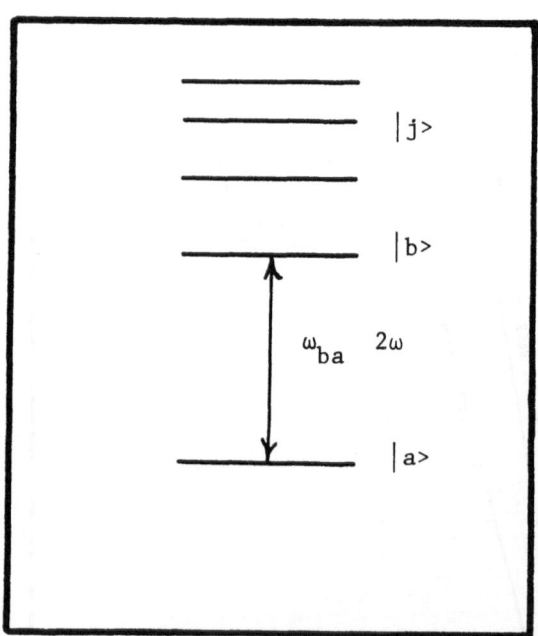

Figure 2.   Schematic energy level diagram of the atomic model used in the calculations

With reference to Fig. 2, a typical atom is described by the state vector

$$|\Psi(t)\rangle = \sum_j C_j(t) \exp(-iE_j t/\hbar) |j\rangle +$$

$$C_a(t) \exp(-iE_a t/\hbar) |a\rangle + C_b(t) \exp(-iE_b t/\hbar) |b\rangle , \qquad (2.1)$$

where the eigenvectors $|j\rangle$ refer to all the higher states of the atomic spectrum.

The atomic evolution is generated by the total Hamiltonian

$$H = E_a |a\rangle\langle a| + E_b |b\rangle\langle b| + \sum_j E_j |j\rangle\langle j| - p\mathcal{E}(x,t) , \qquad (2.2)$$

where p is the dipole moment operator in the direction of polarization of the incident field, and where $\mathcal{E}(x,t)$ is assumed to have the form of a propagating plane wave with a slowly varying envelope and phase:

$$\mathcal{E}(x,t) = \mathcal{E}_o(x,t) \cos(\omega t - kx + \psi(x,t)) , \qquad (2.3)$$

The field carrier frequency ω is set approximately equal to $\frac{1}{2} \omega_{ba} = \frac{1}{2}(E_b-E_a)/\hbar$.

As a result of our assumption concerning the parity of the active levels, the dipole moment operator takes the form

$$p = \sum_j \mu_{aj} |a\rangle\langle j| + \mu_{bj} |b\rangle\langle j| + \text{hermitian adjoint} , \qquad (2.4)$$

The coupling terms μ_{jj}, $|j\rangle\langle j'|$ between intermediate states are ignored.

The exact Schrödinger equation for the state vector (2.1) is equivalent to an infinite hierarchy of coupled amplitude equations involving not only $C_a(t)$ and $C_b(t)$, but also the amplitudes of all the intermediate states $|j\rangle$. As a first step we need to eliminate the intermediate state amplitudes using the slowly varying amplitude approximation and the assumed quasi-resonance condition $2\omega \approx \omega_{ba}$. This program can be accomplished as indicated: we find the formal solution for $C_j(t)$ and eliminate the intermediate state amplitudes from the equations for $C_a(t)$ and $C_b(t)$. The result is a pair of coupled integro-differential equations of the type

$$i\hbar \frac{d}{dt} C_a(t) = - \sum_j \mu_{aj} \mathcal{E}(x,t) \exp(-i\omega_{ja}t) \cdot \frac{i}{\hbar} \int_0^t dt'$$

$$(\mu_{ja} \mathcal{E}(x,t') C_a(t') \exp(i\omega_{ja}t) + \mu_{jb} \mathcal{E}(x,t') C_b(t')$$

$$\exp(i\omega_{jb} t') , \tag{2.5}$$

(A similar equation for $C_b(t)$ follows from Eq. (2.5) upon interchanging the indices a and b with one another.)

The exact problem may now be reduced to manageable form by taking advantage of the slowly varying amplitude approximation. This amounts to replacing the field amplitude $\mathcal{E}_o(x,t')$ and the atomic amplitudes $C_a(t')$ and $C_b(t')$ inside the integral with their values at the upper limit of integration. Finally, the exact integration of the remaining exponential factors can be easily carried out. At this point, for consistency, only the secular terms must be retained.

As a result, the atomic amplitudes $C_a(t)$ and $C_b(t)$ are found to satisfy the coupled equations of motion

$$\frac{d}{dt} C_a(t) = \frac{i}{\hbar} \{ \tfrac{1}{4} k_{aa} \mathcal{E}_o^2 C_a(t) + \tfrac{1}{4} k_{ab} \mathcal{E}_o^2 C_b(t) e^{i\alpha} \}$$

$$\frac{d}{dt} C_b(t) = \frac{i}{\hbar} \{ \tfrac{1}{4} k_{bb} \mathcal{E}_o^2 C_b(t) + \tfrac{1}{4} k_{ab} \mathcal{E}_o^2 C_a(t) e^{-i\alpha} \}, \tag{2.6}$$

The parameters

$$k_{aa} = \frac{2}{\hbar} \sum_j \mu_{ja}^2 \frac{\omega_{ja}}{\omega_{ja}^2 - \omega^2}$$

$$k_{bb} = \frac{2}{\hbar} \sum_j \mu_{jb}^2 \frac{\omega_{jb}}{\omega_{jb}^2 - \omega^2} , \tag{2.7}$$

are the electric susceptibilities of a single atom in the states $|a\rangle$ and $|b\rangle$, respectively, and

$$k_{ab} = \frac{1}{\hbar} \sum_j \frac{\mu_{ja} \mu_{jb}}{\omega_{jb} + \omega} \tag{2.8}$$

is the so-called two-photon gyroelectric ratio.

The phase angle α is defined as

$$\alpha = (2\omega - \omega_{ba})t - 2kx + 2\psi\ (x,t). \qquad (2.9)$$

As a check of the internal consistency of the elimination process, we observe that

$$\frac{d}{dt}\ (|C_a|^2 + |C_b|^2) = 0\ , \qquad (2.10)$$

as one would expect from probability conservation arguments if the virtual states are not populated during the evolution.

The amplitude equations (2.6) bear a considerable similarity to those which govern the evolution of two-level atoms undergoing one-photon transitions[1]. In the two-photon case, however, the field amplitude enters quadratically as one might expect in view of the nature of the atomic transition.

III. FIELD EQUATIONS

The field propagation in the medium is described by the wave equation in one dimension

$$\frac{\partial^2}{\partial x^2}\mathcal{E} - \frac{1}{c^2}\frac{\partial^2}{\partial t^2}\mathcal{E} = \frac{1}{c^2\varepsilon_o}\ \frac{\partial^2}{\partial t^2}P\ , \qquad (3.1)$$

where the source term P is the macroscopic polarization of the sample and ε_o is the vacuum permittivity. A convenient representation for the macroscopic polarization is

$$P(x,t) = P_c\ \cos\ (\omega t - kx + \psi) + P_s\ \sin\ (\omega t - kx + \psi). \qquad (3.2)$$

The in-phase (P_c) and out-of-phase (P_s) components of the polarization will be slowly varying functions of space and time if the carrier frequency of P(x,t) is approximately the same as that of the driving field. In the presence of competing contributions oscillating at frequencies other than ω, the polarization components P_c and P_s will, of course, be no longer slowly varying. This issue will be discussed more at length in Section 4.

Within the slowly varying amplitude and phase approximation, Eq. (3.1) reduces to the following pair of coupled transport equations

$$\mathcal{E}_o(c\ \frac{\partial}{\partial x} + \frac{\partial}{\partial t})\psi = -\ \frac{\omega}{2\varepsilon_o}\ P_c\ ,$$

$$(c\ \frac{\partial}{\partial x} + \frac{\partial}{\partial t})\mathcal{E}_o = -\ \frac{\omega}{2\varepsilon_o}\ P_s\ , \qquad (3.3)$$

which govern the propagation of the field amplitude $\mathcal{E}_0$ and phase ψ.

IV. ATOMIC POLARIZATION

The equations of motion developed in Sections 2 and 3 are made self-consistent by calculating the polarization components P_c and P_s in terms of the quantum mechanical atomic amplitudes. By definition, the macroscopic polarization is given by

$$P = N < \Psi(t) \; |p| \; \Psi(t) > \qquad (4.1)$$

where p is the dipole moment operator (Eq. (2.4)), $\psi(t)>$ the atomic state vector (Eq. (2.1)), and N is the atomic number density. As expected, the total polarization depends on the entire set of atomic amplitudes, i.e.

$$P = N \{ \sum_j \mu_{ja} \, C_a C_j^* \, e^{i\omega_{ja}t} + \mu_{jb} \, C_b C_j^* \, e^{i\omega_{jb}t} + C.C. \} \qquad (4.2)$$

Our immediate task is to eliminate the intermediate state amplitudes C_j consistently with the approximations made in deriving the atomic equations of motion (Eq. (2.6)). First we replace C_j in (Eq. (4.2)) with their exact formal solution, we replace the slowly varying amplitudes $\mathcal{E}_0(x,t')$, $C_a(t')$ and $C_b(t')$ with their values at the upper limit of the time integrations, and carry out the remaining elementary integrals.

The result of the lengthy but simple calculation reveals polarization contributions that oscillate at a frequency ω, as well as terms oscillating at the frequencies ω_{ja}, ω_{jb}, and 3ω. These additional terms oscillating at the "wrong" frequencies open the door to competing effects which may accompany the two-photon absorption or emission processes. A detailed analysis of the third harmonic contribution has shown that it can be eliminated if the incident pulse is circularly polarized[10]. Raman type competitions, on the other hand cannot be ruled out. The question of their importance in this problem remains to be worked out in detail.

In the following development we retain only the polarization components that oscillate at the frequency of the incoming field. In this case, the atomic polarization takes the form

$$P = N \{k_{aa} |C_a|^2 + k_{bb} |C_b|^2 + k_{ab}(C_a C_b^* e^{-i\alpha} + C_a^* C_b e^{i\alpha})\}$$

$$\mathcal{E}_o(x,t)\cos(\omega t - kx + \psi) \tag{4.3}$$

$$+ N k_{ab} i(C_a C_b^* e^{-i\alpha} - C_a^* C_b e^{i\alpha}) \mathcal{E}_o(x,t)\sin(\omega t - kx + \psi)$$

As in the one-photon case, the polarization depends on the expected bilinear combinations of atomic amplitudes (e.g. $C_a C_b^*$). In this case, however, the in-phase component depends also on the population of the active levels through $|C_a|^2$ and $|C_b|^2$.

V.   COUPLED ATOM–FIELD EQUATIONS

By analogy with the Bloch vector representation of the one-photon theory[1,9], it is convenient to introduce the new atomic variables

$$R_1 = i(C_a^* C_b e^{i\alpha} - C_a C_b^* e^{-i\alpha})$$

$$R_2 = -(C_a C_b^* e^{-i\alpha} + C_a^* C_b e^{i\alpha}) \tag{5.1}$$

$$R_3 = |C_b|^2 - |C_a|^2$$

This choice leads to the familiar looking vector form for the atomic equations

$$\dot{R}_1 = \left[\frac{k_{bb}-k_{aa}}{4\hbar} \mathcal{E}_o^2 + (2\omega - \omega_{ba} + 2\frac{\partial \psi}{\partial t})\right] R_2 + \frac{k_{ab}}{2\hbar} \mathcal{E}_o^2 R_3$$

$$\dot{R}_2 = -\left[\frac{k_{bb}-k_{aa}}{4\hbar} \mathcal{E}_o^2 + (2\omega - \omega_{ba} + 2\frac{\partial \psi}{\partial t})\right] R_1 \tag{5.2}$$

$$\dot{R}_3 = -\frac{k_{ab}}{2\hbar} \mathcal{E}_o^2 R_1$$

The field equations take the form

$$(c\frac{\partial}{\partial x} + \frac{\partial}{\partial t}) \mathcal{E}_o^2 = \frac{\omega N k_{ab}}{\varepsilon_o} R_1 \mathcal{E}_o^2$$

$$(c\frac{\partial}{\partial x} + \frac{\partial}{\partial t}) (2\omega - \omega_{ba} + 2\frac{\partial \psi}{\partial t}) = \frac{\omega N k_{ab}}{\varepsilon_o} (\dot{R}_2 - \frac{k_{bb}-k_{aa}}{2k_{ab}} \dot{R}_3) \tag{5.3}$$

It is convenient to introduce the following notations

$$\gamma = \frac{k_{bb}-k_{aa}}{2k_{ab}} \quad , \quad g = \frac{\omega N k_{ab}}{\varepsilon_o}$$

$$\omega_R = \sqrt{1+\gamma^2} \; \frac{k_{ab}}{2\hbar} \; \mathcal{E}_o^2 = \text{Rabi frequency} \qquad (5.4)$$

$$\Omega = 2\omega-\omega_{ba} + 2\frac{\partial\psi}{\partial t}$$

and describe the evolution of the entire system in the travelling reference frame

$$\eta = \frac{x}{c} \quad , \quad \tau = t - \frac{x}{c}$$

Finally, the coupled Schrödinger-Maxwell equations can be cast into the form

$$\frac{\partial R_1}{\partial \tau} = (\frac{\gamma}{\sqrt{1+\gamma^2}} \; \omega_R+\Omega)R_2 + \frac{\omega_R}{\sqrt{1+\gamma^2}} \; R_3$$

$$\frac{\partial R_2}{\partial \tau} = -(\frac{\gamma}{\sqrt{1+\gamma^2}} \; \omega_R+\Omega)R_1$$

$$\frac{\partial R_3}{\partial \tau} = - \frac{\omega_R}{\sqrt{1+\gamma^2}} \; R_1 \qquad\qquad (5.5)$$

$$\frac{\partial \omega_R}{\partial \eta} = g \; \omega_R \; R_1 - \ell\omega_R$$

$$\frac{\partial \Omega}{\partial \eta} = g(\frac{\partial R_2}{\partial \tau} - \gamma\frac{\partial R_3}{\partial \tau})$$

The field damping term $-\ell\omega_R$ has been added phenomenologically to describe the effects of diffraction and other non-resonant scattering losses. The coupled equations (5.5) are valid in the coherent limit, i.e., when atomic relaxation effects can be neglected. In this limit three exact relations can be derived, two in the form of conservation laws, the third providing a link between the field intensity and the detuning variable.

The first relation is a consequence of the probability conservation (Eq. (2.10)) which in terms of the Bloch variables takes the form

$$R_1^2 + R_2^2 + R_3^2 = 1 \tag{5.6}$$

The second, which is valid only for resonant propagation ($\Omega=0$), is a direct consequence of the second and third atomic equations. These can be combined to give

$$R_2 - \gamma R_3 = R_2(o) - \gamma R_3(o) = -\gamma \tag{5.7}$$

The third exact relation follows from the field equations after they have been combined to yield

$$\frac{\partial}{\partial \eta}(\omega_R \Omega) = - \ell \omega_R \Omega \tag{5.8}$$

Equation (5.8) can be integrated at once with the result

$$(\omega_R \Omega)_\eta = (\omega_R \Omega)_{\eta=o} \, e^{-\ell \eta} \tag{5.9}$$

This last result shows that, in the coherent limit, a resonant pulse ($\Omega=0$) can propagate through the active medium always maintaining the resonance condition.

VI. RESONANT COHERENT PROPAGATION. ENERGY EQUATION

The coupled equations (5.5) can be simplified considerably in the case of resonant propagation, where it can be easily shown that the Bloch equations are formally satisfied by[4(a),8]

$$R_1 = \frac{R_3^e}{\sqrt{1+\gamma^2}} \, \sin \sigma$$

$$R_2 = \frac{R_3^e \gamma}{1+\gamma^2} \, (\cos \sigma - 1) \tag{6.1}$$

$$R_3 = \frac{R_3^e}{1+\gamma^2} \, (\cos \sigma + \gamma^2) \, , \quad (R_3^e = \text{equilibrium population})$$

provided the variable $\sigma(\eta,\tau)$ is connected to the Rabi frequency by

$$\frac{\partial\sigma}{\partial\tau} = \omega_R \tag{6.2}$$

or by

$$\sigma(\eta,\tau) = \int_0^\tau d\tau'\, \omega_R(\eta,\tau') \tag{6.3}$$

Thus $\sigma(\eta,\tau)$ is proportional to the accumulated pulse energy up to the local time τ. Clearly the total pulse energy $\Sigma(\eta)$ will be given by

$$\Sigma(\eta) = \lim_{\tau\to\infty} \sigma(\eta,\tau) \tag{6.4}$$

The dynamics of the coupled atom-field system is governed by Eq. (6.1) and by the partial differential equation

$$\frac{\partial^2\sigma}{\partial\eta\,\partial\tau} = \frac{g}{\sqrt{1+\gamma^2}}\,\sin\sigma\,\frac{\partial\sigma}{\partial\tau} - \ell\,\frac{\partial\sigma}{\partial\tau} \tag{6.5}$$

which follows directly from Eq. (6.2) and the field equations. On the surface, Eq. (6.5) is reminiscent of the so-called area equation first proposed by Arecchi and Bonifacio to describe the pulse evolution in a homogeneously broadened laser amplifier under coherent and resonant propagation conditions[9]. Two main differences will be recognized: first of all, the gain term in the Arecchi-Bonifacio area equation has the form $G \sin \sigma$, where G is a gain parameter; secondly, the "area" σ is not a measurable quantity in the one-photon amplifier theory.

A mathematical consequence of the structure of the gain term in our Eq. (6.5) is that the equation can be integrated exactly with respect to the local time with the result

$$\frac{\partial}{\partial\eta}\,\Sigma(\eta) = -\ell\Sigma(\eta) + \frac{g}{\sqrt{1+\gamma^2}}\,(1-\cos\Sigma(\eta)) \tag{6.6}$$

The steady state behavior of the total pulse energy, $\Sigma(\eta)$, is governed by the transcendental equation

$$1-\cos\Sigma(\infty) = \frac{\sqrt{1+\gamma^2}}{g}\,\ell\,\Sigma(\infty) \tag{6.7}$$

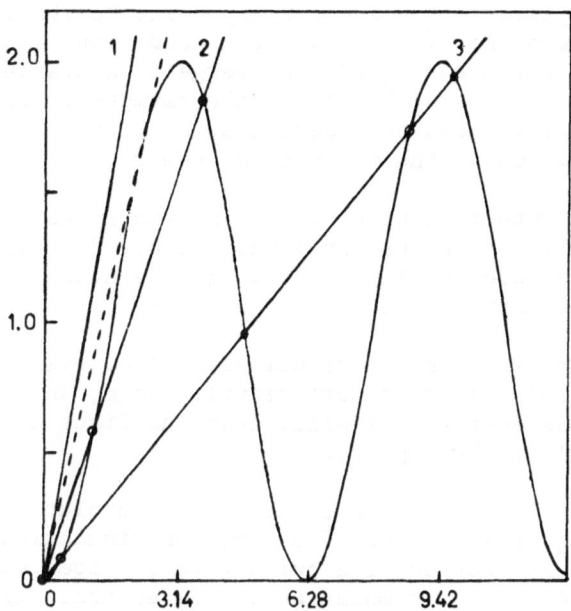

Figure 3. The asymptotic steady state solution of the energy
equation (Eq. (6.7)) correspond to the intercepts of
the straight line $\sqrt{1+\gamma^2}\ \dfrac{\ell}{g}\ \Sigma$ with the curve $1-\cos\Sigma$.
The straight lines 1,2 and 3 have slopes equal to
1, 1/2, 1/5 respectively. The stable solutions are
marked with solid circles. The unstable solutions
are marked with open circles. The critical slope
(dashed line) equals 0.7246.

It is clear from Fig. 3 that non-trivial solutions of Eq. (6.7)
will exist only if the gain to loss ratio exceeds a certain
threshold value and also that, if such threshold is exceeded,
multiple steady state solutions are possible. Which of the
possible multiple solutions is going to be realized physically,
depends on the incident value $\Sigma(o)$.

By inspection of Fig. 3 several quantitative conclusions
can be drawn:

i)   Two simultaneous requirements must be satisfied for the
asymptotic propagation of a pulse having a total energy
$\Sigma(\infty) \neq o$: the parameter $\sqrt{1+\gamma^2}\ \dfrac{\ell}{g}$ must be smaller than
the critical value 0.7246 (the slope of the dashed line
in Fig. 3), and the incident pulse energy must exceed

the numerical value corresponding to the first unstable root of Eq. (6.7). If both conditions are satisfied, the total energy will converge to a stable non-vanishing value. The stability of the asymptotic solution of Eq. (6.7) can be assessed at once by analyzing the sign of $d\Sigma/d\eta$ in the vicinity of the roots.

ii)   The output value of the pulse energy can be larger or smaller than the input value $\Sigma(o)$, depending on the magnitude of $\Sigma(o)$ relative to the nearest stable asymptotic root.

iii) There is no small signal gain, i.e. even if the gain to loss ratio is sufficiently large, an input pulse whose energy is smaller than the first unstable root will not be amplified.

More detailed information on the transient evolution and envelope modulation requires the use of a computer simulation. As we show in the next section, the pulse envelope does not approach a steady state[11]. Our simulation indicates that, if the threshold conditions are met, the propagating pulse narrows indefinitely as the energy approaches its stable asymptotic value. Furthermore, whenever multiple stable solutions are possible, the n-th stable root $\Sigma(\infty)$ corresponds to a pulse envelope that has been split into (n-1) distinct pulses.

VII.   COMPUTER SIMULATIONS. COHERENT PROPAGATION

The set of equations (5.5) has been analyzed with a hybrid computer. In this section we summarize the main results with an eye on the following features of the problem

i)   Validity of the threshold conditions for propagation of simple and multiple pulses.

ii)   Transient evolution of single and multiple pulses.

iii) Approach to steady state of the pulse energy.

The system is assumed excited in a swept excitation mode corresponding to the initial and boundary conditions

$$R_1(\tau=o,\eta) = R_2(\tau=o,\eta) = 0$$
$$R_3(\tau=o,\eta) = 1$$
$$\Omega(\tau,\eta=o) = 0 \qquad\qquad (7.1)$$
$$\omega_R(\tau,\eta=o) = \omega_R(\tau)$$

The input pulse intensity was assigned the shape

$$\omega_R(\tau) = \omega_R^o \sin^2 (\pi \ \tau/\tau_p)$$

where τ_p is the pulse duration from the leading to the trailing edge.

As expected, the computer simulation shows that if the detuning variable Ω is zero at the boundary of the medium, it remains equal to zero during the evolution of the pulse.

The threshold conditions have been checked by choosing a gain to loss ratio smaller than the threshold value (reciprocal slope of the dashed line in Fig. 3) or an incident pulse energy smaller than the first unstable root corresponding to a sufficient large gain to loss ratio in Fig. 3. In both cases the incident pulse is attenuated during the propagation.

The first two examples of propagation above threshold are shown in Figs. 4 and 5. These simulations display the evolution of two input pulses with initial energies $\Sigma(o) = 3$ and $\Sigma(o) = 8$, respectively, which propagate in a medium characterized by a gain to loss ratio equal to 2. Under these conditions, Eq. (6.7) and Fig. 3 predict the propagation of a single pulse. This is confirmed in Figs. 4 and 5.

In the case of Fig. 5, the incident pulse has an energy larger than the expected steady state value $\Sigma(\infty)$. Strong transient modulation and energy loss characterize the pulse evolution. Still, as observed with all the simulations where the threshold conditions were satisfied, peak power amplification occurs even when the pulse experiences a net energy loss. An example of pulse break-up is shown in Fig. 6. The gain to loss ratio equals 5 and the input energy is set equal to 9.8. According to Eq. (6.7) and Fig. 3, the asymptotic value of the pulse energy is predicted to correspond to the third stable solution ($\Sigma(\infty) = o$ is considered to be the first stable root). Two propagating pulses are observed under these conditions. By contrast (Fig. 7), a pulse with a smaller input energy ($\Sigma(o) = 5$) travelling in a medium with the same gain to loss ratio does not exhibit pulse break-up. The predicted asymptotic value of the pulse energy equals the second stable root displayed in Fig. 3 for this value of $g/\ell \ \sqrt{1+\gamma^2}$.

In all the cases shown above, the total pulse energy at every section along the amplifier was read off at the far right of the solid lines representing the evolution of $\sigma(\tau,\eta)$. With due allowance for the fact that some of the simulations had to be terminated before reaching their asymptotic stage, the agreement with the predicted values of $\Sigma(\infty)$ was found to be quite satisfactory.

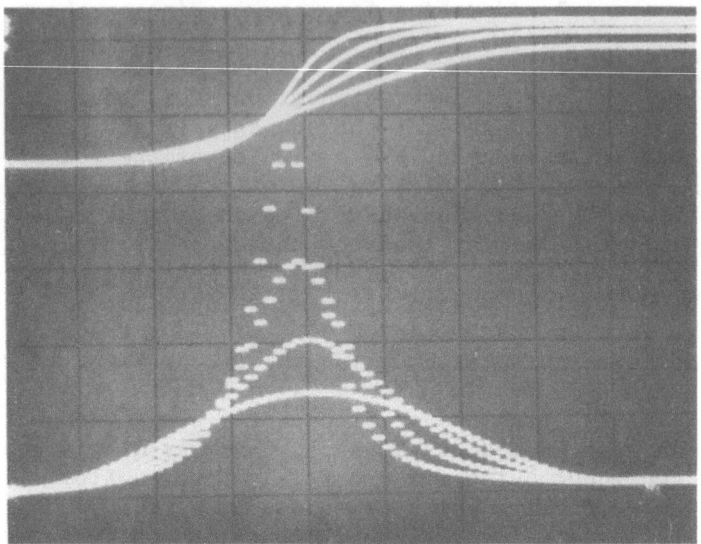

Figure 4.   Computer simulation illustrating the evolution of
            the pulse intensity through the amplifying medium.
            The different dashed curves represent the intensity
            envelope in different sections of the amplifier. The
            solid curves are the corresponding integrated energies
            $\sigma(\eta,\tau)$. The value of $\sigma(\eta,\tau)$ at the far right gives
            the total energy $\Sigma(\eta)$. The horizontal axis is the
            local time axis, with $\tau=0$ (leading edge of the
            pulse) at the far left. The input energy is $\Sigma(o)=3$
            and the gain to loss ratio $g/\ell \sqrt{1+\gamma^2}$ equals 2.

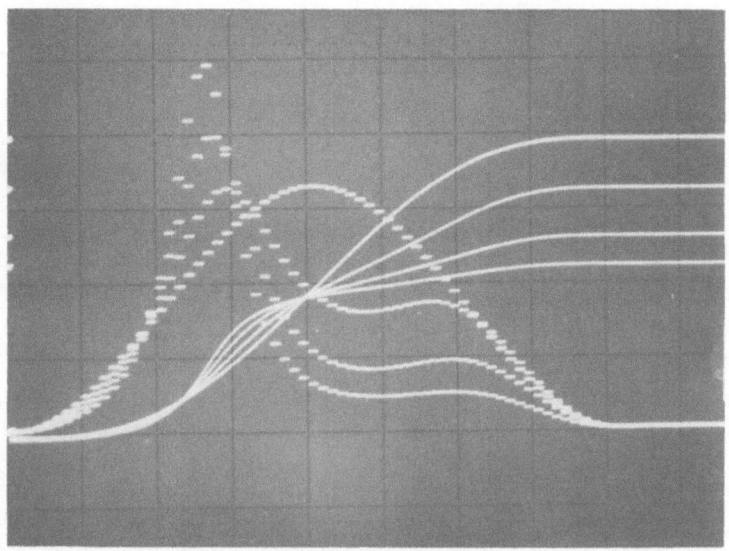

Figure 5. Evolution of a pulse with an initial energy $\Sigma(o)=8$;
the gain to loss ratio equals 2 as in Fig. 4. Note
the transient envelope modulation and pulse energy
loss.

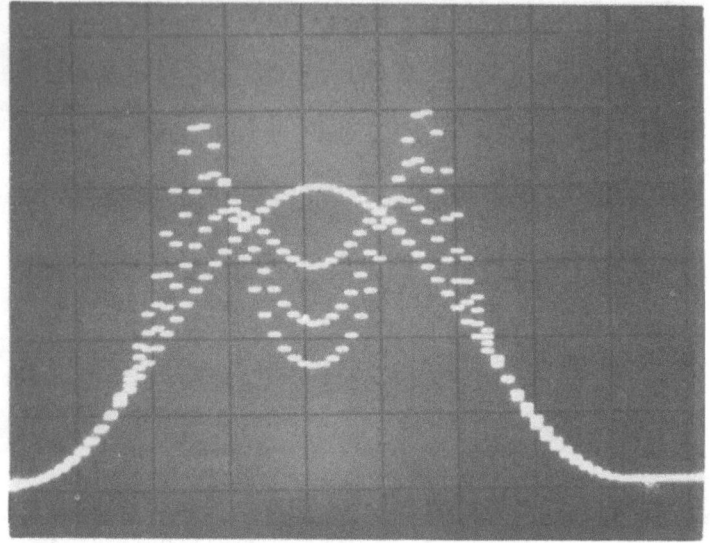

Figure 6. Evolution of a pulse with initial energy $\Sigma(o)=9.8$ in
a medium with a gain to loss ratio equal to 5. The
asymptotic pulse energy is expected to take a value
equal to the third stable root shown in Fig. 3
(straight line #3). Under these conditions, pulse
break-up is observed.

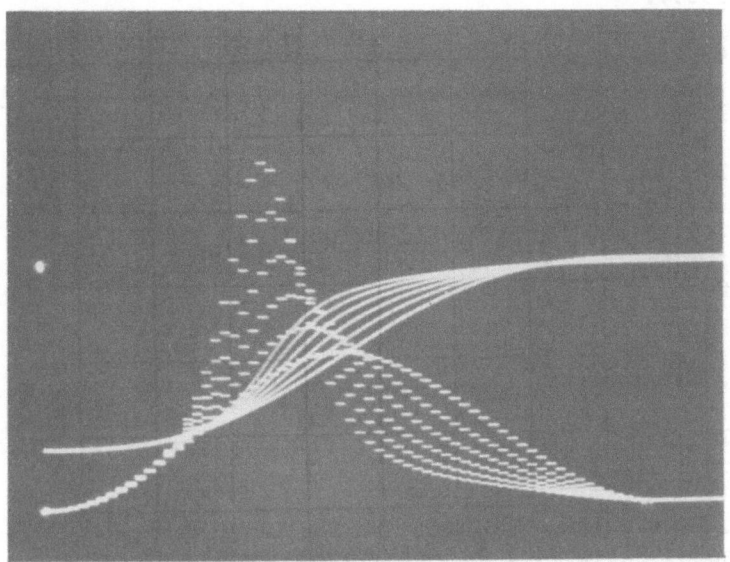

Figure 7.   Propagation of a pulse with initial energy $\Sigma(o)=5$.
The gain to loss ratio is the same as in Fig. 6.
A single peak is expected on the basis of Eq. (6.7)
and Fig. 3.

REFERENCES

(1)   For a nice survey see for example, L. Allen and J.H. Eberly,
Optical Resonance and Two-level Atoms, John Wiley and Sons
(New York), 1976.

(2)   A very extensive literature is available on two-photon
processes. For a critical comparison of different theoretical approaches and additional references see for
example: D. Grischkowski, in Laser Applications to Optics
and Spectroscopy, Vol. II of Physics of Quantum Electronics,
Eds. S.F. Jacobs, M. Sargent III, J.F. Scott, M.O. Scully
(Addison-Wesley 1975), D. Grischkowski, M.M.T. Loy,
P.F. Liao, Phys. Rev. A12, 2514 (1975).

(3)   F. Biraben, B. Cagnac, G. Grynberg, Phys. Rev. Lett. 32
643 (1974),
M.D. Levenson, N. Bloembergen, Phys. Rev. Lett. 32, 645
(1974).

(4)(a) E.M. Belenov, I.A. Poluektov, Sov. Phys. JETP <u>29</u>, 754
        (1969).
        M. Takatsuji, Physica 51, 265 (1971); Phys. Rev. <u>B2</u>,
        340 (1970).
        N. Tan-no, K. Yokoto, H. Inaba, Phys. Rev. Lett. <u>29</u>, 1211
        (1972).

(5)   M. Takatsuji, Phys. Rev. <u>A11</u>, 619 (1975).
        F.H.M. Faisal, J. Phys. B., <u>10</u>, 2003 (1977).
        J.C. Diels, Opt. Quant. Electron <u>8</u> 513 (1976).
        H.P. Yuen, F.Y.F. Chu, Proceedings of the Fourth Rochester
        Conference on Coherence and Quantum Optics, Eds. L. Mandel
        and E. Wolf (to be published).

(6)   R.L. Carman, Phys. Rev. <u>A12</u>, 1048 (1975).

(7)   N. Tan-no, K. Yokoto, H. Inaba, J. Phys. <u>B8</u>, 339 (1975);
        ibid. <u>B8</u>, 349 (1975).

(8)   L.E. Estes, L.M. Narducci, B. Shammas, Lettere al Nuovo
        Cimento, Serie 2, <u>1</u> 175 (1971).

(9)   This logical scheme has a venerable history in laser
        physics. It was implemented in the context of the one-
        photon laser amplifier problem by A.T. Arecchi, R. Bonifacio
        IEEE, Jour. Quant. Electron, <u>QE1</u>, 169 (1965).

(10)   See D. Grischkowski et al. in Ref. 2.

(11)   This result was shown analytically by H.P. Yuen and F.Y.F.
        Chu in Ref. 5.

HIGH RESOLUTION SPECTROSCOPY IN ATOMIC RYDBERG STATES

by S. Haroche

Ecole Normale Supérieure, Laboratoire de Physique

24, rue Lhomond, 75231 PARIS CEDEX 05, France

1. INTRODUCTION

This chapter is intended to be a summary of lectures given at the Summer School, on High Resolution Spectroscopy in Atomic Rydberg States. Only the main lines of the course are presented here and the reader is referred for details to the bibliography given at the end.

The aim of the lectures was to give an overview of various high resolution Doppler free spectroscopy techniques which have been recently developed on Rydberg levels, using the laser as a tool.

One can quite generally say that the development of lasers has induced two major revolutions in atomic spectroscopy :

i) it has provided the possibility of preparing and studying new species, or to excite new transitions very difficult to observe by other means, such as ionic states, molecular levels, exotic atoms, Rydberg states, forbidden transitions ...

ii) it has also allowed to develop new linear and non linear techniques making use of the monochromaticity, intensity and coherence properties of the laser light, in order to achieve ultrahigh resolution well beyond the Doppler limit.

These two types of advantages, independently or in conjunction with each other, have in the recent years widely broadened the scope of Atomic Physics and to a large extent renewed this field of research.

The case of Rydberg states spectroscopy is particularly typical of this new situation. These states may indeed be considered as the obvious "domain" where the lasers have shown their great advantages over conventional light sources on both points i) and ii) above : at first, the high intensity of the laser fields is needed to excite the atoms from ground states to the Rydberg levels, because the corresponding oscillator strengths are very weak. Second, very high resolution is a must in order to study these levels, since the energy structures in these states are usually very small. Almost all known high resolution spectroscopy techniques involving lasers have been indeed applied on these states and can thus be easily compared with each other in the same "context". This is the main reason why I have chosen here to restrict myself to this specific field of spectroscopy, rather than giving scattered examples of Laser Spectroscopy taken from different domains of atomic physics. Another reason is of course that Rydberg states have a big intrinsic interest for spectroscopists, astrophysicists and metrologists...

Before entering into the description of various spectroscopical techniques, it is useful to recall briefly the main characteristics of Rydberg states which might be relevant for the understanding of the remaining of this course.

2. GENERAL PROPERTIES OF ATOMIC RYDBERG STATES

Rydberg states are very excited states of atoms in which a valence electron corresponds to a very high value of the principal quantum number n. The most dramatic property of these states is certainly the very large size of their orbitals, which can easily be of the order of several thousands of angstroems ("biological" size !). As these states are very close to the continuum of positive energy states, they are very sensitive to all kinds of perturbations able to ionize the atoms.

Rydberg states belonging to any atomic species have -for obvious reasons- strong similitudes with the corresponding levels of Hydrogen. In a Rydberg state, the valence electron spends indeed most of its time very far from the remaining of the atom, made of the nucleus and of the "core" of the other electrons. These other electrons shield the charge of the nucleus which thus appears to the valence electron as a single elementary charge similar to the one of the proton. Of course, this is only an approximation and most of the spectroscopic interest of the Rydberg states lies precisely in the small but detectable differences with respect to Hydrogen. The Hydrogen model is however very useful because it is entirely solvable and gives a good "feeling" about the Rydberg states dimensions and main physical characteristics. Let us study it first, before explaining how the departure from this model can be described by introducing the notion of "Quantum Defects".

2.1. The Hydrogen Model ([1])

In Hydrogen, the excited state $|n\ell>$ corresponding to principal quantum number n and angular momentum ℓ has an energy :

$$E = - \frac{R}{n^2} \tag{1}$$

where $R = 109737$ cm^{-1} is the Rydberg constant. Order of magnitude of energy differences between adjacing levels with $\Delta n = 1$ is obtained by deriving (1) with respect to n :

$$\Delta E = \frac{2R}{n^3} \tag{2}$$

ΔE falls typically in the 10 cm^{-1} domain for $n = 30$ (millimetric transitions).

The spatial extension of the atomic orbitals is well represented by the mean value of r :

$$< r > = \frac{1}{2} \left[3n^2 - \ell(\ell + 1) \right] a_0 \tag{3}$$

where a_0 is the Bohr radius. $< r >$ is of the order of 1000 Å for $n = 30$. The mean value of r^2 :

$$< r^2 > = \frac{n^2}{2} \left[5n^2 + 1 - 3\ell(\ell + 1) \right] a_0^2 \tag{4}$$

is roughly proportional to n^4 and gives a good order of magnitude for the geometrical atomic cross section. This n^4 dependence appears also in all physical quantities proportional to $< r^2 >$: rates of electric dipole transitions between closely lying levels, atomic diamagnetism ...

The oscillator strengths linking the ground state to a n, $\ell = 1$ Rydberg state are proportional to $1/n^3$. The lifetime of a given n, $\ell = 1$ state increases on the other hand as n^3. Lifetime of states having not only a large n value, but also large ℓ's are much longer than the lifetimes of $\ell = 1$ states. As a result, the average lifetime of an n manifold (average over all ℓ's) increases much more rapidly than n^3 and is in fact proportional to n^4[5] .

Practically, Rydberg states of hydrogen with n $\sim$ 100 have lifetime in the 10^{-3} - 10^{-2} second range and can be considered as metastable in most spectroscopic experiments.

When taking into account the electron spin, each $\ell \neq 0$ level acquires a structure due to the fine structure perturbation :

$$V_{fs} = \frac{e^2}{2m^2 c^2} < \frac{1}{r^3} > \vec{\ell} . \vec{s} \tag{5}$$

The splitting between the $j = \ell + 1/2$ and $j = \ell - 1/2$ levels is given by :

$$\Delta E_{n\ell} = \frac{R\alpha^2}{n^3} \frac{1}{\ell(\ell+1)} \tag{6}$$

where α is the fine structure constant. This structure is typically of the order of a few Mc/s for $n \sim 20$ and $\ell \sim 2$-3.

When applying an external electric field perturbation, the various levels of an n manifold exhibit a linear Stark effect, with a splitting into n equidistant sublevels. The separation between the extreme level is :

$$(\Delta E)_{Stark} \simeq 3q_e \mathcal{E} \, a_o \, n \, (n-1) \quad (q_e : electron \ charge) \tag{7}$$

and is roughly equal to $2q_e \mathcal{E} |< r >$, that is to the voltage drop across the largest dimension of the Rydberg atom times the electron charge.

If the electric field is increased beyond a critical value approximately given by the formula :

$$E_c(n) \simeq \frac{3.10^8}{n^4} \quad V/cm \tag{8}$$

the various levels of the n manifold are ionized, the valence electron being "pulled" out of the atom. This critical value ($E_c \sim$ 400 V/cm for n = 30) corresponds roughly to an externally applied field whose magnitude matches the average internal field experienced by the valence electron in its Rydberg orbital. One can also say that the external field lowers the potential barrier that prevents the electron from escaping the atom and that $E_c(n)$ corresponds to the critical value for which the top of the barrier matches the energy of the n-manifold. One should note that formula (8) is only an approximate one. When describing in detail the ionization effect, one realizes that various sublevels of a given manifold are in fact ionized in slightly different fields around E_c. This is due to the fact that the different sublevels of the n-manifold are Stark shifted with respect to each other and reach different energies by the time the electric field is large enough to ionize them. Furthermore, for a given sublevel, ionization is not strictly speaking a discontinuuous phenomenon occuring just above the threshold field. In fact, due to the possibility of tunelling accross the energy barrier, the ion production rate becomes appreciable a little bit below the actual threshold E_c and very rapidly increases when the threshold is reached [2].

2.2. Non-Hydrogen Rydberg Atoms : The Quantum Defect Theory

In non - Hydrogen Rydberg states, the departure from the hydrogenic behaviour can be described within the framework of the so-called Quantum Defect Theory. This theory is a quite simple one in

the case of alkali-atom Rydberg states for which the atomic core
has a closed shell structure. In this case, the perturbation due to
the core can be divided into two parts :

i) a penetration contribution ([3]) due to the fact that the valence electron sees, when it penetrates into the core a deeper potential than in Hydrogen

ii) a polarization contribution ([4]) due to the fact that the
outer electron slightly polarizes the core, the induced
polarization reacting back on the valence electron and changing its energy. The former contribution dominates in the
so-called penetrating orbitals ($\ell=0,1,2$ in Na). The latter is
generally much smaller and dominates in non penetrating high
angular momentum states ($\ell > 2$ in Na).

These two types of perturbation can be taken care of by retaining for the non-hydrogen Rydberg states the hydrogen wave function
description, provided one slightly changes the principal quantum
numbers n which appear in the analytical expression of the
wave functions and energies. These numbers cease to be integers and
become equal to $n - \varepsilon_\ell$ where ε_ℓ is the so-called quantum defect
of the $n\ell$ level. ε_ℓ -which to first approximation depends only
on the angular momentum of the level- summarizes most of the effect
of the core perturbation on the valence electron and has a contribution coming from both penetration and polarization of the core.
Introduction of the ε_ℓ correction takes care of the change of the
boundary conditions of the Schrödinger equation due to the presence of the core around the nucleus. Its effect is to pull the nodes
of the wave function toward the nucleus. On the energies, the effect of ε_ℓ is to depress the position of the levels, which become :

$$E_n = - \frac{R}{(n - \varepsilon_\ell)^2} \qquad (9)$$

The quantum defect due to the penetration contribution can in fact
be understood by making a very instructive comparison with the
frequency pulling effect of Laser Physics : in fact, the boundary
conditions at $r = 0$ and $r = \infty$ for the radial Schrödinger equation
of hydrogen are quite similar to those obeyed by the electromagnetic field in a laser cavity. The existence of the atomic core results in an effective change of the potential near $r = 0$ quite similar to the change of refractive index produced by the introduction of a dielectric near one of the mirrors of the laser cavity.
The presence of such a dielectric changes the dispersion formula
linking the wavelength to the frequency   of the electromagnetic
field. In order to still match the boundary conditions which impo-
ses an integer number of wavelengths between the mirrors, the frequency of the cavity has slightly to change and the modes of the
e.m. field are slightly pulled towards the dielectric. The corresponding effect occurs in the Rydberg states : the "frequency" of

the valence electron, i.e. its energy, is slightly modified and the nodes of the wave function are displaced, all these effects being described by the change ε_ℓ of the principal quantum number n.

As a result, the degeneracy of the n-manifold is lifted and the various nℓ states of the same n value take slightly different energies. Accurate measurements of these small energy gaps (typically in the 10^9-10^{10} c/sec range for n $\sim$ 20, ℓ $\sim$ 2-3) allow to study in details the penetration and polarization effects and thus give interesting informations about the geometry and polarizabilities of the alkali ion which constitutes the core of the atom.

For two or more valence electron atoms (for example alkaline-earth), the effect of the open shell core on the Rydberg levels is slightly more subtle to describe. In particular, some Rydberg states might coincide -or nearly coincide- in energy with other levels corresponding to the excitation of two valence electrons to much smaller orbitals (high energy valence states of low quantum numbers). These coincidences might result in strong polarization effects of the core and produce resonant variations of quantum defects along a series of Rydberg levels. These effects are analyzed in, the context of the so-called Multichannel Quantum Defect Theory ([5]) which cannot be described in details in this course.

2.3. Hydrogenic and Non-Hydrogenic Behaviour of Rydberg States

The modification to energies and wave functions due to the quantum defects might, in some cases, result in a very important change of the atomic properties as compared to hydrogen. In other cases, the modification may, on the contrary, be relatively small. Quite generally, all properties which depend on diagonal matrix elements involving only the wave function of a single Rydberg state are almost "hydrogenic". On the other hand, "non-diagonal" characteristics depending on matrix elements between different states might be strongly affected by quantum defects... Among the former characteristics, one can quote the energy of the states -which differ at most by a few per cent from those of hydrogen-the atomic sizes and diamagnetic properties. Among the strongly modified properties, one finds the oscillator strengths, natural lifetimes and atomic polarizability ...

The reason why diagonal contrîbutions are much less affected by quantum defect than the off-diagonal ones is quite clear : the presence of the core results, as we have seen, in a "pulling" of the atomic wave function nodes towards the nucleus. Integral involving the square of a given wave function will not be drastically modified by such a pulling, whereas integrals involving overlap of two wave functions pulled by different amounts, can be strongly modified.

Calculation of these modifications as a function of the respective quantum defects of the various levels involved can be done by several analytical methods, the most popular and simpler one being the Coulomb approximation of Bates and Damgaard ([6]) which amounts to extend hydrogenic formulas involving integer parameters to non integer quantum number ...

Study of these approximations is of course beyond the scope of these short lecture notes.

So far relativistic properties such as fine structures are concerned, the simple quantum defect model is not sufficient to account for the differences between hydrogenic and non hydrogenic Rydberg states. In fact, it is known that some levels (nd levels of Na for example) have an inverted fine structure ([7]) (the highest $j = \ell + s$ level lies below the lowest j one), which cannot be understood at all in a central potential model. To understand this oddity, one has to take into account not only the perturbing effect of the "average" core distribution on the valence electron, but also the correlations between the valence and core electron resulting from exchange contributions ([8]).

Non hydrogenic behaviour might also be very important when studying the response of a Rydberg state to an externally applied electric field ([4]). In low electric fields, such that the electric field perturbation matrix elements are smaller than the small energy differences due to the quantum defects, the Stark effect is quadratic. In "high" electric fields, the external electric field perturbation dominates the "internal" perturbation due to the presence of the core and the Stark effect becomes linear as in hydrogen. Transition between quadratic and linear regions ([4]) occurs for a critical field decreasing as $n^{-5}\ell^{-6}$, typically of the order of 0.1 V/cm for $n = 20$, $\ell = 3$, in Na.

Field ionization might also be different in hydrogen and in other non hydrogen Rydberg levels. The rate of field ionization depends indeed on the way the electric field continuously modifies the energy diagram when it is increased from zero to high values ([9]). The dynamical behaviour of the atom thus depends on a subtle manner on the whole Stark diagram of the atom.

3. GENERAL FEATURES OF HIGH RESOLUTION SPECTROSCOPY IN RYDBERG STATES

Spectroscopy of Rydberg states consists in the investigation of the systematic and sometimes very small differences between hydrogen and the other atoms, that we have described in the previous paragraph. The study of these differences is a good check of atomic structure perturbation theory, provides information on the core of the atom and allows to develop techniques which might have interesting applications (i.e. : isotope separation).

Among the various types of measurements one can make, let us
quote :
- Determination of Δn ≠ 0 energy intervals which usually fall in
  the millimetric domain
- Determination of Δn = 0, Δℓ ≠ 0 intervals which are in the cen-
  timetric domain
- Determination of fine structure intervals, in the MHz range
- Measurements of polarizability
- Measurement of Zeeman diagram and atomic diamagnetism
- Measurement of oscillator strength and atomic lifetimes

For most of these experiments, one has to rely on high resolution spectroscopy techniques allowing to get rid of the optical
Doppler effect which would wash out completely the structure of
the lines. Two different types of methods are quite generally
competing with each other in all fields of atomic spectroscopy:

i) The so-called "Double Resonance" methods which do not resolve the structures in the optical transitions, but which measure
the small energy intervals directly in the radiofrequency range
either by electric or magnetic resonance techniques, or by quantum
beats and level-crossing methods.

ii) The so-called "Optical High Resolution Methods" which rely
on the resonant interaction of atoms with monochromatic and tunable
laser light. These methods are the only ones which provide accurate
measurements of optical transitions frequencies. Among these, one
can quote Laser-beam Spectroscopy, Fluorescence line-narrowing and
Saturation Spectroscopy, Multiphoton Doppler free Spectroscopy ...

Of course, both methods i) and ii) have been applied to a lot
of various energy levels beside Rydberg states. Their adaptation
to Rydberg levels have however implied some specific and important
changes to the detection procedures that we will now describe
briefly.

For ordinary low excited atomic or molecular states, the most
convenient detection techniques are the observation of the fluorescent light emitted by the system interacting with the radiofrequency or laser beam , or the detection of the transmitted radiation. The characteristics of these radiations (frequency and polarization distributions) generally provide the needed information
about the studied resonant processes. These types of detection
have also been developped for Rydberg levels, but they are generally very difficult to implement because the oscillator strength of
the transition linking these levels to low lying states are very
low and the corresponding signals usually very small. One thus
generally prefers to use, when it is possible, non optical techniques of detection which provide a much higher signal to noise ratio.

The most popular one is the selective field ionization method
([10]) developped in atomic-beam type of experiments : at a given
"detection" time, a pulse of electric field of convenient amplitude and direction is applied to the atoms in order to ionize them.
The ion or electron current is collected and amplified in a multiplier and gives rise to an electrical signal which allows to detect
a single atom at a time. Selectivity of the detection results from
the fact that the ionization threshold depends on n (Equ. 8) and
also slightly varies in a given manifold for different m_ℓ substates.
Small changes in the various Rydberg state populations induced by
laser of Rf fields can thus be easily detected. Atomic coherence
between m_ℓ substates precessing at the corresponding Bohr frequency can also be detected as a modulation of the ion current in a
field of convenient direction and amplitude.

A variant of the previous technique is the so-called "thermoionic diode detection" ([11]) which can be applied in atomic cell
experiments : the cell in which the Rydberg atoms are studied
contains electrodes to which a small voltage is applied. When no
Rydberg states are present, the current in this diode is blocked
and a space charge develops into the gas. When a Rydberg atom is
produced, it has a probability of being ionized through collision
with other atoms. The resulting ion and electron modifies the
space charge in the cell and the diode starts to drive some current.
This current is detected, which results in a strong amplification
of the initial ion or electron. This method is a very simple and
elegant one. It has though some drawbacks when compared to the
previous one : perturbation due to the necessary collisions with
the background gas and to the applied voltage, no selectivity for
Rydberg levels of different n and m_ℓ ...

4. "DOUBLE-RESONANCE TECHNIQUES" APPLIED TO RYDBERG STATES

4.1. Ordinary "double resonance":microwave spectroscopy in Rydberg states ([12])([13])

A typical double resonance experiment on Rydberg states is represented on Figure 1. An atomic beam of alkali atom (for example
Na) is irradiated by two synchroneous laser pulses preparing the
atoms in an ns or an nd state by stepwise excitation through a
resonant intermediate p state (Figure 1-a). The atoms thus excited
undergo a microwave or a radiofrequency resonance towards nearby
lying Rydberg levels (for example ns → n'p or nd - n'f transition
with n-n' = 0, ±1, ±2; Figure 1-b). It can be either a single photon, or a multiphoton transition. The resonance can be detected
by a change of the fluorescence pattern of the light reemitted by
the atom ([12]). For example, a nd - n'f transition will result in
the apparition of a n'f - 3d fluorescence transition that can be
detected with adequate filters and photomultipliers. However, this

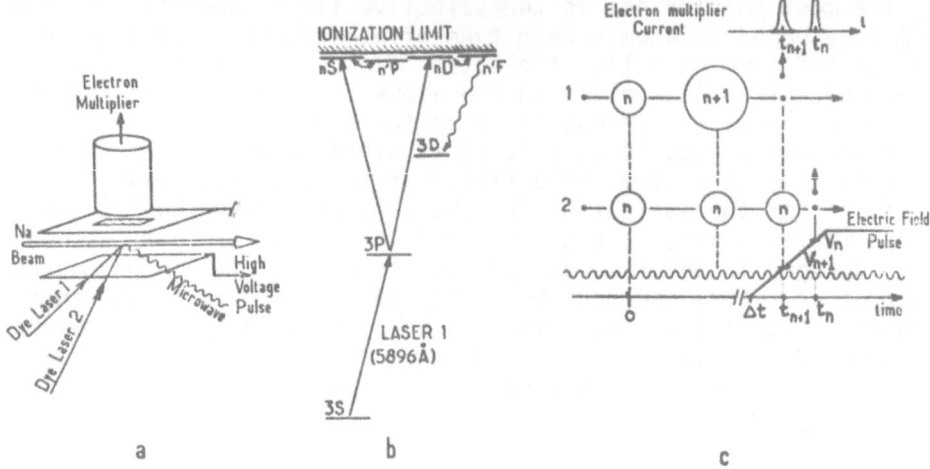

<u>Figure 1</u> : Microwave spectroscopy in Na Rydberg state : scheme of
          experimental set-up (fig. a), of energy diagram showing
          the relevant levels (fig. b) and of the sequence of
          events experienced by the atoms when field ionization
          detection is used (fig. c).

type of detection is less sensitive than the electrical detection
(^{13}) sketched on Figure 1-c in the case of a n → n+1 resonance.
At time ΔT after the laser excitation, a pulse of electric field
smoothly raising in a time of about 1 microsecond is applied to
the atoms. The voltage values corresponding to field ionization
from states n and n+1 are reached at two different times t_n and
t_{n+1}. If the microwave is off resonance, only level n is populated
(directly by the lasers) and one gets a single ion current peak
around t_n. When the microwave is on resonance, level n+1 is also
populated and another peak appears around t_{n+1}. By gating the detection around this latter time, one obtains an ion current signal
measuring directly the rate of microwave transition between n and
n+1. The theoretical resolution of the experiment is very high,
because the microwave can be made almost monochromatic and the
Doppler effect is, for microwave transition, negligible. Present
resolution of the millimetric resonances is of about 1 Mc /sec
(^{13}). Resolution as good as 1 kc/sec should be reached when improving
microwave source stability and frequency counting techniques.
Figure 2 gives an example of spectrum showing both single and double photon microwave transitions. Systematic experiments of this
kind have allowed to measure with high accuracy quantum defects
of s, p, d and f levels, np fine structures, and atomic polarizabilities as large as 10^{10} times those of ground or low lying
excited states of the same atomic varieties (^{13}).

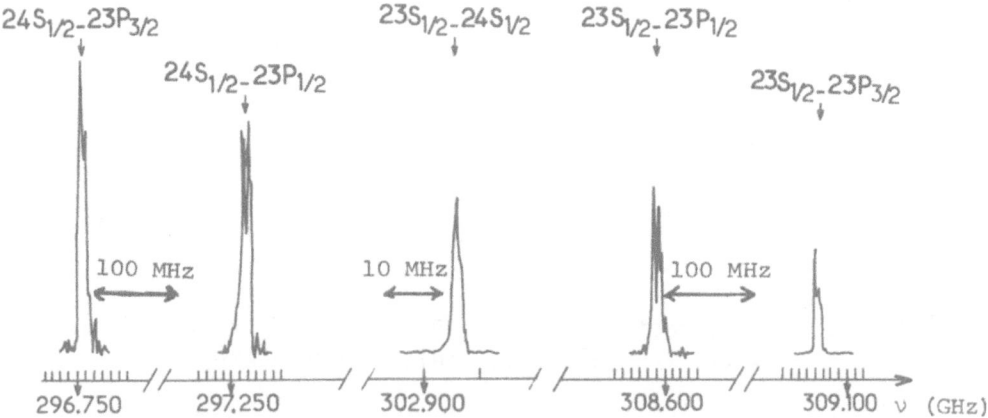

<u>Fig. 2</u> : Single photon 23s-23p, 24s-23p and two-photon 23s-24s
            resonances in Na. (Ref. 13-b).

A most remarkable feature of these experiments is the extremely low power of microwave required to saturate the transitions.
In fact, 10^{-12} watts are enough to induce detectable transitions
in less than 10^{-6} seconds for $\Delta n = 1$ resonance around n = 30. This
extreme sensitivity is due of course to the huge electric dipole
moments of these states which act as giant atomic antennas for
microwaves ! A simple calculation shows that quantum efficiency
equivalent to the one of a good photomultiplier tube can be achieved for the detection of microwaves in such an experiment. Application to practical microwave detection for astrophysical or plasma
diagnostic purposes is presently being considered ([14]).

Due to the extreme sensitivity to electric fields, great care
should be taken in all these experiments to control the field seen
by the atoms ([13]). The effect of an applied field of 1,5V/cm on the
n = 28 ↔ n = 29 resonances is shown on Figure 3 : the pattern of
lines obtained is typical of the linear Stark effect which is already
observed in this small field. The energy separation between the extreme lines, about 5 GHz, i.e., $1,5 \cdot 10^{-5}$ eV, gives a direct measure
of the atomic size of the n ∿ 30 level: a field of 1,5 V/cm corresponds to a drop of $1,5 \cdot 10^{-5}$ volts across the largest atomic dimension
which is thus found to be of the order of 10^{-5} cm, that is, 10^3 Å!

4.2. How to measure radiofrequency intervals without radiofre-
     quency fields : quantum beats and level crossing techniques

We have, in the previous section, described some specific double resonance experiments in the microwave domain. Similar experiments have also been performed with radiofrequency sources (∿ few
Mc/sec) to measure much smaller splitting ([15]) (for example fine
structures). In this latter frequency domain though, the radiofrequency source is not absolutely necessary and it is possible to

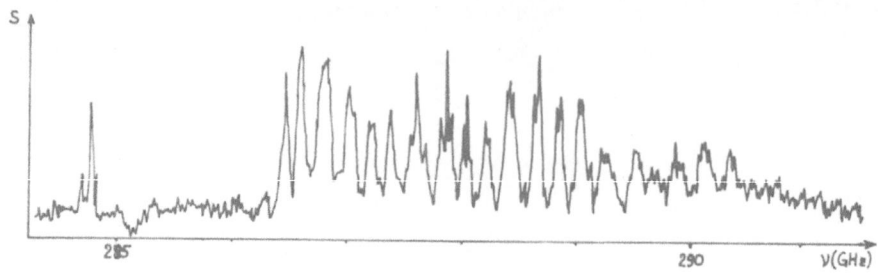

Fig. 3 : Stark effect of the n = 28 - n = 29 resonance in Na
in a 1,5 V/cm electric field   (Ref. 13-a)

measure small energy intervals without even applying on the atoms
any radiofrequency field ! One needs just to prepare the excited
states of interest in a linear superposition of substates having
slightly different energies. This superposition then starts to
precess at the Bohr frequencies corresponding to the energy split-
tings and the detection of this precession provides the spectroscopic signal of interest. This is the principle of atomic coherence experiments, whose simplest example is the quantum beat one
The basic features of the quantum beat phenomenon have been described in detail elsewhere ([16]). It can be in particular shown that
quantum beat is a very clear and nice example of quantum interference effect in atomic fluorescence. We will however not present this aspect of the phenomenon here, but just recall a very
simple description of the quantum beat effect, which will be enough
for understanding its spectroscopic applications to Rydberg states.

Suppose that a laser pulse excitation prepares at time t = 0
the atomic system in a linear superposition of two states |a > and
|b > of different energies:

$$|\psi(0) > \; = C_a \;|a > + \; C_b \;|b >$$ (10)

After excitation, this superposition evolves "freely" according to
the atomic Schrödinger equation. At time t, this superposition of
states has become :

$$|\psi(t) > \; = C_a \; e^{-iE_a t/\hbar} \;|a > + \; C_b \; e^{-iE_b t/\hbar} \;|b >$$ (11)

Suppose that one detects the light intensity emitted by spontaneous
emission by this excited system in a given direction, with a detec-
tion polarization $\vec{\varepsilon}_d$: the detected intensity $I_{\varepsilon_d}(t)$ is, according
to well known dipole-emission rules, proportional to the mean va-
lue of the square of the atomic dipole projection along the polarization $\vec{\varepsilon}_d$:

$$I_{\epsilon_d}(t) \propto \; < \; \psi(t) \; | \; (\vec{D}. \; \vec{\epsilon}_d)^2 \; |\psi(t) \; > \tag{12}$$

($\vec{D}$: electric dipole operator).

Introducing a sum rule over all possible final states $|f>$ in the emission process, Equ. 12 becomes :

$$I_{\epsilon_d}(t) \propto \Sigma_f \; < \; \psi(t) \; |\vec{D}. \; \vec{\epsilon}_d|f \; > \; < \; f|\vec{D}. \; \vec{\epsilon}_d \; |\psi(t) \; > \tag{13}$$

Replacing in Equ. 13 $|\psi(t) >$ by its expression of Equ. 11 and taking into account the finite lifetime $1/\Gamma$ of the excited states, one finally gets :

$$I_{\epsilon_d}(t) = \left[A(\vec{\epsilon}_d) + B(\vec{\epsilon}_d) \cos \nu_{ab} \; t \right] e^{-\Gamma t} \tag{14}$$

with : $\nu_{ab} = \hbar^{-1} (E_a - E_b)$

$$\begin{cases} A(\vec{\epsilon}_d) = C_a^\star \; C_a \; \Sigma_f \; |(\vec{D}.\vec{\epsilon}_d)_{a.f}|^2 + C_b^\star \; C_b \; \Sigma_f \; |(\vec{D}.\vec{\epsilon}_d)_{bf}|^2 \\[2mm] B(\vec{\epsilon}_d) = 2\mathcal{R}e \; C_a^\star \; C_b \; \Sigma_f \; (\vec{D}.\vec{\epsilon}_d)_{af} \; (\vec{D}.\vec{\epsilon}_d)_{fb} \end{cases} \tag{15}$$

The transient fluorescence signal from the excited state thus appears as the sum of two different terms : an exponentially decaying term, $A(\vec{\epsilon}_d) \; e^{-\Gamma t}$, proportional to the probabilities $C_a^\star C_a$ and $C_b^\star C_b$ of finding the atom in states a and b following the laser excitation, and a decaying term modulated at frequency ν_{ab}, $B(\vec{\epsilon}_d) \cos \nu_{ab} \; t \; e^{-\Gamma t}$, which is proportional to the "atomic coherence" $C_a^\star C_b$ between states a and b. This modulation on top of the transient fluorescence signal is precisely called quantum beat. By measuring its frequency, one gets directly a determination of the energy splitting $E_a - E_b$. The physical interpretation of quantum beats is quite clear : the laser pulse excitation prepares the atomic system in a non equilibrium state which is not an energy eigenstate. The square of the atomic dipole in this state is non stationary and modulated at the eigenfrequency ν_{ab}. The emission of the dipoles is thus modulated at the same frequency and the atomic system behaves as a lighthouse which emits in a given direction bursts of lights with the periodicity $2\pi/\nu_{ab}$.

Note that the beating corresponds to the precession of the atomic coherence of each atom. It is <u>not</u> an interference effect coming from the emission of different atoms. Hence, there is almost no Doppler effect and an ensemble of atoms in a thermal sample will exhibit a beat modulation without any Doppler spread : quantum beat detection is thus a Doppler free high resolution technique in Spectroscopy.

Another important remark should be made : coefficients $A(\vec{\epsilon}_d)$ and $B(\vec{\epsilon}_d)$ in (14) depend on the polarization of the detection (and also on the polarization $\vec{\epsilon}_1$ of the light excitation) : quantum beat is a polarization dependent effect. One case demonstrates the important sum rule :

$$\sum_{\vec{\epsilon}_d} B(\vec{\epsilon}_d) = 0 \qquad (16)$$

which means that the total intensity emitted by the excited levels decays exponentially without beats : the beats appear only on polarized components of the light intensity. Equ. (16) also shows that for different detection polarizations ϵ_d^1, ϵ_d^2 ..., the beats should appear in phase opposition [17].

Figure 4 gives an example of fine structure quantum beat detections in the nD level of Na (n ranging from 9 to 16) ([7]).

The slowing down of the beat pattern clearly reveals the continuous decrease of the fine structure as one climbs up in the energy ladder of the Rydberg level.

The perturbation of these beat patterns in an electric field has also been studied ([18]) and has allowed to assign a negative sign for the fine structure in these levels.

These quantum beat experiments have thus been of great interest for the theoretical understanding of the fine structure interactions in Rydberg states.

For more excited levels, the fluorescence detection becomes very difficult. Variant of the quantum beat technique using electric detection can then be developped. The method is then based on the fact that various substates differing from each other by their orientation with respect to an externally applied electric field are ionized -as we have seen- for slightly different field amplitudes. Thus by adjusting the field amplitude and direction, one can detect a well defined state superposition whose precession entails a modulation of the ion current as a function of the delay between the exciting laser pulse and the detection electric pulse. The method has been successfully applied to detect fine structure beats in higher D states of sodium ([19]).

To conclude this analysis of quantum beat spectroscopy, let us summarize the advantages of the method :

- it is simple (there is no scanning or searching of the line)
- it does not add any perturbation on the studied system (the signal is detected when the atoms are in the "dark" after the pulse excitation)
- it allows to study very small structures (the closer they are to zero, the better it is !).

Of course the technique has also some limitation as compared to "true" double resonance methods. In particular, it does not allow -(due to laser pulse finite duration and detection frequency cut off)- to detect too large energy structures. This drawback

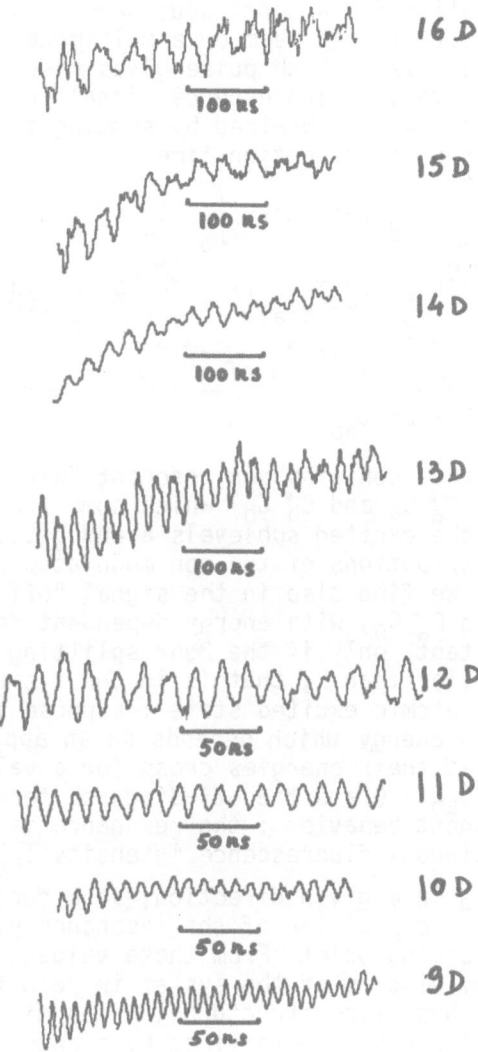

<u>Fig. 4</u> : Fine structure quantum beats in the n = 9 to n = 16D
           levels of Na   (Ref. [7]).

can however to some extent be overcome by using a related type of
atomic coherence experiment, namely the level crossing method [20].
Suppose now that, instead of exciting the atoms with a single pulse
of duration T, one irradiates the system with a random succession
of short pulses, each of them having the duration T : this is a
possible description for a continuous wave (c.w.) broadband light
beam (e.g. a spectral lamp beam or a multimode c.w. laser beam of
spectral width $\Delta = 1/T$). Each pulse gives rise to its own beat
pattern and the average fluorescence signal in a given direction,
with polarization $\vec{\epsilon}_d$, is obtained by summing the contribution
of all pulses prior to detection time

$$
\overline{I_{\epsilon_d}} = \int_{-\infty}^{t} I_{\epsilon_d}(t-\tau)\ d\tau = \int_{0}^{\infty} I_{\epsilon_d}(\tau')\ d\tau'
$$

$$
= \frac{C_a^{\star} C_a}{\Gamma} \sum_f |(\vec{D}.\vec{\epsilon}_d)_{af}|^2 + \frac{C_b^{\star} C_b}{\Gamma} \sum_F |(\vec{D}.\vec{\epsilon}_d)_{bf}|^2
$$

$$
+ \frac{C_a^{\star} C_b \sum_f (\vec{D}.\vec{\epsilon}_d)_{af} (\vec{D}.\vec{\epsilon}_d)_{fb}}{\Gamma + 2i\pi\nu_{ab}} + \text{c.c.} \tag{17}
$$

The signal thus contains non resonant "diagonal terms" (terms
proportional to $C_a^{\star} C_a$ and $C_b^{\star} C_b$) which come from light "incoherently" emitted by the excited sublevels a and b ... and which corres-
pond to the contributions of the non modulated part of each elementary signal. We find also in the signal "off-diagonal" terms
(proportional to $C_a^{\star} C_b$) with energy dependent denominators. These
terms are important only if the Bohr splitting ν_{ab} is of the
order of or smaller than Γ, that is in the vicinity of a level
crossing in the atomic excited state : suppose that level $|a>$
and $|b>$ have an energy which depends on an applied magnetic
field H_0 and that their energies cross for a value $\overline{H}_0$ of H_0 :
for this value $\nu_{ab} = 0$ and the off-diagonal term in (17)
exhibits a resonant behavior : the resonance is detected by monitoring the continuous fluorescence intensity $\overline{I_{\epsilon_d}}$ emitted with

polarization ϵ_d in a given direction, as a function of the applied
magnetic field. The position of the resonance yields the value $\overline{H}_0$
of the level-crossing point. From these values, one can deduce
the whole energy diagram of the system in zero field and thus
measure fine or hyperfine structures, which might be much larger
than the one which can be determined by quantum beats. Systematic
investigations of such level-crossing resonances have been made
in the Rydberg states of various alkalis and have provided very
interesting spectroscopic informations [21].

For all the experiments described in this section, the only
characteristics of laser that are needed are the high intensity
and -for quantum beats- the pulse operation. These are only
quantitative advantages over classical spectral lamps.

All the techniques described here have indeed already been applied in the prelaser age on ground and low lying states using classical lamp excitation. The laser has thus "only" allowed to extend and adapt these already known techniques to very excited states (point i) of the "laser revolution" mentionned in the introduction).

We now turn to methods in which not only the quantitative but also the "qualitative" changes introduced by laser in spectroscopy are made use of (point ii) of the "laser revolution" described in the introduction).

5. HIGH RESOLUTION OPTICAL TECHNIQUES APPLIED TO RYDBERG STATES

5.1. Laser - Atomic Beam Spectroscopy

We turn now to methods allowing to measure directly with high resolution optical frequencies ν_{opt}. In this case, one has to overcome directly the large optical Doppler effect which usually gives each absorption or emission line a width of about 1 GHz (10^9 c/sec) in a gas. The simplest method is the one which consists in irradiating an atomic beam with a monochromatic tunable single mode laser crossing the beam at right angle. In this way, the first order Doppler effect in absorption is almost cancelled (one observes only the residual Doppler effect due to the divergence of the beam). In this case, the ultimate width of the spectral lines will depend on four parameters : the residual Doppler effect $(\Delta\nu_D)_{res} = k\Delta\nu_D$ where k is the collimation ratio of the atomic beam, the natural atomic excited state line width $\Delta\nu_W$ equal to the reciprocal of the excited state lifetime, the "instrumental" transit time width $\Delta\nu_{inst} = 1/\tau_0$ where τ_0 is the time of flight of the atoms across the laser beam, and the laser linewidth $\Delta\nu_l$. The largest of these four parameters will limit the observable spectral width. In general $\Delta\nu_L$, $\Delta\nu_{inst}$, $\Delta\nu_D$ and $\Delta\nu_W$ can be of the order of or smaller than a few MHz and the laser-atomic beam experiment can easily achieve a resolution of about 10^8-10^9.

This type of experiment have been recently performed to investigate Rydberg states of alkali atoms. The detection of the resonance is generally achieved by the field ionization technique : an ion current is observed each time the laser radiation frequency coïncides with a transition between the atomic ground state and a Rydberg level. Figure 5-a shows a typical spectrum, corresponding to the series of 5s-np transitions of Rb ([22]). The lines are here detected with a "moderate" resolution, basically limited by the linewidth of the frequency-doubled laser beam used for the experiment ($\Delta\nu_L \sim 10^9$ c/sec). Improvement of the laser quality has resulted in a dramatic increase in resolution ([23]), as shown in Fig. 5-b which represents a small portion of the above spectrum corresponding to a blow up of the 5s-54p transition : the line-

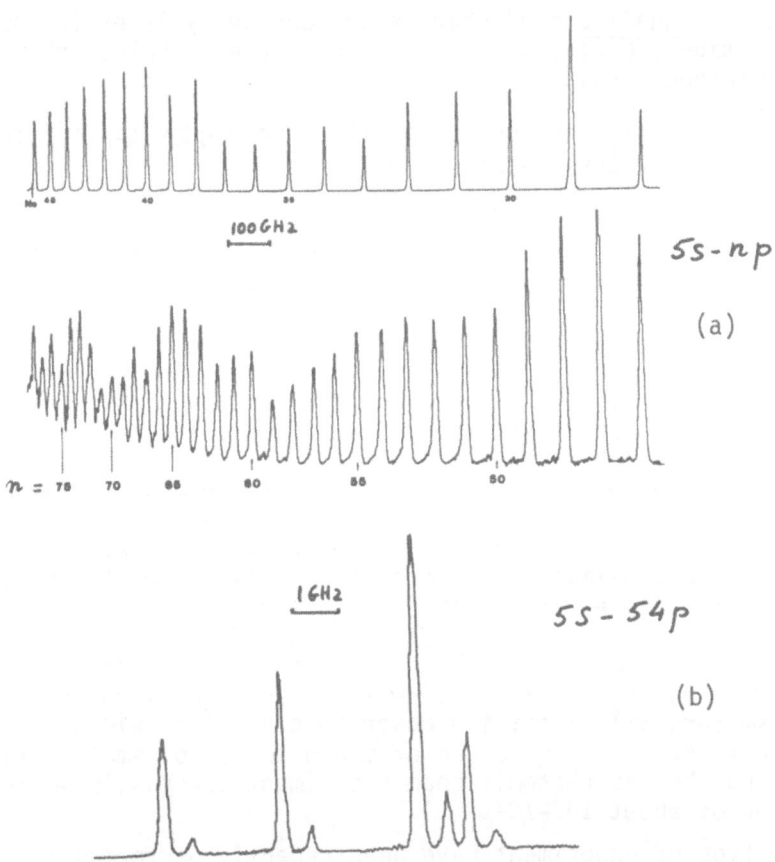

<u>Fig. 5</u> : Spectrum of the 5S-np resonance in Rubidium : broad scan
with moderate resolution (Fig. a) and high resolution scan
showing details of the 5S-54p resonance (Fig. b)
(Ref. [22] and [23])

width of the laser is now $\Delta\nu_L \sim 20$ Mc/sec and the broad peak of Fig. 6-b is resolved into eight lines resulting from the combined effects of the ground state hyperfine structure, the excited fine structure and the presence in the beam of two isotopes of Rb (^{85}Rb and ^{87}Rb) (residual line width of 150 MHz due to low atomic beam collimation).

Another very interesting type of detection of laser-beam resonances in Rydberg states has been developed. It consists of deflecting the produced excited state by applying a gradient of electric field to the beam : the huge polarizability of the level results in a large force bending the atomic trajectories out of the initial beam path. This electric deflection of the beam can then be detected by placing adequate atomic detectors in the right direction. The method has been used to detect 6s-np transition in Cs ([24]).

At last, one might add that laser frequency tuning can, in some experiments, be replaced by the Stark tuning of the atomic line, the laser frequency being kept constant ([25]).

5.2. Fluorescence line-narrowing ([26]) and saturation spectroscopy ([27])

The use of an atomic beam can be avoided and an ordinary resonance cell with randomly distributed atomic velocities can be used without loss of resolution if one can take advantage of the fact that the monochromatic laser beam preferentially interacts with a well defined velocity group in the Maxwellian Doppler profile of the velocity distribution. The laser beam of frequency ν_L and wave vector $\vec{k}_L$ is indeed coupled only with atoms having a velocity $\vec{v}$ such that the resonant condition

$$\nu_L - \vec{k}_L \, \vec{v} = \nu_0$$

is fulfilled (ν_0 being an atomic transition frequency).

The velocity group thus excited will eventually reemit in the backward direction e.m. radiation with the frequency

$$\nu'_0 = \nu_0 - \vec{k}_L \, \vec{v} = 2\nu_0 - \nu_L \ .$$

The spectrum of backward fluorescence thus only depends on the atomic and laser frequencies, and not on atomic velocities. Analyzing this spectrum with a Fabry-Perot interferometer allows in principle to study small structures buried in the Doppler profile. Instead of observing the fluorescence spectrum of the backward scattered light, one can alternatively analyze as a function of ν_L the absorption of a weak probe beam of the same frequency propagating in the $-\vec{k}_0$ direction. This is the well known saturation spectroscopy method, described in details in various references ([27]).

In both fluorescence line narrowing and saturation spectroscopy, a relatively large number of atoms is needed, since the techniques make in fact use only of a small percentage of the available atoms, those whose velocity matches the required condition. Furthermore, these techniques need relatively large oscillator strengths since the former asks for large enough fluorescence yields and the latter implies strong non linear interaction with the laser light. For all these reasons, these methods are not well suited for Rydberg states studies and have in fact, to my knowledge, never been used in this context. We have quoted them here only for sake of completeness. We will end this survey by the description of another laser atomic-cell technique which has actually been used in Rydberg states studies : the multiphoton-Doppler free method.

5.3. Multiphoton Doppler-Free Spectroscopy ([28]) in Rydberg States

The principle of the method is quite simple. Atoms in a cell are interacting with two counterpropagating beams of the same frequency. In the atom rest frame, these beams are seen respectively at frequencies

$$\nu_+ = \nu_L - \vec{k}_L \vec{v}$$

and $$\nu_- = \nu_L + \vec{k}_L \vec{v}.$$

Two photon absorption involving a photon from each beam will be resonant when the velocity independent condition

$$\nu_+ + \nu_- = 2\nu_L = \nu_0$$

is fulfilled.

The two-photon line is thus not affected by the Doppler effect and all atoms, whatever their velocity, participate to the resonance. Of course, large light intensities are needed to induce these second-order transitions. However, this drawback can, to some extent, be overcome by chosing transitions for which there is an intermediate level near the middle of the energy gap, which considerably increases the transition probability ([28]). Another way of getting large signals is to use two lasers of slightly different frequencies in order to nearly match, with one of the lasers, the frequency of an intermediate transition ([29]). The fact that all atoms participate to the signal and also the possibility of using very sensitive electrical detection is sometimes a great advantage over the methods described in the previous subsection and allow to detect resonances towards very excited Rydberg states, even if the corresponding transition rates are very small. For example, 5s-nD transitions in Rb, with n as large as 75, have been observed using the thermoionic diode detection method, with a resolution good enough to measure the small fine structure splitting of the Rydberg series of levels ([30]). Other two photon studies in K have also been performed ([31]).

6. CONCLUSION

I have given in this chapter several examples of high resolution laser spectroscopy experiments performed on alkali atom Rydberg states. I hope to have shown that these experiments provide interesting informations about the physical properties of these states that have been summarized at the beginning of this chapter.

Development of some of these experiments might certainly be also of great interest for practical applications : new infrared and microwave detectors, isotope separation schemes, may be new types of atomic clocks, could be developped using Rydberg states high resolution spectroscopy techniques.

On a more theoretical level, alkali-atom Rydberg states spectroscopy could provide a better determination for the Rydberg constant. It could also allow the investigation of interesting light-matter interaction processes involving the "dressing" of weakly bound electrons by strong electromagnetic fields ([32]).

Of course the present investigations are by no means limited to the simple case of one-electron atoms. A great lot of experiments is now carried on, using related techniques on Rydberg spectra of two or more electron atoms, and more generally in all kinds of weakly bound states of atomic or molecular species. The reader who wishes to have a broader view of this new field of atomic physics can consult the proceedings of a Colloquium on highly excited states recently held in Aussois, France, by the CNRS ([33]).

REFERENCES

([1])   H. Bethe, E. Salpeter, Quantum Mechanics of one and two-electron atoms. Academic Press, New York (1957)

([2])   D.S. Bailey, J.R. Hiskes and A.C. Rivière, Nucl. Fusion, 5, 41 (1965)

([3])   See for example J.C. Slater, Quantum Theory of Matter, Mc Grac Hill (1951)

([4])   R. Freeman and D. Kleppner, Phys. Rev. A, 14, 1614 (1976)

([5])   M.J. Seaton, Proc. Phys. Soc., 88, 801 (1966)
        K.T. Lu, U. Fano, Phys. Rev. A 2, 81 (1970)

([6])   Bates and   Damgaard, Philos. Trans. Roy. Soc. London, 242, 101 (1949)

([7])   C. Fabre, M. Gross and S. Haroche, Opt. Commun. 13, 393 (1975)

([8])   R.M.Sternheimer, J.E. Rodgers, T. Lee, T.P. Das, Phys. Rev. A, 14, 1595 (1976)
        E. Luc-Koenig, Phys. Rev. A, 13, 2114 (1976)

([9])   T.F. Gallagher, L.M. Humphrey, W.E. Cooke, R.M. Hill and S.A. Edelstein, Phys. Rev. A, 16, 1098 (1977)

([10]) T.W. Ducas, M.G. Littman, R.R. Freeman, D. Kleppner, Phys. Rev. Letters, 21, 279 (1975)

([11]) S.M. Curry, C.B. Collins, M.Y. Mirza, D. Popescu and I. Popescu, Opt. Commun. 16, 251 (1976)

(12) T.F. Gallagher, R.M. Hill, S.A. Edelstein, Phys. Rev. A, 13,
     1448 (1976)
     W.E. Cooke, T.F. Gallagher, R.M. Hill and S.A. Edelstein,
     Phys. Rev. A, 16, 1141 (1977)
(13) a. C. Fabre, P. Goy and S. Haroche, J. Phys. B, Letters, 10,
        L-183 (1977)
     b. C. Fabre, S. Haroche, P. Goy, Phys. Rev. A (to be published)
        (1978)
(14) D. Kleppner and T.W. Ducas, B.A.P.S., 21, 600 (1976)
(15) T.F. Gallagher, L.M. Humphrey, R.M. Hill, W.E. Cooke and
     S.A. Edelstein, Phys. Rev. A, 15, 1937 (1977)
(16) S. Haroche, in "High Resolution Laser Spectroscopy", K. Shimoda
     editor, Springer Verlag (1976)
(17) S. Haroche, M. Gross and M. Silverman, Phys. Rev. Letters, 33,
     1063 (1974)
(18) C. Fabre, S. Haroche, Opt. Commun. 15, 254 (1975)
(19) H. Walther, private communication
(20) F.D. Colegrove, P.A. Franken, R.R. Lewis, R.H. Sands, Phys.
     Rev. Letters, 3, 420 (1959)
(21) S. Svanberg, Proceedings of TICOLS (3rd Int. Conf. on Laser
     Spectroscopy), J. Hall editor, Springer Verlag (1977)
(22) D.H. Tuan, S. Liberman and J. Pinard, Opt. Commun. 18, 533
     (1976)
(23) J. Pinard and S. Liberman, Opt. Commun. 20, 344 (1977)
(24) A.F.J. Van Raan, G. Baum and W. Raith, J. Phys. B, Letters, 9,
     L-349 (1976)
(25) T.W. Ducas, M.L. Zimmerman, Phys. Rev. A, 15, 1523 (1977)
(26) We give here a simple description of fluorescence line narrow-
     ing in a two level system. For similar effects in the three-
     level system, see for example M.S. Feld, A. Javan, Phys. Rev.
     177, 540 (1969)
(27) P.H. Lee and M.L. Skolnick, Appl. Phys. Letters, 10, 303 (1967);
     V.S. Letokhov, JETP (Sov. Phys.) Letters, 6, 101 (1967);
     E.V. Baklanov and V.P. Chebotaev, JETP (Sov. Phys.), 33, 300,
     (1971)
     S. Haroche and F. Hartmann, Phys. Rev. A, 6, 1280 (1972)
(28) L.S. Vasilenko, V.P. Chebotaev and A.V. Shishaev, JETP (Sov.
     Phys.) Letters, 12, 161 (1970)
     B. Cagnac, G. Grynberg and F. Biraben, J. de Phys. 34, 845
     (1973)
(29) R.T. Hawkins, W.T. Hill, F.V. Kowalski, A.L. Schawlow and
     S. Svanberg, Phys. Rev. 15, 967 (1977)
(30) K.C. Harvey and B.P. Stoicheff, Phys. Rev. Lett. 38, 537 (1977)
(31) C.D. Harper, S.E. Wheatley, M.D. Levenson, J.O.S.A., 67, 5
     (1977)
(32) P. Avan, C. Cohen-Tannoudji, J. Dupont-Roc, C. Fabre, J. de
     Phys. 37, 993 (1976)
(33) Etats atomiques et moléculaires couplés à un continuum
     Atomes et Molécules hautement excités. Colloque International
     du C.N.R.S., Aussois (1977), to be published.

THE METHOD OF SEPARATED FIELDS IN OPTICS

V.P. Chebotayev

Institute of Semiconductor Physics

pr. Nauki, 13, Novosibirsk, 90, USSR

Three methods are of great importance for super-high resolution spectroscopy and in obtaining supernarrow resonances. These are as follows: the method of saturated absorption, two-photon spectroscopy, and the method of separated optical fields. The first two methods are well known and developed. Recent progress in developing these methods has been discussed in [1]. The narrowest resonances and high resolution were obtained by the method of saturated absorption. It is based on the use of a Lamb dip [2] in an absorption line. The use of absorption saturation resonances has been begun in 1967 with investigations of He-Ne lasers with a Ne absorbent which were independently carried out in Perkin Elmer laboratories [3] and in our institute[4].

Recent data of investigations of absorption saturation resonances in methane which are the narrowest should be presented before describing the method of separated fields which is new for optical spectroscopy. It will be helpful in explaining the reasons which made us begin developing the method of separated fields in optics. To obtain narrow resonances requires to use a telescopic beam expander in order to increase the time of particles-field interaction. The authors of [5, 6] used a telescopic beam expander outside a cavity. We used it inside a cavity [7, 8]. The experimental arrangement is shown in Fig. 1. The beam diameter is 14 mm and the length 5 m. Investigations of narrow resonances require special laser spectrometers with a narrow radi-

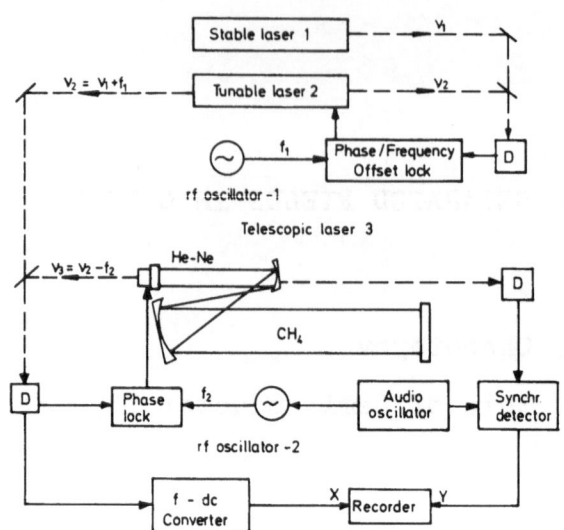

Fig. 1. The experimental arrangement

ation line whose frequencies can be tuned and kept with a high accuracy within the given spectral region to store a signal. The spectrometer consists of a reference frequency-stabilized laser with a narrow radiation line of 7 Hz and of a tunable laser. The frequency of the tunable laser is locked to that of the stable laser with the aid of an electron system of phase offset lock. The frequency difference is equal to the frequency of a radio-frequency oscillator with an accuracy of better than 1 Hz. Tuning the frequency of the radio-frequency oscillator permits to vary the tunable laser frequency within the wide limits. The linewidth of the tunable laser is equal to that of the stable laser. We used this spectrometer in investigations of resonances of 1 kHz wide in methane.

Fig. 2 a, b shows the typical record of the magnetic hyperfine structure of the $F_2^{(2)}$ line of methane. Three main components of the structure can be easily seen. At the methane pressure of about 20 μtorr we observed line splitting of the hyperfine structure due to the recoil effect (Fig. 2b). The similar results were obtained in [6]. In the course of experiments we found out an appreciable difference in the widths of the principal hyperfine components. The strongest transition F = 8→7 had a larger width as compared with the two others. This difference in widths, as has been shown in

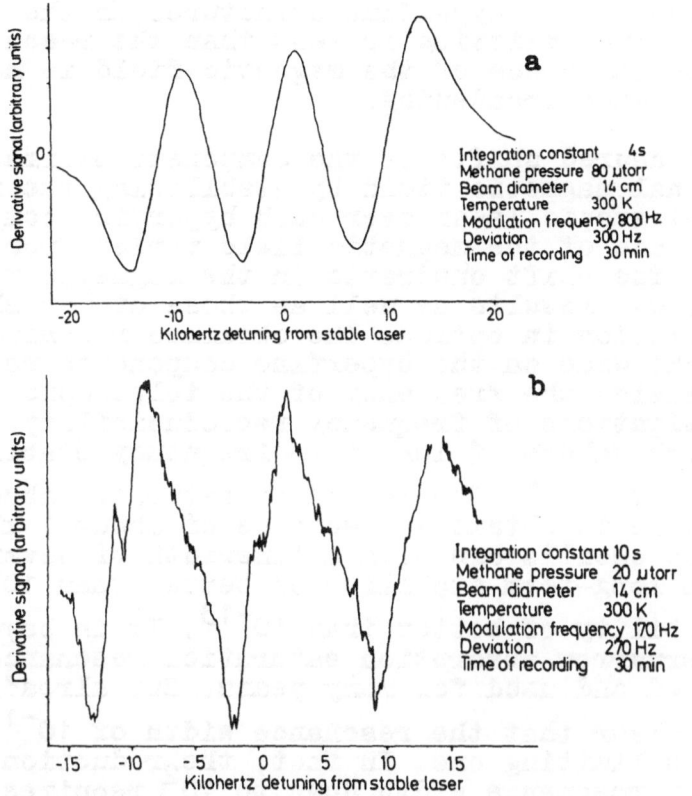

Fig. 2a and b. Record of the magnetic hyperfine structure (a) and of the recoil effect (b) of the $F_2^{(2)}$ line of the ν_3 band of methane. Methane pressure in the cell is a) 80 μtorr; b) 20 μtorr

additional experiments, was caused by the action of the non-cancelled Earth magnetic field. A high intensity of resonances on the hyperfine components and the achieved high resolution of about 1 kHz permitted us to explore for the first time an anomalous Zeeman effect in the region of weak fields at the $F_2^{(2)}$ transition of methane.

Fig. 3 shows the record of the hyperfine structure of methane in a longitudinal magnetic field of 5 Gauss. It is seen that the intense $F = 8 \rightarrow 7$ component is more split than the others. This difference in splitting is associated with the difference in g - factors at the transitions of the hyperfine structure. In the case where the line splitting is less than the resonance width, the influence of the magnetic field is manifested in resonance broadening.

We measured shifts of the component maxima in the longitudinal magnetic field by stabilizing the frequency of the telescopic laser over each hyperfine component. The magnitude of the magnetic field varied from 0 to 5 oersteds. The shift quadratic in the magnetic field was observed. Our results as well as those of [6] show the best resolution in optics. The obtained resonances of about 1 kHz wide on the hyperfine components were used in stabilizing the frequency of the telescopic laser and in investigations of frequency reproducibility. We obtained high values of long-term frequency stability at the level of 10^{-14}. The method of saturated absorption permitted us to obtain a resonance of about 1 kHz wide, to produce oscillators with a linewidth of several Hz, to obtain long-term stability of better than 10^{-14} and reproducibility of better than 10^{-13}. It is beyond doubt that supernarrow absorption saturation resonances will be explored and used for many years. But already now it is quite clear that the resonance width of 10^{-11} is close to a limiting one. In fact, the reduction of a saturation resonance width down to 10^2 requires the production of telescopic expanders of a beam of more than 2 m in diameter (for CH_4). It is impossible under

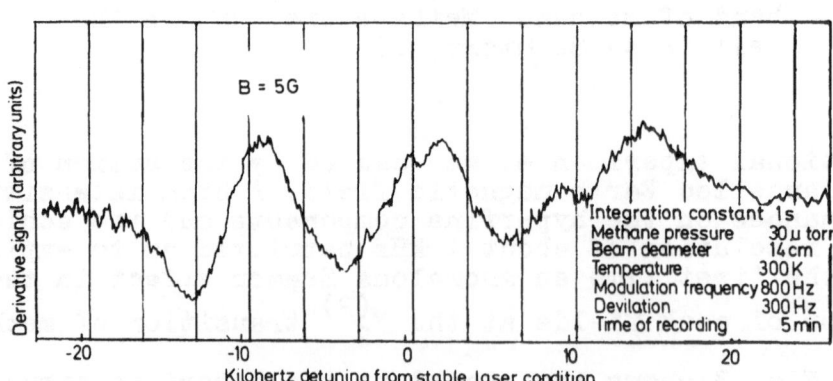

Fig. 3. The record of the magnetic hyperfine structure of the $F_2^{(2)}$ line of methane in the longitudinal magnetic field

laboratory conditions. So further progress in this
sphere should be associated with the methods based on
the other physical ways. These are two-photon resonances
in the field of oppositely traveling waves and resonances which are due to coherence transfer of particle
states in spacially separated fields. The two methods
are based on coherence effects.

The method of two-photon Doppler-free spectroscopy
has become one of the most effective methods of atomic
and molecular spectroscopy. A great number of works is
devoted to this method [1, 9]. Recently, in 1975-76, we
grounded the method of separated optical fields [10, 11].
A number of experiments carried out by different scientific groups was reported this year [12-15]. The first
experimental results and theoretical analysis have shown
that the method of separated fields can become one of
the most effective methods of laser spectroscopy of super-
high resolution. The method of separated fields is well-
known in the microwave range. The method was developed
by Ramsey [16] in the 50-60th, and up to now it is considered to be one of the most effective methods in obtaining narrow resonances.

The idea of the method as used in the radio-frequency range can be explained with the aid of Fig. 4. A
beam of particles resonantly interacts with two coherent
fields E_1 and E_2. The fields are produced in two cavi-
ties which are $L = T/u$ away from each other (u is an
atomic velocity, T is the flight time between the fields).
Coherence between the two fields is due to excitation
of both fields with the same oscillator. Our approach
is valid for any harmonic oscillator which is excited
with a periodic force whose frequency coincides with the
oscillator frequency. However, for definiteness we shall
consider a two-level atom. The field at frequency ω in
the first cavity $E_1 e^{-i\omega t}$ + c.c. and in the second
$E_2 e^{-i\omega t + i\varphi}$ + c.c., where φ is the phase difference between the fields E_1 and E_2. The equations for probability
amplitudes are

$$\dot{b} = \frac{i}{\hbar} d_{21} E \, e^{-i\Omega t} a$$

$$\dot{a} = \frac{i}{\hbar} d_{12} \overset{*}{E} \, e^{i\Omega t} b \qquad\qquad (1)$$

b and a are the probability amplitudes of a particle
on the level 2 and 1, respectively, d_{21} is the matrix
element of the transition.

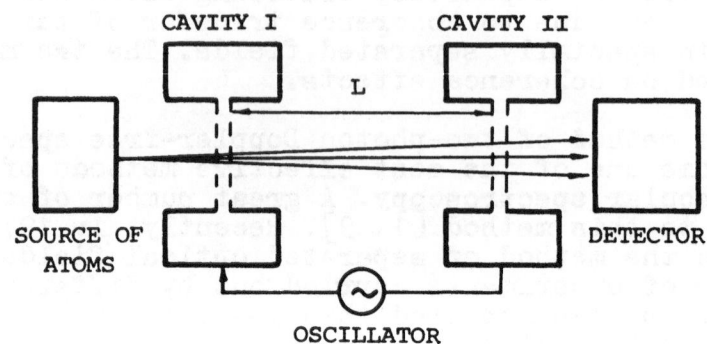

**Fig. 4. The schematic diagram of
the method of separated fields
in a microwave range**

We shall consider weak fields in order to use the theory
of perturbation $d_{21} E/\hbar \ll 1/\tau$. Frequency detunings
$\Omega = \omega - \omega_{12}$ will be assumed small so that $\Omega\tau \ll 1$,
where τ is the particle flight time in the cavity, ω_{12}
is the transition frequency. Then after the interaction
with the first field the probability amplitude on level
2 is

$$ b = G_1\tau, \quad a = 1, \quad G_1 = \frac{i d_{21} E}{\hbar} $$

As we neglected damping of the levels, the probability
amplitude is not changed in the region of drift. It has
a dipole moment at the transition frequency

$$ d(t) = d_{21} b a^x e^{-i\omega_{21} t} + c.c. $$

In the second cavity a particle with a dipole moment
d(t) will absorb or emit energy at the interaction with
the field. The part of energy which is absorbed by the
particle at the interaction with two fields is

$$W = 2\text{Re}\left[-i\omega\int_{T}^{T+\tau} d(t)E_2(t)dt\right]$$

$$= 2\hbar\omega_{21}\tau^2 d_{21}^2 \frac{|E_1||E_2|}{\hbar^2}\cos(\Omega T - \varphi) \tag{2}$$

where φ is the phase difference between fields E_1 and
E_2. The absorbed energy is a periodic function of detuning Ω of the flight time T between the fields. Real
beams of particles have velocity distribution:

$$f(v) = \frac{4v^2}{\sqrt{\pi}\,u^3}\,e^{-(v/u)^2}$$

After averaging an interference term over velocities
we obtain

$$\Delta I = \frac{|E_1||E_2|\,d_{21}^2\,\tau^2 I_o}{\hbar^2}\,F(\Omega T) \tag{3}$$

where I_o is the particle flow on the lower level, F is
the function typical for the method of separated fields

$$F(\Omega T) = 2\int_0^\infty ye^{-y^2}\cos\frac{\Omega T}{y}\,dy \tag{4}$$

At $\varphi = 0$ the function $F(\Omega T)$ has a maximum at $\Omega = 0$
and a characteristic width which is inverse to the particle flight time between the fields (Fig. 5). For long-
-lived particles the flight time can be $\sim 10^{-2}$ s (L =
10^2 cm) and it is possible to obtain resonances with a
width of several tens Hz. In the optical band these resonances correspond to relative widths of 10^{-13} to 10^{-14}.
However, in employing this method in the optical band
we encountered difficulties which seemed to be insuper-
able. These difficulties were connected with the influence of a Doppler effect and with a short wavelength as
compared with the field occupied region. In the optical
region it is necessary to realize not only coherence
transfer of individual particles but also transfer of
macroscopic polarization of a medium. This is the principal difference between the methods in the microwave

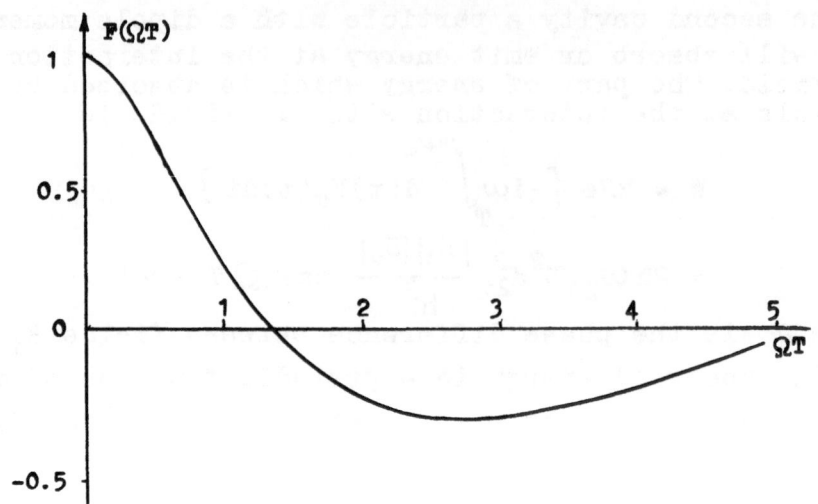

Fig. 5. The function $F(\Omega T)$ describing
a resonance shape in separated fields

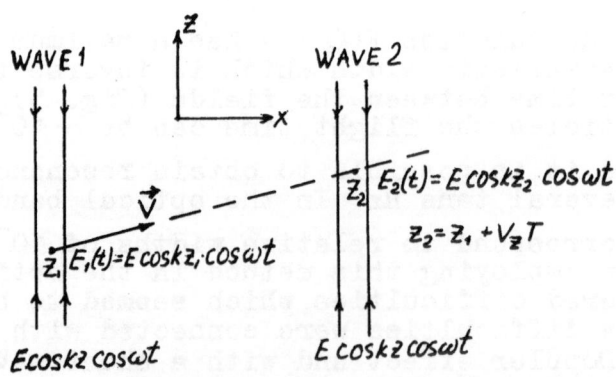

Fig. 6. Explanation to the interaction
of two separated optical fields with
moving particles

and optical ranges. Let the position of fields be chosen
such that the field phases coincide in the plane perpendicular to the beams (Fig. 6). We assume light beams to
be so narrow that phase retardation can be neglected at
the interaction of a particle with the field. An energy
absorbed by the particle with the velocity projection
V_z is

$$W = 2\hbar\omega \frac{d_{21}^2 |E_1| |E_2|}{\hbar^2} \tau^2 \cos(\Omega T - kV_z T) \tag{5}$$

Note that here the Doppler phase angle $\varphi = \Delta z k =$
$= kV_z T$ plays the role of the phase difference between
the fields in the z_1 and z_2 points. Thus, the absorbed
energy is a rapidly oscillating velocity function. If
a characteristic width in the velocity distribution is
such that $\Delta V_z \cdot kT >> 1$, then it is natural that a mean
value of the absorbed energy is zero at velocity averaging. It is easy to estimate divergence Θ of a particle
beam at which a resonance in separated fields can be
observed. As $T \simeq L/u$, $\Theta < \lambda/L$. For the distance $L = 10$ cm
and $\lambda = 10^{-4}$ μm we must have a beam divergence of 10^5.
It is clear that production of such beams is impractible.

New methods proposed in [10-11] for two- and multilevel
atoms are based on elimination of the influence of the
Doppler effect at the expense of nonlinear particle-
-fields interaction. Of great importance for atomic
spectroscopy will be two-photon resonances in separated
fields. The first experiments in pulsed fields were carried out in [13, 14] and in spacially separated fields
in [15]. Of great importance for molecular spectroscopy
is the method of separated fields that requires the use
of the three light beam system. The resonant field interacting with a gas produces macroscopic medium polarization. Motion of particles results in that the medium
polarization is destroyed outside the region of interaction. This makes impossible the realization of the method
of separated fields for two widely separated fields. As
has been shown in [10], macroscopic polarization is
transferred at large distances after the interaction
with two fields. Let us consider the physical idea which
explains polarization transfer and appearance of a resonance in separated fields. It is based on a phase jump
of an atomic dipole moment at the interaction with the
field. After the interaction with the first standing-
-wave field the particle has a dipole moment (in the

first order of perturbation theory):

$$d_1(t) = d_0(z_1)e^{-i\omega_{21}t} + \text{c.c.}$$

where $d_0(z_1) = d_{12}b_1 = d_{12}G_1 e^{ikz_1}$, z_1 is the particle coordinate in the first field (here we assume that $\Omega\tau \ll 1$). It can be easily shown that after the interaction with the second standing-wave field the particle dipole moment in the third order of perturbation theory is

$$d_2(t) = d_1(t) + \left\{ d_{12} \left[\frac{\tau}{2} G_2 e^{-i\Omega T}(e^{ikz_2} + e^{-ikz_2}) - b_1 \frac{|G_2|^2}{4}\tau^2 \right.\right.$$
$$\left.\cdot(2 + e^{2ikz_2} + e^{-2ikz_2}) - b_1^* \frac{G_2^2}{4}\tau^2 e^{-2i\Omega T}(2 + e^{2ikz_2} + e^{-2ikz_2}) \right]$$
$$\left. \cdot e^{-i\omega_{21}t} + \text{c.c.} \right\} + \text{terms containing } G_1^2, G_2^3 \qquad (6)$$

The addition of interest that is quadratic in field E_2 contains two terms. The first term is connected with the two-photon process of photon emission and absorption. The term with no change of the dipole moment phase is due to two-photon emission and absorption of photons from unidirectional waves. Additions to the dipole moments with the phase jump $\pm 2kz_2$ are connected with the two-photon process of emission and absorption of photons from oppositely traveling waves. The second term is due to another process. It corresponds to the one-quantum processes, transition of a particle from an upper level (emission) and transition of a particle from a lower level (absorption). In this process the phase jump is equal to $\pm 2kz_2 - 2z_1 - 2\Omega T$ and due to the interaction with traveling waves. Running ahead we note that macroscopic polarization transfer is due to the phase jump $-2kz_2$ and $-2\Omega T + 2k(z_2 - z_1)$. These phase jumps permit the appearance of oppositely traveling polarization waves. If in the first beam the particle interacts with the standing-wave field in the z point, the field-quadratic dipole moment of the particle after the interaction with the second field can be represented as

$$d_2(t) \sim A\cos kz_1 (1 + \cos 2kz_2)|G|^2\tau^2(1 + e^{-2i\Omega T})$$
$$\cdot e^{-i\omega_{21}t} + \text{c.c.} \simeq B\cos k(2z_2 - z_1)(1 + e^{-2i\Omega T})$$
$$\cdot e^{i\omega_{21}t} + \text{c.c.} \qquad (7)$$

where A and B are constants.

Now suppose that there is the third standing wave $E_3 \cos kz_3 \cos \omega t$ at the distance x away from the first field. Then the fiels-particle interaction is determined by the phase difference between the dipole moment and the field. The spacial phase difference is determined by the expression $k(z_3 - 2z_2 + z_1)$. At the distance x the phase difference is equal to $\Delta u = V_z(x - 2L/u)$. It is zero and independent of V_z at $x = 2L$ as $z_3 = 2z_2 - z_1$. This means that irrespective of V_z all the particles turn out to be synchronous with the third field due to the phase jump at the interaction with the second field at $x = 2L$. Of importance for the resonance in the line center is the time phase between the field and the particle dipole moment. Thus, the frequency dependence is due to the interaction of particle with standing-wave fields. The interaction with traveling waves produces polarization which is independent of frequency detuning. The particle-absorbed energy in the third beam is [10]:

$$W = \hbar\omega \frac{|G_2|^2}{2} \tau^4 |G_1| |G_3| \cos^2 \Omega T \cdot \cos(\varphi_1 + \varphi_3 - 2\varphi_2) \quad (8)$$

where φ_i is the ith field phase.

The resonance shape in arbitrary fields was analysed in [17]. Fig. 7 shows the resonance shape after velocity averaging. The authors of the work [12] observed absorption resonances in three separated fields. Observation was performed on short-lived atomic levels of neon with the aid of rapid beams. So a relative resonance width was rather large (about 60 MHz). The effective use of absorption resonances in molecular systems is difficult due to very low absorption. Worthy of notice is the new phenomenon which is of interest for molecular spectroscopy, coherent radiation in separated fields [18]. The method of separated fields is usually associated with narrow absorption resonances which are due to spacial coherence transfer in the atomic state. Macroscopic polarization at the distance 2L away from the first beam is responsible for the resonance with a width that is inverse to the flight time. At the same time, in accordance with Maxwell's equations, macroscopic polarization produces coherent radiation. The scheme for observing this phenomenon is given in Fig. 8. Two standing waves interact with a rarefied gas. Continuous coherent radi-

A curve number	ΓT_0	$G_{opt}\tau_0$	$\dfrac{\alpha(0)}{\alpha_0}\cdot\dfrac{L}{\omega}10^2$
1	0.1	0.8	3.63
2	0.5	0.96	1.56
3	1	1.09	0.63

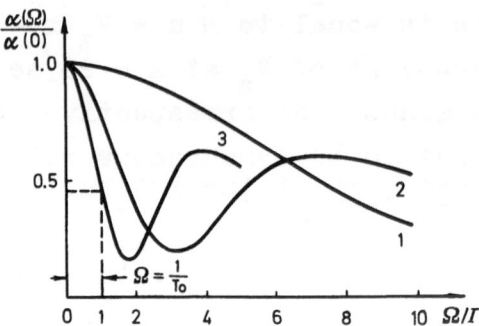

Fig. 7. The shape of the nonlinear Ramsey resonance. $\alpha(\Omega)$ represents the addition to an absorption coefficient due to the interaction of atoms with three rays; α_0 is an unsaturated absorption coefficient; $\alpha(0)/\alpha_0$ is a relative resonance amplitude; Γ is the homogeneous half-width of a transition line

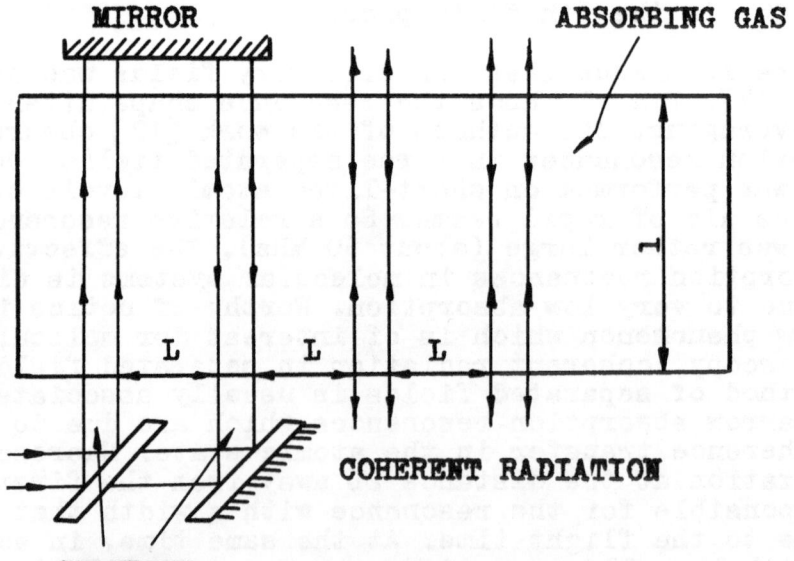

Fig. 8. The scheme for observing coherent radiation in separated fields

ation arises at the distances 2L, 3L and so on away
from the first field. A resonance intensity has a sharp
maximum with a width which is inverse to the particle
flight time. It is evident that physically similar processes can be observed in time separated fields (pulsed
fields). It should be noted that the other phenomena
are rather well known, such as: superradiance and quantum echo which use time separated fields and coherence
conservation in the state of particles. However, there
is an appreciable difference between basic ideas underlying such phenomena as absorption resonances in separated fields and quantum echo. So no wonder that these
methods were independently developed and treated. The
described phenomenon (coherent radiation in separated
fields - CRSF) has a common nature with an absorption
resonance. At the same time some CRSF properties are
closely related to properties of superradiance, coherent
resonant scattering, and quantum echo. The medium polarization in the z point is

$$P(x,z,t) = -d_{12}G_1 |G_2|^2 \frac{L^3}{4}\left[e^{i\Omega x/u} + e^{i\Omega(k-2L)/u}\right]$$

$$\cdot \int dV_z f(V_z)\cos\left[kz - \frac{kV_z}{u}(x-2L) + 2\varphi_2 - \varphi_1\right] \quad (9)$$

$f(V_z)$ is the function of velocity distribution of
particles. The medium polarization arises at the field
frequency at the distance x = 2L in a narrow region of
the space $\Delta x = (\Delta V_z/u)k$. V_z is a characteristic width
of the function of velocity distribution of particles.
Velocity averaging gives density of particles N if
$k \cdot \Delta V_z \cdot \tau \ll 1$. The polarization P(x, z, t) produces two
oppositely traveling waves. Their amplitudes are

$$E_{\pm} = -2\pi ik\ell d_{12}G_1 |G_2|^2 \frac{\tau^3}{8}(e^{2i\Omega T} + 1)$$

$$e^{\pm i(2\varphi_2 - \varphi_1)} \cdot N \quad (10)$$

where N is the particle density, ℓ is the
absorption cell length. The amplitude phase
of the first term depends on the frequency detuning Ω. This term has the same nature as an absorption
resonance. It is associated with the two-quantum process
of emission and absorption of photons from oppositely
traveling waves. The phase jump $\mp 2kz_2$ takes place in
this process that permits the appearance of a spacial
polarization harmonic. Its time phase turns out to be
dependent on the frequency detuning Ω. The second term

in (10) is due to the interaction with traveling waves.
In this case the phase jump $\pm 2kz_2 - 2i\Omega T$ takes place.

The radiation phase which is due to this process is
independent of the detuning Ω . This process appears to
be closely related to the phenomenon of quantum echo in
its nature. Note that coherent radiation arises at the
distances 3L, 4L and so on only in the line center. At
these distances coherent radiation arises only at the
interaction of particles with standing-wave fields.

The principal CRSF properties should be noted:
    1. Coherent radiation arises at the field frequency
but not at the transition frequency.
    2. The radiation intensity is proportional to the
square of density of absorbing particles.
    3. The radiation intensity has a sharp maximum in
the center of line with a width which is inverse to the
particle flight time between the fields.
    4. At the action of the field with a wide radiation
line coherent radiation has a narrow line whose width
is inverse to the particle flight time between the fields.
So this system can operate as a filter with a very narrow transmission band.
    5. If the first field is standing-wave, coherent
radiation arises in two opposite directions. If the first
field is traveling-wave, radiation arises in two directions at the distance 2L. An intensity and phase of radiation in the same direction are independent of the detuning frequency   . Radiation in the opposite direction
arises only in the line center. This fact indicates two
different processes.
    6. Polarization is transferred by atoms whose velo-
city is about 10^5 s^{-1} cm. So coherent radiation is delayed relative to inducing radiation. Thus, the system
operates as a coherent delay line. The delay time T
depends on the distance between the fields.

The first experiments on observation of CRSF were
carried out in spacially separated and pulsed fields
in the Institute of Semiconductor Physics (Novosibirsk,
USSR) [19-20] . The scheme for observing coherent radiation in spacially separated fields is given in Fig. 9.
It was observed in a methane absorption cell with the
aid of a laser spectrometer at λ = 3.39 μm. The spectrometer consisted of a frequency-stabilized He-Ne laser
with a narrow radiation line of 10 Hz, a tunable laser
3 and an auxiliary heterodyne-laser 2. The principle of
operation of the spectrometer was described in detail
in [8]. Radiation of the tunable laser 3 with a line-

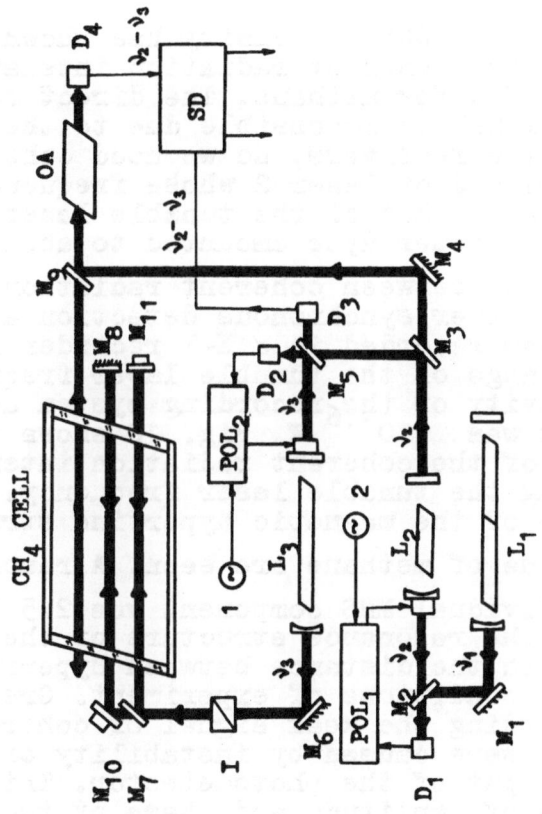

Fig.9. The scheme for observing CRSF in methane at λ = 3.39 µm.
L_1 – stabilized laser, L_2 – frequency-shifted laser, L_3 – tunable laser, POL_1, POL_2 – phase offset lock systems, D_1, D_2, D_3, D_4 – detectors, SD – system of coherent synchronous detection, OA – He-Ne optical amplifier, 1,2 – r.f.oscillators, $M_1 - M_9$ – mirrors, M_{10}, M_{11} – mirrors fixed on piezoceramics to cancel phases, I – optical insulator.

width of 10 Hz was directed into an absorption cell
where two standing waves were formed with the aid of
mirrors. The tunable laser frequency was scanned near
the center of a methane absorption line. The light beam
diameter was about 1 cm. The distance between beams was
3.5 cm. The absorption cell length was 115 cm. The methane pressure in the cell amounted to about 10^{-4} Torr.
The coherent radiation which arose at the distance of
3.5 cm away from the second beam was recorded.

An estimation of the CRSF intensity has shown that
under our conditions the coherent radiation intensity
amounts to about 10^{-15} W for methane. The direct recording of such a weak signal is impossible due to the absence of high-sensitive receivers, so we used coherent
heterodyning with the aid of laser 2 whose frequency was
tuned by 1 MHz away from that of the tunable laser. The
radiation power of the heterodyne amounted to about

10^{-3} W. The beat signal between coherent radiation and
radiation of laser 2 after synchronous detection at the
frequency of 1 MHz was recorded in a X-Y recorder in
dependence on the change of the tunable laser frequency.
The limiting sensitivity of the recording system achieved in the experiment was 10^{-16} W. Fig. 10 shows the
experimental record of the coherent radiation intensity
in methane at varying the tunable laser frequency. Three
principal components of the magnetic hyperfine structure
(MHS) of the $F_2^{(2)}$ line of methane are seen. A resonance
half-width of an individual MHS component was 2.5 kHz.
A typical period of the resonance structure of the observed CRSF coincides with the distance between hyperfine
components under the conditions of experiment. Great
difficulties in recording the weak signal of coherent
radiation in methane were caused by instability of optical signals at the input of the photodetector. This resulted in the change of amplitude and phase of the recorded radiation signal and, hence, in distortion of
the shape of the recorded resonances. Standing waves
were tuned by adjusting the mirrors fixed on piezocera-
mic elements. Thus, already in the first experiments we
could resolve the methane MHS without complicated telescopic systems.

The action of spacially separated fields in the
coordinate system of moving atoms is equivalent to that
of pulsed fields. Under pulsed conditions coherent radiation in the field of standing waves connected with the
above-mentioned phenomena was observed in [20].

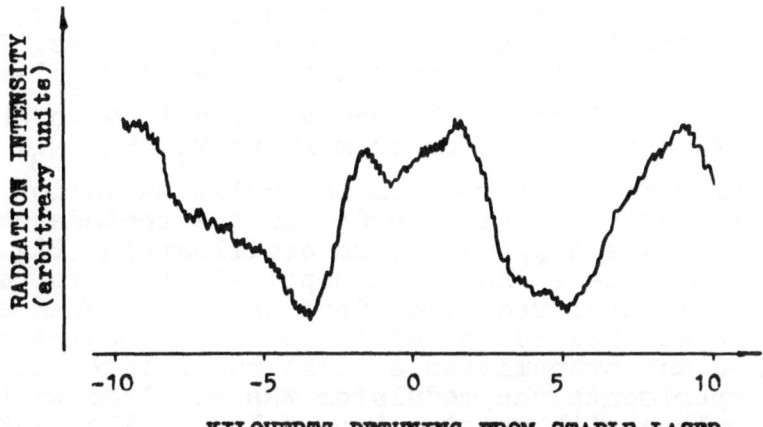

Fig. 10. The dependence of the coherent radiation signal amplitude in methane on frequency

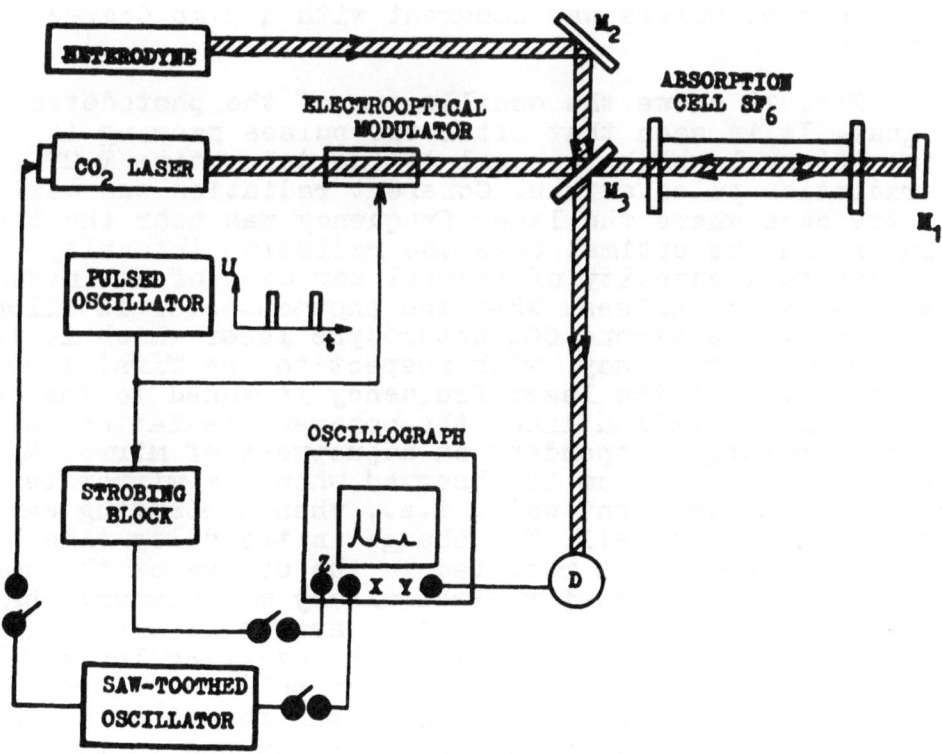

Fig. 11. The scheme for observing coherent
radiation in SF_6 in a standing-wave field

Coherent radiation arises in the intervals of 2T, 3T...,
where T is the interval between exciting pulses. The
experimental arrangement consisted of a CW CO_2 laser,
heterodyne, electrooptical modulator, cell with absorbing gas SF_6 and the system of mirrors M_1, M_2, M_3.

A standing wave was formed in the cell. Radiation from
the absorption cell was recorded in the photodetector
whose signal was supplied to an oscillograph (Fig. 11).
In the normal state the electrooptical modulator did
not transmit laser radiation into the cell. When a voltage pulse was fed to the electrooptical modulator, it
was opened and transmitted a light pulse into the cell.
In our experiments the modulator was supplied with a
pair of pulses with varying time delay T. The light pulse
duration   and delay T could be varied from 0.1 to several microseconds, and the pulse amplitudes were equal.
Estimations have shown that under our conditions phase
retardation from pulse to pulse associated with the laser
frequency instability did not exceed $3 \cdot 10^{-2}$ rad. Thus,
each pair of pulses was coherent with a high degree of
accuracy.

Fig. 12 shows the oscillogram of the photodetector
signal. It is seen that after two pulses passing in the
time interval which is equal to the delay time T,2T...,
a radiation pulse arises. Coherent radiation was maximum
in the case where the laser frequency was near the line
center. In the optimal case the radiation intensity
amounts to a quantity of several per cent of an intensity of exciting pulses. When the photodetector is illuminated with the second CO_2 heterodyne laser which is tuned
by 5 MHz in frequency with respect to the first laser,
beats arise. If the laser frequency is tuned to the center of an absorption line, the coherent radiation intensity is strongly dependent on adjustment of mirror M.
The radiation maximum is observed when the mirror is
normal to an incident wave, i.e., when a standing wave
is formed in the cell. The change in the delay time between exciting pulses resulted in the change of the coherent radiation intensity. Preliminary measurements have
shown that the coherent radiation arises at the frequency
of an exciting laser. Tuning of the exciting laser frequency resulted in the change of the pulse amplitude of
coherent radiation. Fig. 13 shows the record of the coherent radiation pulse intensity dependence on the laser
frequency. In this case X-plates of the oscillograph
were supplied with the same voltage as piezoceramics
which was used in scanning the laser frequency. The

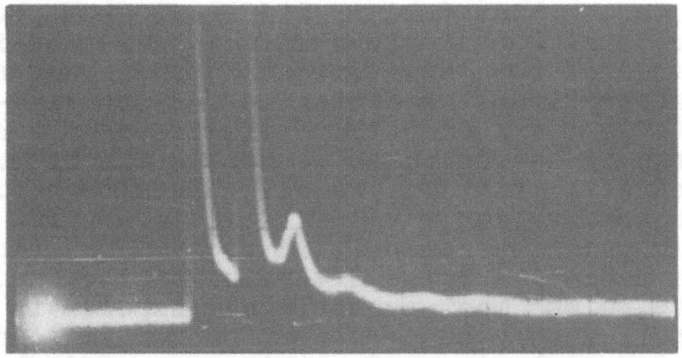

Fig. 12. The oscillogram of coherent
radiation in SF_6. Pressure is 10^{-2}Torr.
Distance between pulses is 0.4 μsec,
pulse duration is 0.2 μsec

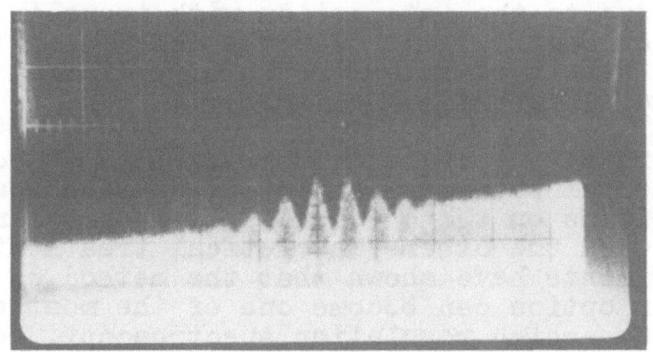

Fig. 13. The dependence of coherent
radiation intensity on detuning frequency
from the line center. T = 0.4 μsec,
τ = 0.2 μsec. Distance between maxima is ~ 2 MHz

Y-input was supplied with the signal of the photodetector. An electron system of strobing illuminated only the
pulse of interest. It is seen that the coherent radiation intensity oscillates depending on laser detuning
from the center of an absorption line. The total oscillation period occurs at tuning of the laser frequency
by a value of $\Omega \sim 1/2$ T. Far from the line center at
the detuning by a value of $\Omega \gg 1/\tau$ the coherent radiation intensity is independent of the laser frequency.

The discovered phenomenon has a number of characteristic features which make it different from phenomena known in optics. Oscillations of the radiation intensity at the change of the laser frequency are observed only near the absorption line center in the standing-wave field. Outside the center the mechanism of formation of the coherent radiation pulse is due to the other process, interaction with traveling waves. Molecules having interacted with a traveling wave of the second pulse produce macroscopic polarization only in the time 2T. The formation of the radiation pulse after the action of two traveling-wave pulses most probably corresponds in its nature to a well-known phenomenon of photon echo. Thus, there are two mechanisms of elimination of dephasing of oscillators due to thermal motion. The first mechanism is associated with the two-photon interaction with a standing wave. Note once more that it is the process that is associated with the appearance of resonances in spacially separated optical fields [7]. The second mechanism is due to the interaction with traveling waves. The radiation oscillations near the absorption line center are the result of interference of radiation associated with two different mechanisms.

An attractive feature of CRSF is that it combines such fundamental phenomena as intense resonances in separated fields on the one hand and superradiance and quantum echo on the other. Theoretical treatment and first experiments have shown that the method of separated fields in optics can become one of the most effective methods of superhigh resolution spectroscopy. We will not discuss various spectroscopic applications in detail but must note the effectiveness of the method in investigations of collisions. Narrow resonances can be used in producing optical frequency standards. New possibilities can appear with the production of optical delay lines.

In conclusion I would like to express my thanks to Drs. S.N.Bagayev, Ye.V.Baklanov, B.Ya.Dubetsky, and L.S.Vasilenko for collaboration.

R e f e r e n c e s

1. V.S.Letokhov, V.P.Chebotayev: Nonlinear Laser
   Spectroscopy (Springer-Verlag, Berlin, Heidelberg,
   New York 1977)

2. W.E.Lamb, Jr.: Phys.Rev. 134A, 1429 (1964)

3. P.H.Lee, M.L.Skolnick: Appl.Phys.Lett. 10, 373
   (1967)

4. V.N.Lisitsyn, V.P.Chebotayev: Zh.Eksp. i Teor.Fiz.
   54, 419 (1968)

5. J.L.Hall, C.Borde: Phys.Rev.Lett. 30, 1101 (1973)

6. J.L.Hall, C.Borde, K.Uehara: Phys.Rev.Lett. 37,
   1339 (1976)

7. V.P.Chebotayev: Laser Spectroscopy. Proc. 2nd
   International Conference, Megeve, France, June
   1975 (Springer-Verlag, Berlin, Heidelberg, New
   York) p. 150

8. S.N.Bagayev et al.: Appl.Phys. 13, 291 (1977)

9. N.Bloembergen, M.D.Levenson: Doppler-Free Two-
   -Photon Absorption Spectroscopy. Topics in Applied
   Physics vol. 13 (Springer-Verlag, Berlin,
   Heidelberg,New York 1976) p. 315

10. Ye.V.Baklanov, B.Ya.Dubetsky, V.P.Chebotayev:
    Appl.Phys. 9, 171 (1976)

11. Ye.V.Baklanov, B.Ya.Dubetsky, V.P.Chebotayev:
    Appl.Phys. 11, 201 (1976)

12. J.C.Bergquist, S.H.Lee, J.L.Hall: Phys.Rev.Lett.
    38, 159 (1977).

13. M.M.Salour, C.Cohen-Tannoudji: Phys.Rev.Lett.
    38, 757 (1977).

14. R.Teets, J.Eckstein, T.W.Hänsch: Phys.Rev.Lett.
    38, 760 (1977)

15. V.P.Chebotayev, A.V.Shishayev, B.Ya.Yurshin,
    L.S.Vasilenko: Appl.Phys. in press

16. N.F.Ramsey: Molecular Beams (Oxford Univ. Press, New York, London 1956)

17. B.Ya.Dubetsky: Sov.J.Quantum Electron. $\underline{6}$, 682 (1976)

18. V.P.Chebotayev: Report at 5th Vavilov Conference on Nonlinear Optics, Novosibirsk, USSR, June 1977 (Appl.Phys. in press)

19. S.N.Bagayev, V.P.Chebotayev, A.S.Dychkov: Appl.Phys. in press

20. V.P.Chebotayev, N.M.Dyuba, M.I.Skvortsov, L.S. Vasilenko: Appl.Phys. in press.

COHERENCE IN BEAM-FOIL SPECTROSCOPY

H.J. Andrä

Institut für Atom- und Festkörperphysik

Freie Universität Berlin, 1000 Berlin 33, W.-Germany

Among all the interrelated coherence effects in atomic spectroscopy the quantum beat (interference) |1-4| phenomenon is conceptually the most basic one from which the other methods like the zero- |1,5| or high field-|6| level crossing can be derived. For this reason and reasons related to the experimental potentials of BFS I choose mainly quantum beats for the discussion of coherence in FBS. For the physical interpretation of QBs we choose a four level system (i,1,2,f) in Fig. 1a with two nearly degenerate levels 1 and 2 separated by an energy interval $\hbar\omega_{21} = \hbar(\omega_2-\omega_1)$. We assume that an ensemble of such atoms is irradiated by a light pulse of length δt at $t=0$ and that the reemitted (scattered) light intensity is observed as a function of time t through a detection time window Δt centered at t. For $\Delta t, \delta t$ being δ-function type time windows this light scattering process through levels 1 and 2 can be described by probability amplitudes A_1 and A_2

(1a) $\quad A_1 = \text{const.} <f,t,k\varepsilon|V_{k\varepsilon}|1,t,0>e^{(-i\omega_1-\gamma_1^2/2)t} <1,0,0|V_{k'\varepsilon'}|i,0,k'\varepsilon'>$

(1b) $\quad A_2 = \text{const.} <f,t,k\varepsilon|V_{k\varepsilon}|2,t,0\ e^{(-i\omega_2-\gamma_2/2)t} <2,0,0\ V_{k'\varepsilon'}\ i,0,k'\varepsilon'>$

with the notation |atomic state, time, field state>. If these two paths are in principle indistinguishable a general postulate of quantum mechanics requires for the probability $P(f,t,k\varepsilon)$ of detecting a $k\varepsilon$-photon and finding the atom in its final state |f>

$$P(f,t,k\varepsilon) \propto |A_1+A_2|^2 \propto I(t,k\varepsilon) \propto$$

(2) $\quad \left| \sum_{n=1,2} \underbrace{<f,t,k\varepsilon|V_{k\varepsilon}|n,t,0><n,0,0|V_{k'\varepsilon'}|i,0,k'\varepsilon'>}_{B_n} e^{(-i\omega_n-\gamma_n/2)t} \right|^2$

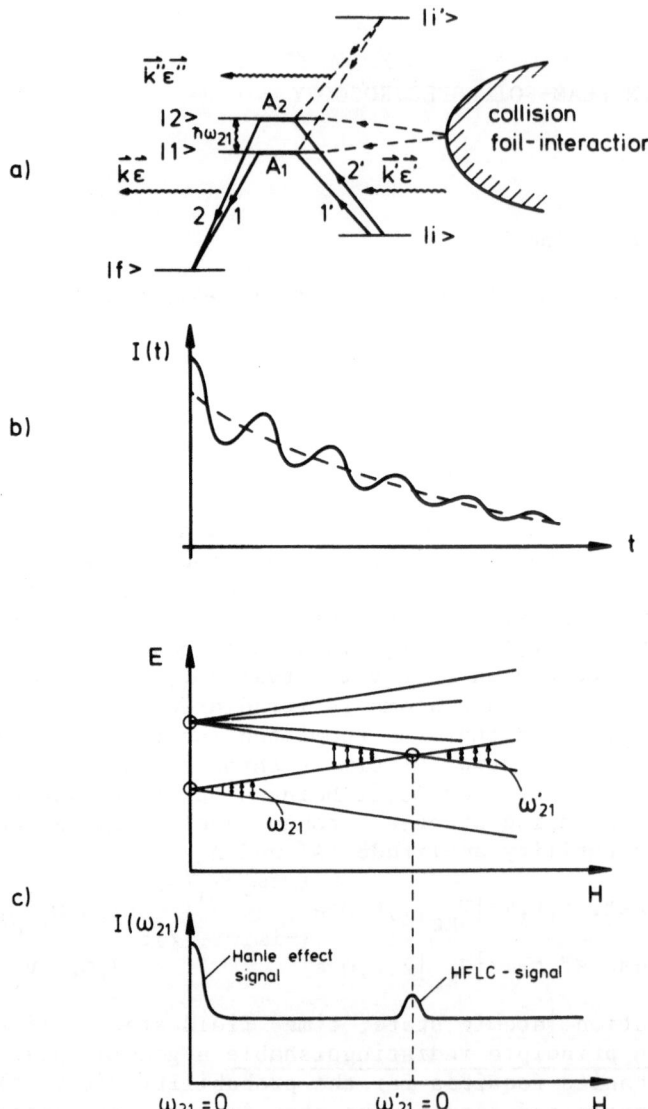

Fig. 1:   Physical interpretation of quantum beats and level
          crossings. See text for explanation.

which becomes with $B_n=B_n^*$ and $\gamma_1=\gamma_2=\gamma$

(3) $I(t,k\varepsilon) \propto \left|B_1^2+B_2^2+2B_1B_2 \cos\omega_{21}t\right| e^{-\gamma t}$.

Hence, one observes an exponential decay with a superimposed quantum beat (QB) in Fig. 1b, the frequency of which directly represents the excited level separation and where it is the emission probability of <u>each single</u> atom which oscillates.

The two paths in Fig. 1a are in principle indistinguishable if the transitions 1',2' and 1,2 are allowed for photons of equal direction and polarization in the excitation (k'ε') and detection (kε) processes respectively, if the transitions 1',2' and 1,2 are energetically indistinguishable due to the time energy uncertainty relation requiring δt, $\Delta t \ll 2/\omega_{21}$, and if the atoms are left in the same final state $|f\rangle$.

If the detection is spread out in time from t=0 to t=∞ one obtains at once the levelcrossing signal as a function of the level separation ω_{21} in Fig. 1c.

(4) $I(\omega_{21},k\varepsilon) = \int_0^\infty I(t,k\varepsilon)dt \propto \dfrac{2B_1B_2\gamma}{\gamma^2+\omega_{21}^2}$.

This short sketch of quantum interference does of course not only rely on the pulsed light excitation. It equally well applies to (k"ε")photon-(kε)photon coincidence experiments or to collisional ion-foil, -tilted foil, or- inclined surface excitation of the two levels 1 and 2 as long as the condition of principally indistinguishable paths is guaranteed. (For a quantum electrodynamic discussion see ref. 7,8.)

For a discussion of these collisional excitations it is advisable to rewrite the QB intensity formula

(5) $I(t,k\varepsilon) \propto \left|\langle f,t,k\varepsilon|V_{k\varepsilon}|\Psi_i(t)\rangle\right|^2$

with the coherent superposition state

$|\Psi_i(t)\rangle = \displaystyle\sum_{n=1,2} |n,t,0\rangle\langle n,0,0|V_{k'\varepsilon'}|i,0,k'\varepsilon'\rangle\, e^{(-i\omega_n-\gamma_n/2)t}$

(6) $= \displaystyle\sum_{n=1,2} c_{ni} |n,t,0\rangle\, e^{(-\omega_n-\gamma_n/2)t}$

where the coefficients c_{ni} are not any more predictable as for light excitation as long as the ion-solid interaction mechanism is not understood. Instead one can only use symmetry arguments depending on the excitation geometry for the deriviation of some general properties of the coefficients c_{ni} and one is essentially forced to determine them by experiments.

With the introduction of the coherent superposition state we
can now say that coherence exists and QBs will occur if such a
superposition state in the eigenstate representation is produced
and if at least two components can decay to the same final state.
Rewriting equ.(5) in simplified form

(7) $I(t,k\varepsilon) \propto <f|V_{k\varepsilon}|\Psi_i(t)><\Psi_i(t)|V_{k\varepsilon}^+|f>$

and expanding it in terms of eigenbasis sets

(8) $I(t,k\varepsilon) \propto \sum_{n,n'} <f|V_{k\varepsilon}|n>\underbrace{<n|\Psi_i(t)><\Psi_i(t)|n'>}_{\rho_{nn'}t)}<n'|V_{k\varepsilon}^+|f>$

one can equally well say that coherence exists and QBs will occur
if off-diagonal density matrix elements in the eigenstates representation are produced by the excitation process:

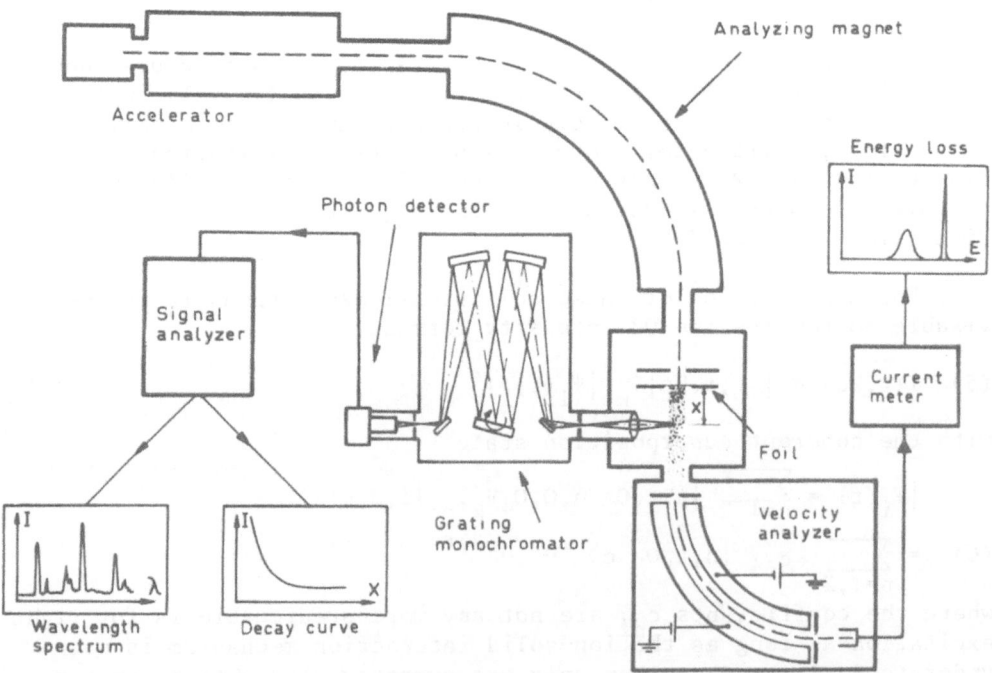

Fig. 2:   Experimental arrangement for studies of atomic spectra,
intensity decay curves and quantum beats with the beam-
foil technique.

$$(9) \quad \rho(t) = \begin{pmatrix} c_1 c_1^* \; e^{-\gamma_1 t} & c_1 c_2^* \; e^{(-\omega_{21}-(\gamma_1+\gamma_2)/2)t} \\ c_1^* c_2 \; e^{(i\omega_{21}-(\gamma_1+\gamma_2)2)t} & c_2 c_2^* \end{pmatrix}$$

In spite of the little knowledge on the c_{ni} coefficients from collisional ion-solid excitation mechanisms, beam-foil spectroscopy (BFS) |9-12| is ideally suited for QB measurements for reasons which become apparent by a short introduction to BFS (see Fig. 2): The ions to be studied are generated in an ion source, accelerated to energies ranging from 10keV to several 100MeV, and are then momentum analyzed in a magnet to yield monoenergetic, isotopically pure beams of 0.1-10µA. This beam is collimated and passed through a target which may consist of a thin self-supporting solid foil (carbon foils of 1-20µg/cm^2 are used most often), a differentially pumped gas target of $\leq$0.1µg/cm^2 at p $\leq$1mbar, or a crossed leaser beam in resonance with a transition of the ions. In the beginning I shall discuss only foil targets. When traversing the foil the ions experience many elastic and inelastic collisions so that the beam emerges from the foil in a highly excited state, consisting generally of ions in various stages of ionization and excitation. This excitation energy is spontaneously released by the ions downstream from the foil in the form of a rich photon (electron) spectrum ranging from the optical into the X-ray region. The foil excited beam thus represents a fast moving spectroscopic source which emits in a high vacuum environment of p $\leq$ 10^{-5}mbar at average velocities v of the order of 10^{8}cm/s and is therefore ideally suited for spectral and transient studies of the photon emission.

For measurements of the transient emission of a specific spectral line a spectrometer is set on this line and its output intensity is recorded as a function of distance x from the foil. This is simply achieved by mechanically moving the foil along the beam axis and feeding the detector pulses into a multiscaler whose channel sweep is synchronized to the x-displacement of the foil. In order to reduce the influence of beam fluctuations the time each channel is measured is usually normalized to the charge collected in a faraday cup or to the "white" photons observed by a detector at fixed distance from the foil. By observing a beam section Δx = 0.01cm, a time resolution of the order of $\Delta t = \Delta x/v' \simeq 10^{10}$s is easily achieved and the point t=0 is defined to within $\leq$10^{-14}s due to the final state formation within $\leq$ 10nm at the exit of the foil. For the time scale calibration of transient measurements the average velocity v of the foil transmitted beam has to be determined with the best possible accuracy.

Without going into any further detail I list the relevant features of BFS for QB-measurements:

1) Unselective foil or gas excitation of all levels of a given charge state – sometimes causing cascade problems.

2) Any charge state of any element may be obtained by choice of beam energy. Limitations are set by the presently available energies of $\leq 10 \text{MeV/amu}$.

3) Collisional perturbation of the excited levels is negligible at a residual gas pressure of $p \leq 2 \cdot 10^{-6}$ mbar.

4) Due to a beam density of 10^5–10^6 cm^{-3} interionic fields, imprisonment of radiation and stimulated emission can be fully reglected.

5) Low excited beam luminosity requires single photon counting.

6) Excellent time resolution of $\Delta t = 10^{-10}$–10^{-11}s with a definition of t=0 to within $\delta t \leq 10^{-14}$s.

7) Time scale calibration to within $\pm 10^{-3}$ in the ns-range by excited beam energy measurement.

8) Excellent differential and integral transient detection linearity.

9) Close lying levels are coherently excited due to the short excited state formation of $\leq 10^{-14}$s.

10) Depending on the excitation geometry collisional excitation may yield nonisotropic excitation – necessary for the observation of QBs.

11) Selective laser excitation eliminates the cascade problems yields maximum nonisotropy of the excited levels and conserves all the other excellent properties of FBS.

With these features in mind which offer unprecedented experimental conditions for the study of QBs we return now to coherent phenomena by stating: The general method of preparing superposition states or off-diagonal density matrix elements in the eigenstate representation is the application of a sudden perturbation to the atomic system which breaks the symmetry of the pre-existing hamiltonian. This includes not only the collisional (pulsed) excitation but also the sudden change of external fields or the sudden change of the atomic environment when a fast atom is leaving a foil surface.

A realistic excited two level system for discussing QBs within the frame work given so far can be approximated by the hydrogen n=2 Lamb-shift splitting $2s_{1/2}$–$2p_{1/2}$. We neglect the $2p_{3/2}$-level, hyperfine structure, and the m-sublevels and use the simple notation in Fig. 3a. Due to the parity selection rule for atomic E1-transitions only the $|2p>$-state can decay to the $|1s>$ ground state. First we prepare a beam in the $|2s>$ state by waiting after a gas excitation until the $|2p>$-state has decayed away completely. If one enters with this beam at t=0 all of a sudden a region with electric field F_0 as shown in Fig. 3b such that the sudden perturbation can be applied, then one has to expand the initial field-free eigenstates {a} in terms of the new eigenstates in the field {n}. The new eigenstates in the field are linear combinations of the field-free ones so that both can decay to the <u>same final</u> $|1s>$-ground state

(10)
$$|s'> = C_{ss} |2s> + C_{ps}|2p> \qquad <s'|2s> = C_{ss}$$
$$|p'> = C_{sp} |2s> + C_{pp}|2p> \qquad <p'|2s> = C_{sp}$$

due to their $|2p>$-admixtures. Since only the 2s-level was populated only the C_{ss} and C_{sp} coefficients are important and a particularly simple expression is obtained

(11) $\quad I(t) \propto |\displaystyle\sum_{n=s',p'} <f|V|n> e^{(-i\omega_n - \gamma_n/2)t} <n|2s>|^2$

which yields

(12) $\quad I(t) \propto A\, e^{-\gamma_{s'}t} + B\, e^{-\gamma_{p'}t} + C\, e^{-(\gamma_{s'}+\gamma_{p'})t/2} \cos(\omega_{s'}-\omega_{p'})t$

where all variables depend on the field strength.
One has thus obtained a situation where a sudden perturbation which breaks the symmetry of the preexisting field-free hamiltonian has produced a coherent superposition state in the field-eigenstate representation, which gives rise to the QB-pattern in Fig. 3c $|13|$.

If the exciting foil is placed into the plane of the left electrode both the $|2s>$ and $|2p>$ states are simultaneously but <u>incoherently</u> excited. (The possible s-p coherence is neglected here.) Without a field one has a diagonal density matrix in the field free eigenstate representation, where p_1 and p_2 give the populations of the 2s- and 2p-states respectively. In order to obtain the density matrix in the field one assumes that the field is suddenly switched on right after the 2s- and 2p-state formation. This sudden perturbation determines the transformation matrix U.

(13) $\quad \rho\{a\} = \begin{array}{c} \\ 2s \\ 2p \end{array}\overset{\begin{array}{cc} 2s & 2p \end{array}}{\begin{pmatrix} p_1 & 0 \\ 0 & p_2 \end{pmatrix}} \xrightarrow[\text{field}]{U\,\rho\{a\}\,U^+ = \rho\{n\}} \rho\{n\} = \begin{array}{c} \\ s' \\ p' \end{array}\overset{\begin{array}{cc} s' & p \end{array}}{\begin{pmatrix} p_1' & x \\ x & p' \end{pmatrix}_2}$

As a result off-diagonal density matrix elements in the field-eigenstate representation occur if and only if $p_1 \neq p_2$. Hence, QBs become observable only if $p_1 \neq p_2$, that is if the excitation density matrix is not proportional to the unit matrix. The experimental result $|14|$ in Fig. 4 exhibits a large enough difference of the p_1 and p_2 for a clean observation of the field induced Ly_α-Stark beats. It is interesting to note that for this particular case the field induced coherence between different parity states leads to an inten-

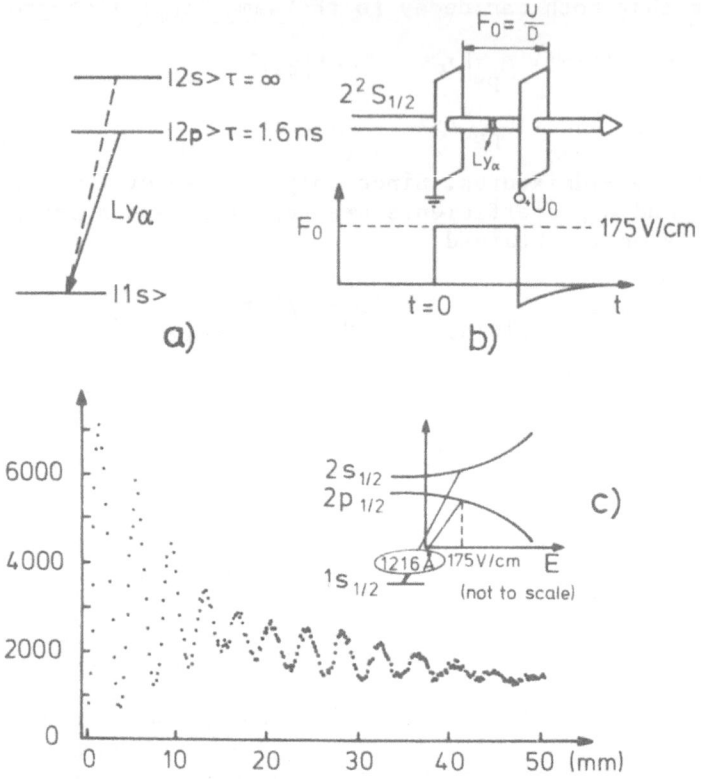

Fig. 3: Stark beats in HI-Ly$_\alpha$ observation.
a) Simplified levelscheme. b) Field electrodes, electric
field F_o, and direction of detection. c) Field dependent
level splitting and observed quantum beats.

sity pulsation of the spatially isotropic Ly$_\alpha$-radiation. It should
also be noted that for a full description of this experiment a
16 x 16 density matrix is actually necessary when all finestructure
and hyperfinestructure components are included which do account also
for the deviation of Fig. 3c from a single frequency beat pattern.

So far we have discussed only QBs involving different parity
states which can be interpreted as time dependent pulsations of the
transition rate of the atom. From now on we concentrate on coherence
phenomena between equal parity states where $\gamma_n = \gamma_{n'} = \gamma$ holds, which
become observable only when the nonisotropic radiation pattern
(unequal initial population of different m-substates) of the atoms
periodically changes its shape relative to a given quantization
axis, thus yielding intensity oscillations for fixed directions of
detection. Hence, an initial nonisotropic excitation must take
place in order to make interference phenomena observable at all.

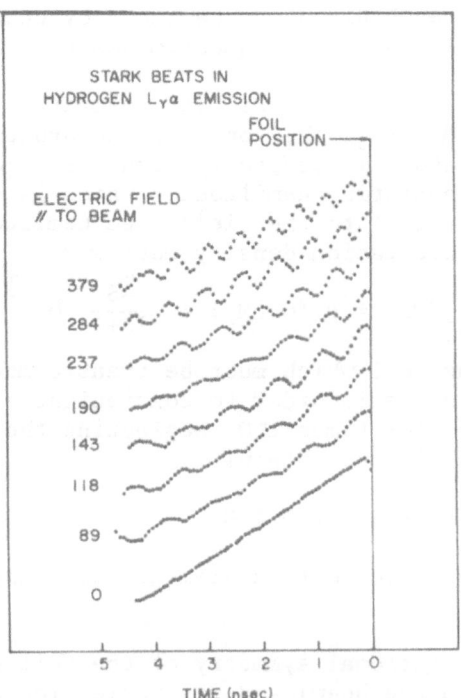

Fig. 4: Plots of log (Ly$_\alpha$-intensity) versus time in an electric
         field parallel to the beam.The parameter is field strength
         in V/cm.

         For their description one must therefore include all m-sub-
states and one needs a general theory for a many level system. To
this end one can adopt the theory of angular correlations |15| or
one can proceed with equ.(8) with the inclusion of all initial and
final states.

(8') $I(t,k\varepsilon) \propto \sum_{\substack{nn' \\ i,f}} <f|V_{k\varepsilon}|n><n|\Psi_i(t)><\Psi_i(t)|n'><n'|V_{k\varepsilon}^+|f>$

(14) $I(t,k\varepsilon) \propto \sum_{nn'} <n'|D_{k\varepsilon}|n><n|\rho(0)|n'> e^{|-i\omega_{nn'}-\gamma/2|t}$

As before one can deduce from equ.(14) that off-diagonal density
matrix elements in the eigenstate representation are necessary in
order to observe quantum beats. The way these off-diagonal density

matrix elements are obtained can be shown by choosing the example of the HI-n=2-^{2}P-finestructure quantum beats:

The Coulomb-interacting foil environment, which acts in the uncoupled base $\{LM_L, SM_s, IM_I|$ only on the orbital angular momentum and leaves S and I isotropic (as long as LS-coupling holds), represents the preexisting hamiltonian which is suddenly changed at the exit of the foil to the field free hamiltonian of the atom. One thus has the excitation density matrix

$$(15) \quad \rho(0,LM_L,SM_s,IM_I) = \rho_L(0,LM_L) \otimes \frac{1_S}{2S+1} \otimes \frac{1_I}{2I+1}$$

at the exit of the foil which must be transformed into the field free coupled eigenbase in order to convieniently describe the evolution of the atom for times t>0. Neglecting the hyperfine interaction in our example one obtains

$$(16) \quad \rho(0,JM_J) = U \, \rho(0,LM_L,SM_s) \, U^{-1}$$

where the transformation matrix elements are simply Clebsch-Gordan coefficients.

Due to the rotational symmetry of the foil excitation (foil normal parallel to the beam) and its reflection symmetry with respect to a plane containing the beam axis (quantization axis) two further conditions are imposed on the $\rho(0,LM_L)$ density matrix elements $|16|$:

$$(17) \quad <LM_L|\rho(0)| \, L'M'_L> = e^{-i(M_L-M'_L)\rho} <LM_L|\rho(0)|L'M'_L>$$

$$(18) \quad <LM_L|\rho(0)| \, L'M'_L> = (-1)^{-M_L-M'_L} <L-M_L|\rho(0)|L'-M'_L> \; .$$

The first one allows only diagonal density matrix elements with the additional requirement $\rho_{MM}=\rho_{-M-M}$ due to the second condition.

For our example the incoherent excitation density matrix $\rho(0,LM,SM_s)$ can therefore be described by only two parameters $\sigma_o = \sigma_{oo}$ and $\sigma_1 = \rho_{11} = \rho_{-1-1}$

(19)

$$\varrho(0, LM_L SM_S) =$$

		M_L	1		0		-1	
M_L	M_S		$1/2$	$-1/2$	$1/2$	$-1/2$	$1/2$	$-1/2$
1	$1/2$		$\sigma_1/2$	0	0	0	0	0
	$-1/2$		0	$\sigma_1/2$	0	0	0	0
0	$1/2$		0	0	$\sigma_0/2$	0	0	0
	$-1/2$		0	0	0	$\sigma_0/2$	0	0
-1	$1/2$		0	0	0	0	$\sigma_1/2$	0
	$-1/2$		0	0	0	0	0	$\sigma_1/2$

which yield after transformation (16) off-diagonal density matrix elements in the eigenstate representation only if $\sigma_0 \neq \sigma_1$:

(20)

$\varrho(0, JM_J) =$

J $J M_J$	3/2 3/2	3/2 1/2	1/2 1/2	3/2 $-1/2$	1/2 $-1/2$	3/2 $-3/2$
3/2 3/2	$\sigma_1/2$	0	0	0	0	0
3/2 1/2	0	$(\sigma_1+2\sigma_0)/6$	$(\sigma_1-\sigma_0)/\sqrt{18}$	0	0	0
1/2 1/2	0	$(\sigma_1-\sigma_0)/\sqrt{18}$	$(2\sigma_1+\sigma_0)/6$	0	0	0
3/2−1/2	0	0	0	$(\sigma_1+2\sigma_0)/6$	$(\sigma_0-\sigma_1)/\sqrt{18}$	0
1/2−1/2	0	0	0	$(\sigma_0-\sigma_1)/\sqrt{18}$	$(2\sigma_1+\sigma_0)/6$	0
3/2−3/2	0	0	0	0	0	$\sigma_1/2$

This requirement is equivalent to the former statement for the observation of the HI-$2s_{1/2}$-$2p_{1/2}$ beats that the excitation density matrix must <u>not</u> be proportional to the unit matrix. It means for our example that an initial unequal population of M_L-substates is necessary which is responsible for the time dependent nonisotropic angular emission pattern to be observed as quantum beats. The experimental arrangement in Fig. 5a (without polarization detection) is therefore sufficient for the observation of the HI-n=2-$p_{1/2}$-$p_{3/2}$ finestructure beat-frequency in Ly_α-emission after foil excitation |17| shown in Fig. 5b. The way the angular Ly_α-intensity distribution changes with time is schematically shown in Fig. 5c, from which it is also obvious that an initial nonisotropic M_L-population is necessary for the observation of such zero field quantum beats. Since the observed 2^2p-finestructure beat frequency was already known with high precision this example serves here only as proof that the foil excitation can lead to nonisotropic M_L-populations, a fact for which symmetry arguments alone are not sufficient.

As application of this zero field quantum beat spectroscopy I show the result obtained after foil excitation of $^7Li^+$ in Fig. 6. The upper part shows the oscillatory part I_{osc} of the LiII-2^3S - 2^3P - 548.5 nm intensity, i.e. after subtraction of the non-oscillating exponential intensity decay (including cascades). It has been measured with a time resolution of $\Delta t \approx$ 40 ps at 300 keV $^7Li^+$-beam energy and the time scale is calibrated to $1.5 \cdot 10^{-3}$ accuracy by beam-velocity measurement behind the foil |18|. I_{osc} clearly exhibits the superposition of several frequencies and a fourier transform of I_{osc} (lower part of Fig. 6) allows to extract 6 clearly resolved frequencies which are in excellent agreement with a recent calculation |19| as indicated in the center of Fig.6.

The zero field quantum beat spectroscopy after foil excitation is thus a good tool for fine- and hyperfine structure measurements in atomic physics. Unfortunately, however, great difficulties have

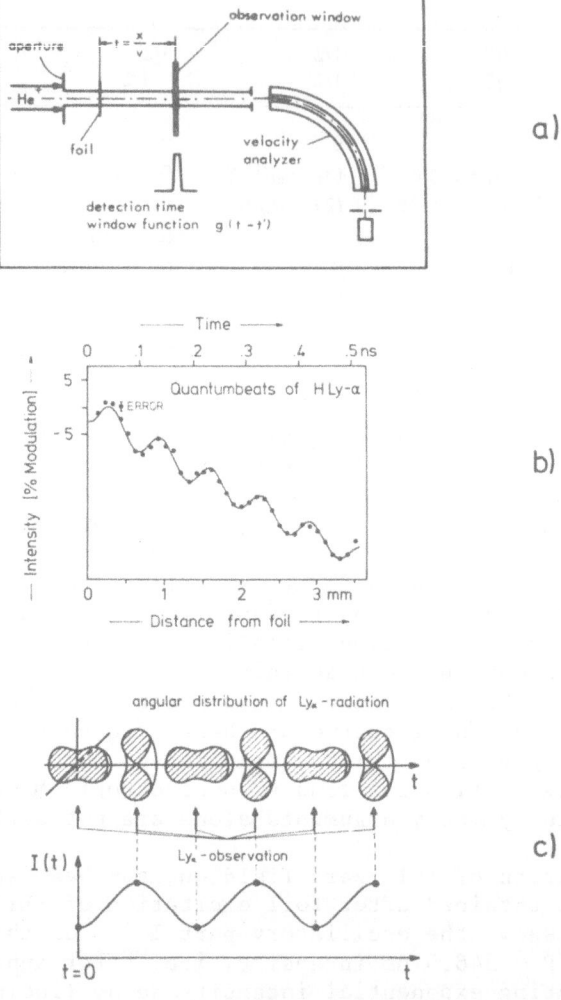

Fig. 5: a) Geometry for the observation of zero-field QBs after
foil excitation. b) Zero-field HI 2p $^2P_{1/2}$ - $^2P_{3/2}$ -fs-QBs
at 10969 MHz observed in H Ly_α-emission without polarization. The solid line is a fit to the data. c) Schematic
view of the transient change of the angular distribution of
the total Ly_α-emission in the zero-field QB experiment of
(b). Initial M_L = ±1-state population is assumed.

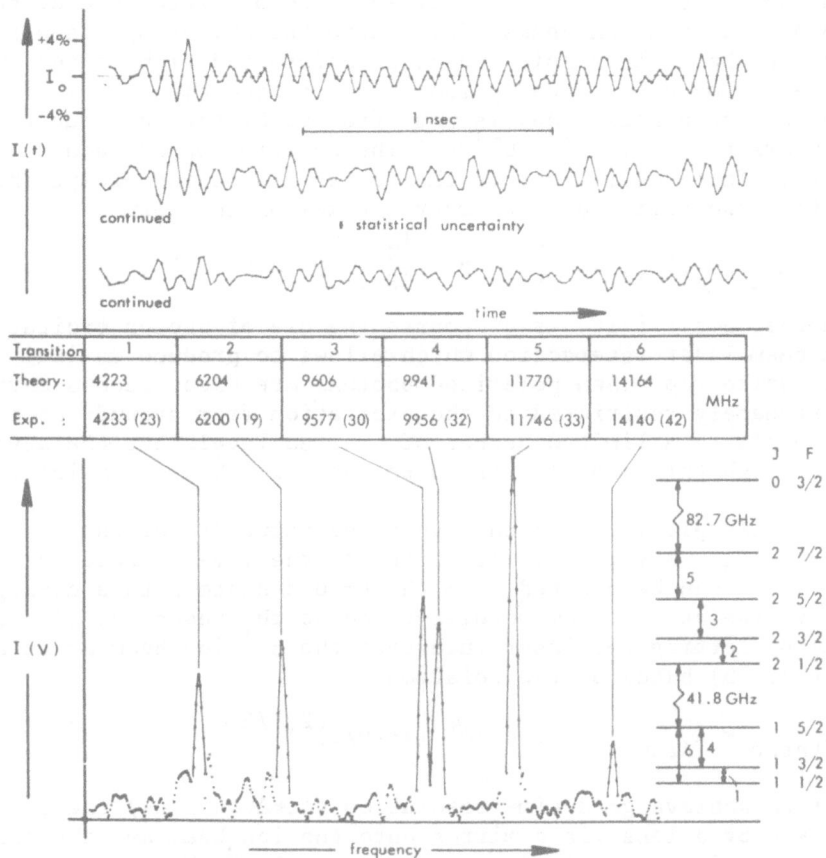

Fig. 6: ^{7}LiII-2p ^{3}P (fs)-hfs measurements with the zero-field QB method after foil excitation. The upper part shows the beat pattern in the 2s ^{3}S - 2p ^{3}P - 548.5 nm emission after subtraction of the non-oscillating intensity and continued twice. The full line connects the data points only. The lower part shows the fourier transform of these beats with the zero point of the frequency being suppressed. In the center six experimentally determined frequencies as indicated in the level diagram are compared with theory.

occured in obtaining sufficient anisotropy for ions and atoms beyond Z = 4 at medium beam energies. In order to make further use of the excellent qualities of the fast beam for such measurements other mechanisms for the creation of anisotropy into the system to be studied have to be introduced.

One such possibility is the recently demonstrated use of nuc-
lear spin polarized ion beams |20|, where the anisotropy is already
existent in the nuclear spin subsystem before and right after the
foil interaction adds a nearly isotropic excited electron shell.
(The nuclear spin anisotropy is not affected by the foil due to the
short interaction time of $<10^{-14}$s.) The description of such an experiment is fully equivalent to the former discussion except that
the initial excitation density matrix takes on the form

$$(21) \quad \rho(0, LM_L, SM_S, IM_I) = \frac{1_L}{2L+1} \otimes \frac{1_S}{2S+1} \otimes \rho(0, IM_I) \ .$$

Another possibility is of course the use of photon excitation
via ion beam laser interaction which allows to produce maximum
possible anisotropy when polarized photons are used. Such a method
is unfortunately restricted to the excitation from ground- or metastable levels to a limited number of excited levels but the advantages outweigh this problem for those cases which are feasible.

As an example I choose the hyperfine structure of the
^{137}BaII-6p $^2P_{3/2}$ level |21|. According to the level scheme in Fig.
7a this level can be excited from the ground state with a cw-argon-
ion laser line at 454.4 nm Doppler-tuned to the resonance line at
455.4 nm by letting the laser intersect the Ba$^+$-ion beam at an angle
ϑ (see Fig. 7b) based on the relation

$$(22) \quad \lambda_{Laser} = \lambda_{Atom} \ (1-v/c \ \cos\vartheta)/(1-(v/c)^2)^{1/2}.$$

In order to achieve good time resolution ($\delta t \approx \Delta t \leqslant 0.5$ ns) the laser
is focussed by a lens via a mirror onto the ion beam and the obser-
vation window is also tilted to the same angle ϑ. For such a geometry
one can calculate the excitation probability per ion of 1.6% when a
laser power density of 43 mW/(mm^{2}xc), an effective Doppler broadened
absorption line width of 4 GHz, and an interaction time of 0.5 ns is
assumed. Hence, with a beam of 5µA Ba$^+$ one expects $5 \cdot 10^{11}$ ions per
second to be excited such that the detected count rates will be
high enough ($>25 \cdot 10^3$) for precise experiments with the geometry in
Fig. 7b.

In a more detailed analysis of the level scheme in Fig. 8 one
finds the ground state split by 8.04 GHz, whereas the excited states
have a maximum separation of 659 MHz only. When Doppler tuning the
laser resonance by changing ϑ in Fig. 6b and recording the 455.4 nm
fluorescence emission a partially resolved double peak structure as
shown in Fig. 8 is observed. It clearly indicates that selective
excitation from either one of the ground state hyperfine components
F=1 or F=2 is achieved with a resolution of ∿4-5 GHz with a free
running argon-ion laser. As a consequence of the selection rules
one has then only coherent excitation of either the upper three or
the lower three hyperfine components of the excited level and

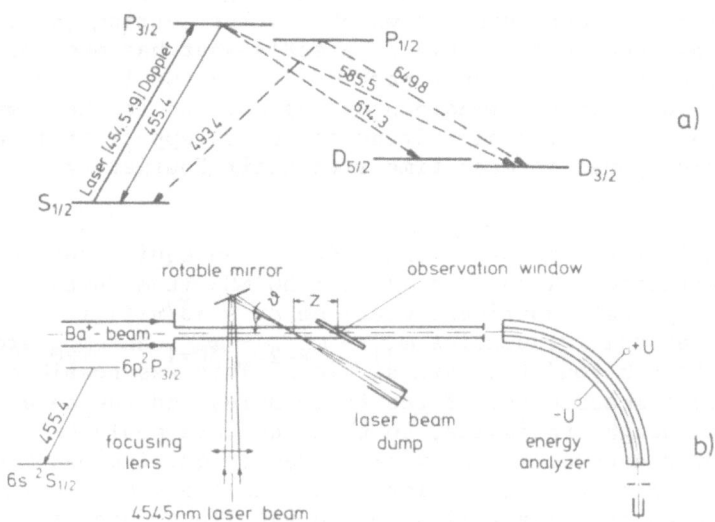

Fig. 7: a) Diagram of the 5 lowest BaII energy levels. b) Schematic
view of the experimental set up used for laser excitation
of a fast Ba⁺ beam with a Doppler tuned fixed frequency
argon ion laser. The observation window is moved along the
beam axis relative to the fixed excitation region.

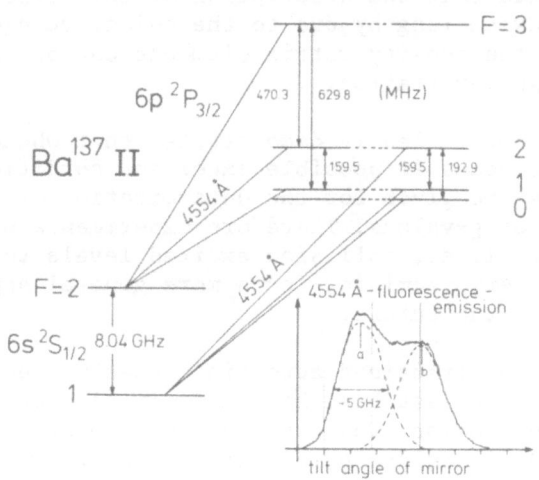

Fig. 8: Level scheme of Ba [137] II indicating the selective laser excitation from F=2 groundstate (a) to the upper three and
from F=1 groundstate (b) to the lower three hfs components
of the excited $6p^2P_{3/2}$ level. (a) and (b) correspond to the
angular settings of ϑ shown in the lower part of the figure.

expects therefore only beats among the upper three or the lower
three components depending on which angular setting (a or b in Fig.
8) of ϑ has been chosen. This is exactly what has been observed in
the experiment: In the upper part of Fig. 9 two low frequency beats
from the lower three hfs-components appear and in the lower part one
low frequency plus two high frequency beats appear from the upper
three hfs-components, each time with their Fourier transforms inserted.

Recalling the natural line width of an equivalent double re-
sonance experiment with $\Delta\nu = 1/\pi \cdot \tau = 50$ MHz (the "natural" width of
an intensity beat experiment would be $\Delta\nu = \sqrt{3}/\pi \cdot \tau$) it is important
to note that the frequencies $\omega_{21} = \omega_{F=2} - \omega_{F=1}$ and ω_{20} are actually
resolved here beyond the natural width. This was possible due to the
simultaneous measurement of the frequencies and the decay up to 7τ
with good enough statistics, so that the data could be divided by
the measured exponential to yield undamped quantum beats up to 7τ,
which by fourier transformation result in the good resolution shown.
Such a procedure fully relies on the excellent detection linearity
of FB-experiments, which makes FBL-QB-experiments superior to pulsed
laser excited QBs detected with presently known electronic time resolution. From the measured frequencies one can readily deduce the
hyperfine coupling constants with significant improvement in accuracy over earlier work.

I want to add that the description of this experiment with
equ.(14) is somewhat lengthy due to the selective excitation condition although the density matrix elements can be calculated explicitly for light excitation.

This measurement clearly demonstrates that whenever laser excitation of fast beams is possible excellent conditions for quantum
beat experiments are given for the determination of fine- and hyperfine structures or g-values. Therefore experiments with laser excitation from short lived, collision excited levels to higher excited
levels |23| are very promising for a more general application of
this technique in the future.

Apart from the transient zero field quantum beat spectroscopy
discussed so far the fast beam has recently also be turned into a
tool for high resolution fluorescence and saturated absorption
spectroscopy. The linewidth $\Delta\nu$ of the Doppler tuned, partially resolved resonance curve in Fig. 8 is approximately given by the addition of the contributions from velocity straggling of the beam $\Delta v/v$,
the ion beam divergence $\Delta\gamma$, the laser divergence (convergence) $\Delta\vartheta$,
the time-frequency uncertainty $\nu \sin\vartheta/\pi\Delta x$ (Δx is the geometrical
diameter of the laser beam), the laser bandwidth σ, and the natural
width Γ

(23) $\Delta\nu = \sigma + \Gamma + \nu \cdot v/c \left| \Delta v/v \cos\vartheta + (\Delta\gamma + \Delta\vartheta) \sin\vartheta \right| + \nu \sin /\pi \cdot \Delta x$.

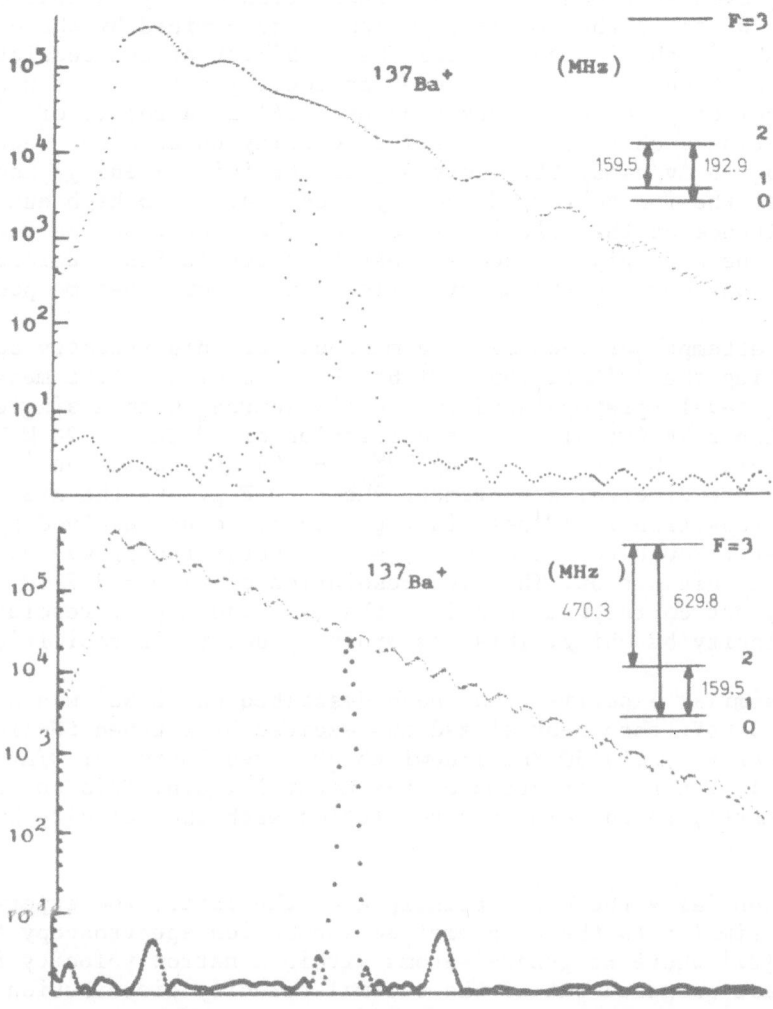

Fig. 9:   Observed zero-field QBs in the 455.4 nm emission of ^{137}BaII
          and their fourier transforms as calculated after deviding
          the data by an exponential and then subtracting the con-
          stant background. The inserted level schemes indicate the
          measured beat frequencies. Please note that the time and
          frequency scales are different in the top and bottom part
          of the figure.

It is important to note that the last three terms can be eliminated
for $\vartheta \to 0$, i.e. in parallel super-imposed beam (PSIB) geometry, such
that the resolution becomes $\nu/\Delta\nu = c/\Delta v$ when σ and Γ can be neglected.

At such favourable conditions the resolution simply depends on the velocity width of the ion beam, which is determined by the energy spread ΔU_o of the ion source and the stability of the acceleration voltage U. Under the assumption of an ideally stable U it was shown that a velocity bunching effect occurs |24| as a result of the acceleration such that the relative velocity uncertainty reduces according to $\Delta v/v = \Delta U_o/2U$, where ΔU_o is the initial energy uncertainty at the ion source. Hence, by acceleration to high energies the influence of the velocity spread of the source on the resolution can be strongly reduced so that high resolution fluorescence spectroscopy down to the natural linewidths should become possible.

An attempt has been made to make use of this velocity bunching by exciting the ^{131}XeII [5p^4 ^{3}P] 6p ^{4}P$^o_{5/2}$ level from the metastable 5d ^{4}F$_{7/2}$ level (prepopulated in the ion source) with a single-mode CW dye laser at 605 nm after acceleration of ^{131}Xe$^+$ to 20 keV and by monitoring the 6s ^{4}P$_{5/2}$ - 6p ^{4}P$^o_{5/2}$ - 529 nm transition |25|. With the experimental arrangement shown in Fig. 10a the HFS of the 605 nm transition in ^{131}XeII in Fig. 10b could be resolved by tuning the laser through the resonances, since energy tuning was not possible with this set up. The high resolution of $v/\Delta v = 3.7 \cdot 10^6$ did, however, not correspond at all to the expected higher resolution with velocity bunching. This was probably due to instabilities of U.

A similar experiment has been described where Na$^+$ was accelerated to 5 keV, then neutralized and excited by a tuned CW-single mode laser to yield 30 MHz linewidth or a resolution of $v/\Delta v = 2.5 \, 10^7$(!) after subtraction of the natural width. This corresponds approximately to the expected resolution with the velocity bunching effect.

Essentially the basic principle of the latter two experiments is very similar to the Doppler-free saturation spectroscopy techniques |27| where in general atoms within a narrow velocity interval Δv are prepared out of the thermal velocity distribution by a first laser beam and are then probed by a second laser beam. Here the narrow velocity interval Δv is prepared by electromagnetic ion optical velocity means which is then probed by a laser beam. A very elegant combination of both techniques was first introduced by Dufay et al. |28|. An improved version |29| of their experiment for the measurement of the ^{137}BaII-6p ^{2}P$_{3/2}$ hfs is shown in Fig. 11. A single mode argon ion laser beam at 454.5 nm was superimposed on a high energy Ba$^+$ beam. With the acceleration voltage HV the beam energy was Doppler tuned through the resonance at ∿251.5 keV (with a width of ∿1.2 keV for ^{138}Ba$^+$ due to instabilities of HV) and was then kept fixed above the resonance at about 254 keV as controlled by spectrometer I, which accepts the resonance fluorescence at 455.4 nm. In zone A the beam was decelerated (Doppler switched) into resonance by the potential V' such that according to the level scheme of ^{138}Ba$^+$ in Fig. 7a population was pumped out of the ground

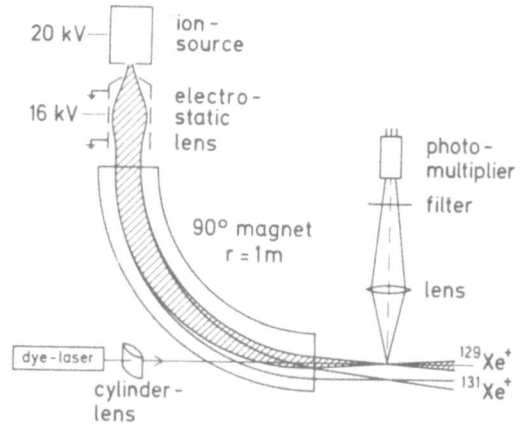

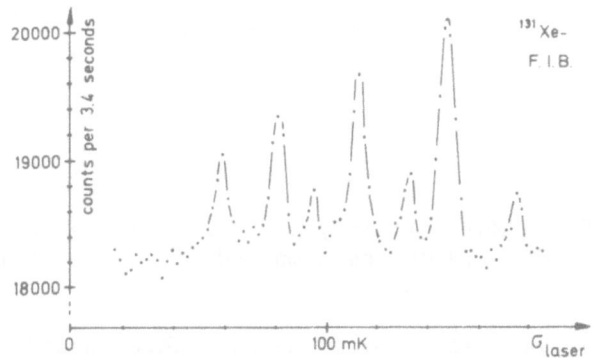

Fig. 10:   a) Experimental arrangement for high resolution fluores-
           cence spectroscopy in superimposed beam geometry with a
           scanning dye-laser where the velocity selection can be
           achieved by a 90°-magnet. b) Hyperfine structure of the
           ^{131}XeII-605 nm as observed with the set-up in (a).
           Further details see text.

state through the $6p\ ^2P_{3/2}$ level into the $5d\ ^2D_{3/2,5/2}$ levels for a
very narrow velocity class of ions in resonance with the laser, i.e.
a very narrow hole was burned into the broad velocity distribution
of the ground state ions. After reacceleration out of zone A the
beam was again decelerated by a potential V' into zone B in order
to probe this hole by observing a minimum (Lamb dip) in the resonance fluoescence detected through spectrometer II when ΔV in zone
A was zero. By variation of ΔV the hole could thus be scanned with

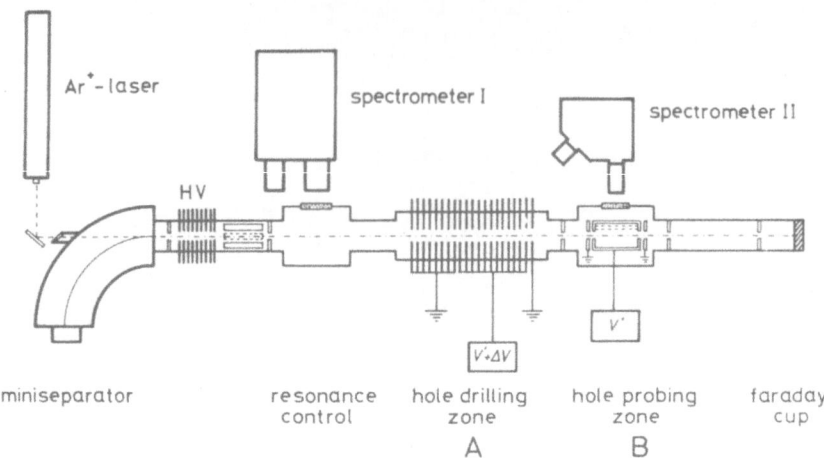

Fig. 11: Experimental arrangement for Doppler tuned "in flight
          Lamb dip spectroscopy". See test for explanation.

a dispersion of 2.5 MHz/Volt to yield for ^{138}Ba$^+$ a power broadened
($\sim$3mW) dip-width of $\sim$110 MHz as compared to the natural width of
50 MHz.

When ^{137}Ba$^+$ was used simulataneously three holes, corresponding
to the F'= 1,2,3 or 0,1,2, states of the 6p ^{2}P$_{3/2}$ level respectively,
could be burned into the F=2 or F=1 ground state velocity distributions separated by 3.08 kV (see Fig. 8). By choosing F=2 the corresponding three holes were probed in zone B with the three equivalent
transitions to yield the central fluorescence-dip in Fig. 12a for
ΔV=0. When ΔV was swept the three holes shifted across the three
probing transitions as indicated in Fig. 12, which explains the
whole dip-structure observed in Fig. 12b |30|. From such results
the authors could extract the hfs-splittings of the ^{135,137}BaII-6p
^{2}P$_{3/2}$ level in excellent agreement with the fast beam-laser QB results at about half the accuracy of the QB experiment.

The great advantage of this "in flight Lamb dip spectroscopy"
technique is not only the ease with which the dip-structure can be
scanned by variation of a rather small voltage at a few MHz/volt
but also its important independence of laser- and HV-instabilities.

Its application will be, however, more limited than the above
mentioned velocity experiment since it relies on the three level
situation and can in principle only approach half the resolution.
The combination of its beautiful Doppler-switching technique with

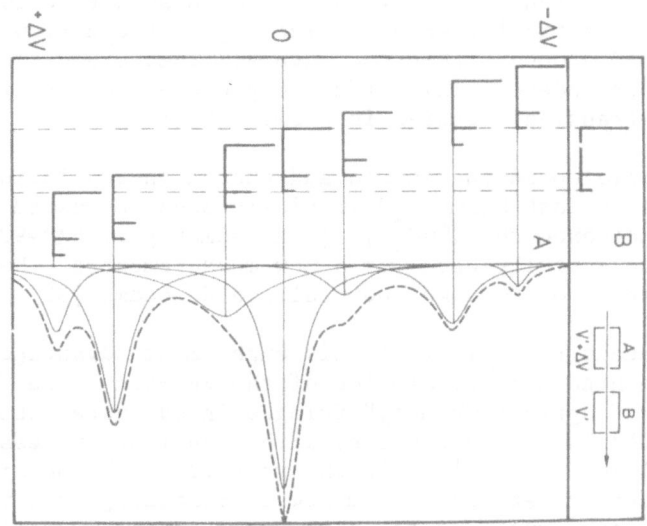

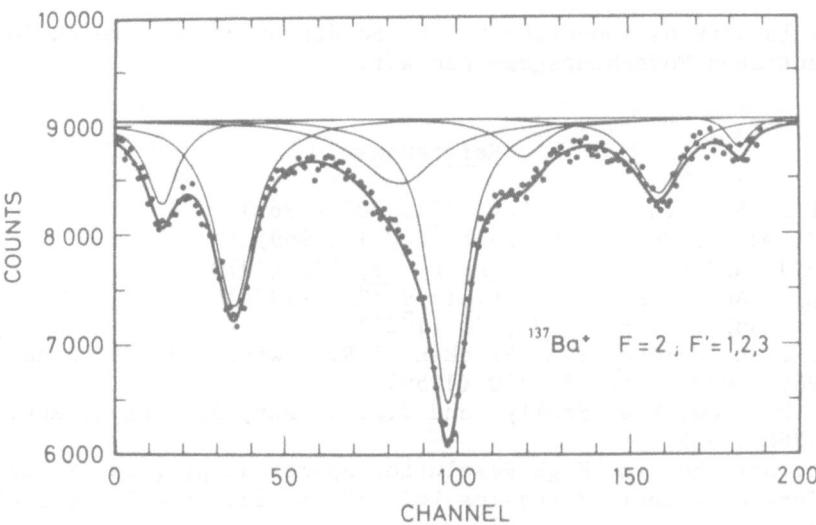

Fig. 12:   a) Reconstruction of the Lamb-dip structure of the
^{137}BaII-6s ^{2}S$_{1/2}$(F=2)-6p ^{2}P$_{3/2}$(F'=1,2,3) transitions by
shifting the three holes burned in the region A through
the three equivalent probing transitions in region B by
variation of ΔV. b) Experimental result of a Doppler
tuned "in flight Lamb dip" measurement of the
^{137}BaII-6p ^{2}P$_{3/2}$ hfs with the set up of Fig. 11. The
result can be fully explained with the reconstruction
in Fig. 12a. The left-right asymmetry of the data is
explained in ref. 30.

velocity bunched ion beams may therefore be an attractive tool for the future in particular when considering that a narrow laser line can be shifted around within the natural linewidth of the same atom at any time dependence. This offers unprecedented conditions for fundamental transient studies in atomic physics.

For completeness one should add that high resolution Lamb-dip spectroscopy on fast beams had initially been suggested for other angles of incidence than $\vartheta = 0°$ |31|. A geometry with $\vartheta = 90°$ could indeed be successfully used for a new measurement of the relativistic Doppler shift on 5-50 keV neutralized Ne beams |32|.

In a recent refinement of this experiment advantage could be taken of the high time resolution of the FBL-technique for the observation of optical "Ramsey" fringes in saturated absorption with separated laser fields, i.e. with transient successive, phase correlated laser pulses |33| in the rest frame of the atoms. Since this technique suffers less from power broadening than normal Lamb-dip spectroscopy improvements in ultimate spectral resolution can be expected from it.

This work as supported by the Sonderforschungsbereich 161 der Deutschen Forschungsgemeinschaft.

References

|1| P.A. Franken, Phys. Rev. 121, 508 (1961)

|2| J. Macek, Phys. Rev. Lett. 23, 1 (1969)

|3| H.J. Andrä, Phys. Rev. Lett. 25, 325 (1970)

|4| H.J. Andrä, Physica Scripta 9, 257 (1974)

|5| W. Hanle, Z. Phys. 30, 93 (1924)

|6| F.D. Colegrove, P.A. Franken, R.R. Lewis, and R.H. Sands, Phys. Rev. Lett. 3, 420 (1959)

|7| W.W. Chow, M.O. Scully, and J.O. Stoner, Jr., Phys. Rev. A11, 1380 (1975)

|8| S. Haroche, in "High Resolution Spectroscopy", ed. K. Shimoda, Topics in Applied Physics Vol. 13, Springer-Verlag (1976), p. 253

|9| L. Kay, Phys. Lett. 5, 36 (1963)

|10| "Beam-Foil Spectroscopy", ed. S. Bashkin, in "Topics in Current Physics", 1, Springer-Verlag, Heidelberg (1976)

|11| H.G. Berry, Rep. Progr. Phys. 40, 155 (1977)

|12| H.J. Andrä, in "Progress in Atomic Spectroscopy", eds. W. Hanle and H. Kleinpoppen, Plenum Press, to be published

|13| A. Gaupp, Diploma Thesis, Freie Universität Berlin, unpublished. See Ref. 4

|14| H.J. Andrä, Phys. Rev. A2, 2200 (1970)

|15| R.M. Steffen and H. Frauenfelder, in "Perturbed Angular Corre-
      lations", eds. E. Karlsson, E. Matthias, and K. Siegbahn,
      North-Holland, Amsterdam (1964), p. 1.

|16| J. Macek, Phys. Rev. $\underline{A1}$, 618 (1970)

|17| P. Dobberstein, H.J. Andrä, W. Wittmann, and H.H. Bukow,
      Z. Phys. $\underline{257}$, 272 (1972)

|18| W. Wittmann, Dissertation, Freie Universität Berlin, to be
      published

|19| A.N. Jette, T. Lee, and T.P. Das, Phys. Rev. $\underline{A9}$, 2337 (1974)

|20| H.J. Andrä, H.J. Plöhn, A. Gaupp, and R. Fröhling, Z. Phys.
      $\underline{A281}$, 15 (1977)

|21| M. Kraus, Diploma Thesis, Freie Universität Berlin, unpublished.
      See ref. 22

|22| H.J. Andrä, in "Beam-Foil Spectroscopy", Vol.2, eds. I.A.
      Sellin and D.J. Pegg, Plenum Press (1976) p. 835

|23| D. Schulze-Hagenest, H. Harde, W. Brand, and W. Demtröder,
      Z. Phys. $\underline{A282}$, 149 (1977)

|24| S.L. Kaufmann, Opt. Comm. $\underline{17}$, 309 (1976)

|25| Th. Meier, H. Hülmermann, and H. Wagner, Opt. Comm. $\underline{20}$, 397
      (1977)

|26| K.R. Anton, S.L. Kaufmann, W. Klempt, R. Neugart, E.W. Otten,
      and B. Schinzler, Verhandl. der Deutschen Phys. Gesellsch.
      2/1977, p. 506

|27| V.S. Letokhov, in ref. 8, p. 95

|28| M. Dufay, M. Carré, M.L. Gaillard, G. Meunier, H. Winter,
      and A. Zgainski, Phys. Rev. Lett. $\underline{37}$, 1678 (1976)

|29| H. Winter and M.L. Gaillard, to be published in J. Phys. B
      (1977)

|30| F. Beguin, M.L. Gaillard, H. Winter and G. Meunier, to be
      published in J. de Phys. (1977)

|31| W. Chow, A.D. Maio, and M.O. Scully, Nucl. Instr. Meth. $\underline{110}$,
      469 (1973)

|32| J.J. Snyder and J.L.Hall, in "Lecture Notes in Physics",
      Vol. 43, eds. S. Haroche, J.C. Pebay-Peyroula, T.W. Hänsch,
      and S.E. Harris, Springer, Berlin (1975), p. 6

|33| J.C. Bergquist, S.A. Lee, and J.L. Hall, Phys. Rev. Lett. $\underline{38}$,
      159 (1977).

MAGNETIC RESONANCES IN SODIUM VAPOURS ORIENTED

BY A C.W. DYE LASER

Gerardo Alzetta

Istituto di Fisica, Università Pisa
and
LAFAM, C.N.R., Pisa, Italy

Dye lasers employing rhodamine 6G are very good sources for the optical pumping of sodium vapours owing to the coincidence of the maximum of emission of the rhodamine with the sodium absorption D-lines. The high intensity and coherence of the emitted radiation allow to easily evidence many effects which usually are rather difficult to be shown by the use of ordinary lamps. We have taken advantage also of the fact that the human eye has its highest sensitivity in this spectral region to set up an experimental apparatus which allows a direct and immediate observation of these effects.

Let us briefly recall how an optical pumping cycle of the Kastler type works in producing orientation of atoms along a beam of a dye laser, tuned to the D_1-line and crossing a cell containing the vapour. The ground state and 3P1/2 excited state of sodium are shown in Fig.1 toghether with allowed transitions to and from 3P1/2 states under illumination with σ^+ circularly polarized radiation. As can be seen from this diagram, owing to the selection rule $\Delta m = +1$ the state +2 cannot be excited to the P1/2 state and an increase of the population of this level is obtained owing to the allowed decay of P states into this level.

If relaxation processes can be disregarded, the absorption

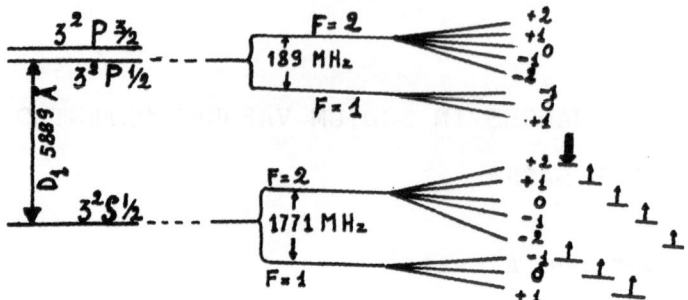

FIG.1. H.f.structure of S1/2 ground state and P1/2 excited
state of sodium. At the right, the ground state
splitting in a weak magnetic field and the σ^+ optical
pumping cycle.

of resonant radiation by the vapour and its fluorescence
will come to an end when all the levels with quantum numbers different from +2 are completely depleted. The atoms
populate the +2 state and the vapour becomes completely
transparent. The relaxation processes, usually due to
collisions with the cell walls, can be strongly reduced
if some buffer gas (He,Ne,etc.) is introduced into the
cell and special coating of the wall is used.
It is worthwhile to point out the influence of a weak
magnetic field which interacts with the magnetic moment
of the oriented atoms. If the field is applied in the
same direction of the pumping beam and the splitting of
the energy levels is small compared with the width of
the D-line the pumping process follows the above gives
scheme.   If the magnetic field direction does not coincide with the beam axis, a precession of the spins around
the magnetic field takes place and the mixing of the states will give rise to a continuous absorption of the irra-

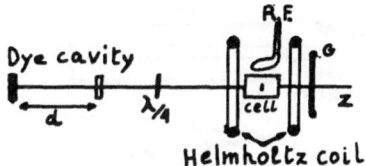

FIG.2. Optical pumping in an
inhomogeneous magnetic field.

diating light with a weak orientation of the vapour and
a permanent fluorescence along the beam.
Now it is easily realized that the optical pumping process
is by no means affected if a magnetic field, parallel to
the direction of the beam, has a weak axial gradient along
the beam owing to the fact that the direction and not the
value of the magnetic field is of importance in fixing
the quantisation axis. In this case while optical pumping
will lead to such an orientation of the vapour as the one
obtained in a uniform magnetic field, the Zeeman splitting
of the ground state sublevels results in a function of z.
The following experimental set-up can be used (Fig.2). A
beam of a dye laser, tuned to the sodium D-line, is circularly polarized by a quarter-wave plate and crosses a cell
heated up to 130°C, containing saturated Na vapour and
some Torr of neon as buffer gas. The cell is placed at
the centre of a pair of Helmholtz coils. A gradient of
the magnetic field along the beam is obtained by superpo-
sing to the uniform magnetic field the field given by
an additional coil (gradient coil G) coaxial with the
Helmholtz coils. A smaller coil can produce a r.f. field
on the cell, perpendicular to the direction of the pumping-
-beam axis.

Magnetic resonances due to radiofrequency fields.
If the vapour is irradiated by a radiofrequency field,
resonances will occur only in the regions of the cell
where the magnetic field and the frequency of the r.f.
field fulfil the resonance condition:

$$h \nu_{RF} = g \beta H_z$$

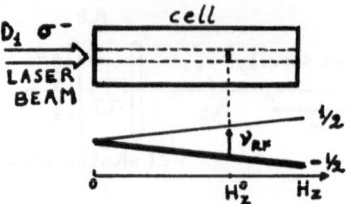

FIG.3. Example of level splitting in a linearly increasing field and position of an r.f. resonance between the
two differently populated levels.

where g is the Landè factor and β is the Bohr magneton
(see Fig. 3). Therefore the decrease of transparency
and the fluorescence of the vapour, induced by the radiofrequency transition, will appear only in correspondence
of the surface on which the magnetic field has the resonance value. If we observe laterally the cell the resonances will appear therefore as bright lines or spots
due to the induced fluorescence, as can be seen in the
pictures of Fig. 4 a and b. In Fig. 4a the resonance
due to an r.f. field of about 7 MHz is observed at the
center of the cell where the field has the value $H_z^o =$
10,1 gauss. If the polarization of the light beam is
changed into σ^- by turning the $\lambda/4$ plate, the spot is
seen in a different place where the field is 9,8 gauss,
Fig. 4b. In this case the - 2 level results to be
populated by the pumping process and the unequal spacing
of the Zeeman sublevels, due to the Back-Goudsmith effect,
changes the position of the resonance. The presence of
a single spot shows that resonances between the others
sublevels are missing and that complete transfer of population to +2, or -2, level has been achieved. By increasing the r.f. power, many photon transitions and
power broadening can be observed as reported in Fig. 5.
The method of the inhomogeneous magnetic field is very
convenient for the observation of a wide class of phenomena which are magnetic field dependent as Zeeman and
hyperfine light shifts due to virtual transitions and

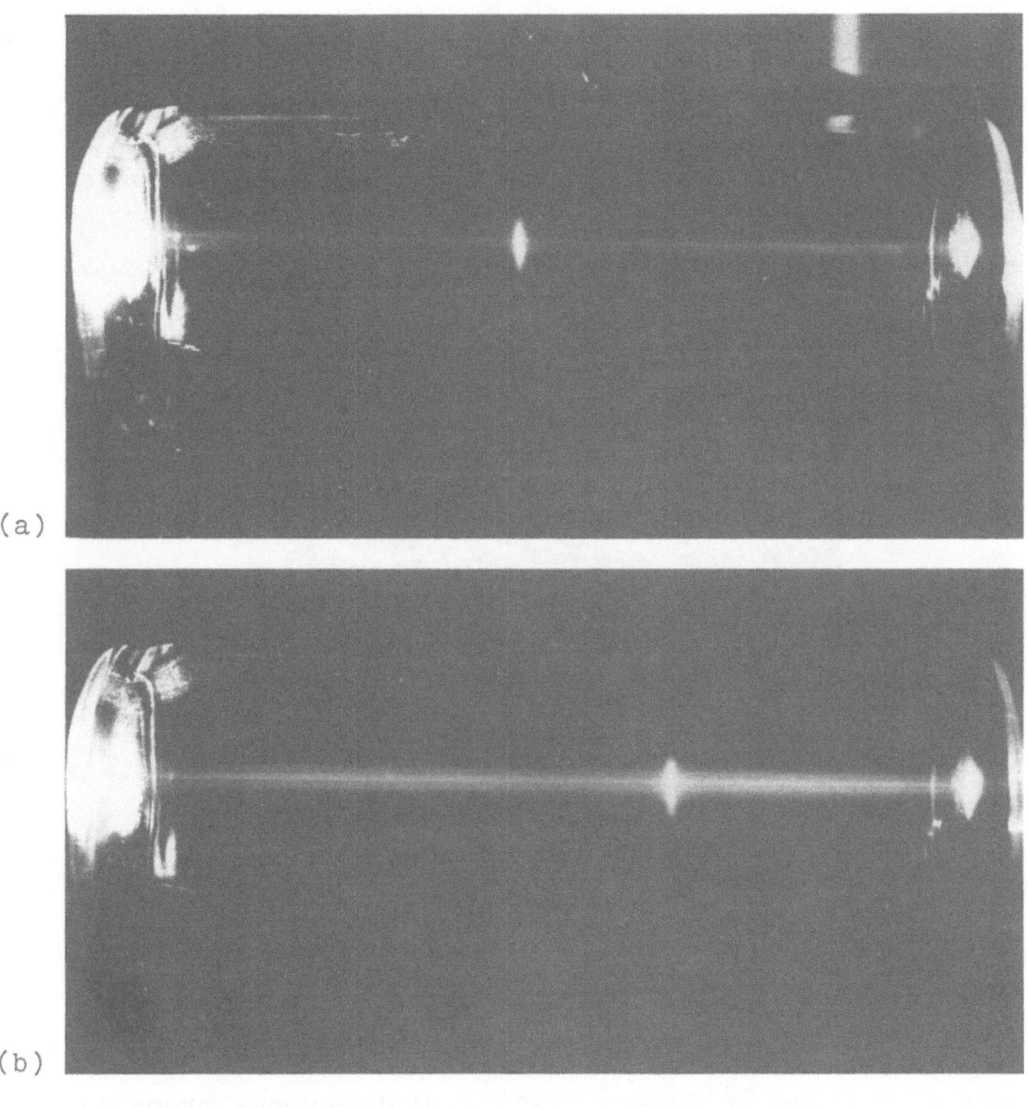

FIG.4. 7MH$_Z$ r.f.transition as seen inside a cell containing oriented Na. The magnetic field increases along the patn of the light beam from right to left.
a)(2,-2;2,-1)transition with left polarized laser radiation;
b)(2,+2;2,+1)transition with right polarized radiation.

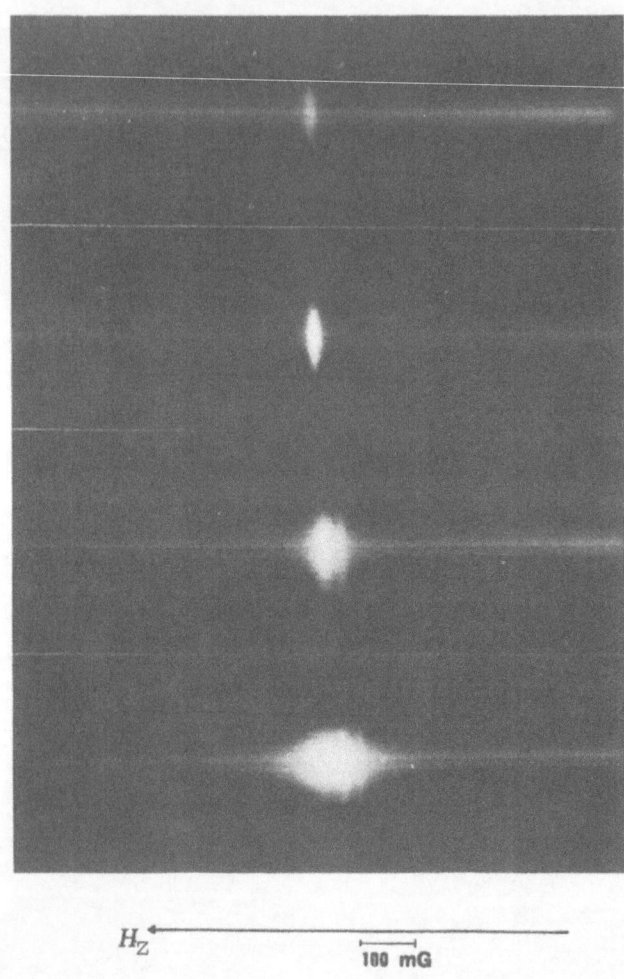

FIG.5. Pattern of (2,-2; 2,-1) radiofrequency transition for different power levels.   Power increases from top to bottom.   In the third and forth picture two and three photon transitions are seen at the right of the power broadened main line.

nonlinear effects due to the presence of many r.f. fields.
An example is reported in Fig. 6 where the pumped vapour
is submitted simultaneously to two strong fields oscillating at hyperfine and Zeeman frequencies.   In Fig. 6a
the   fluorescence signal due to the $(2,+2;1,+1)$ h.f. resonance is observed for a frequency of 1798 MHz at the
position where the static field has the value of 12,5
gauss.   If the vapour is also submitted to a strong r.f.
field of lower frequency which induces transitions between
the Zeeman sublevels, besides the power broadened many-
-photon transitions (of which only the four-photon
$(2,+2;2,-2)$ transition is clearly visible) another weak
resonance is visible on the left of the h.f. one. This
is a two-photon hyperfine transition connecting the $(2,+2)$
and $(1,0)$ levels via the intermediate $(1,+1)$ level (Fig.6b).
If the position of the Zeeman pattern is displaced nearer
to the h.f. resonance other multiple quantum transitions
appear, connecting the $(2,+2)$ and $(1,-1)$ levels via the
$(1,+1)$ and $(1,0)$ intermediate levels.

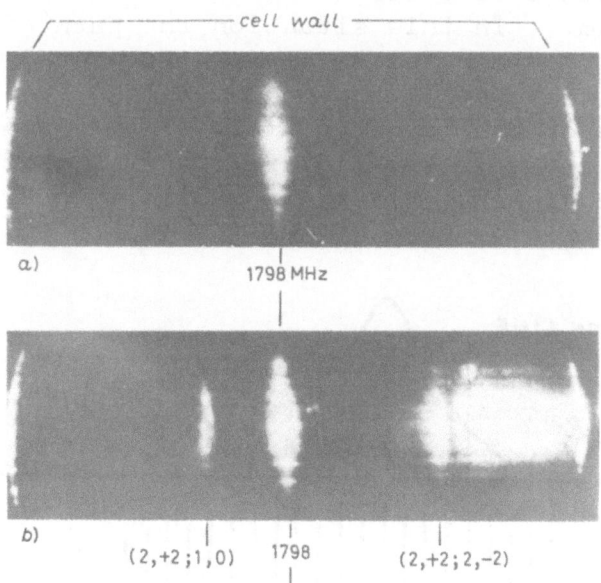

FIG. 6. Nonlinear effects involving hyperfine and Zeeman
transitions. a) $(2,+2;1,+1)$ h.f. resonance at 1798 MHz;b)
New line $(2,+2;1,0)$ due to a two photon transition induced
by the h.f. field and an r.f. field at 10 MHz.

The presence of these lines has been discussed in a paper by Di Giacomo and Feo(1), where the position, the intensity and the width of the lines have been calculated by taking into account simultaneously the optical pumping and the interaction with the radiofrequency fields. The calculations are performed under the hypotesis of broad excitation and in the absence of buffer gas, but they are, at least qualitatively, valid also for our experimental conditions. The experimental observations are indeed in good agreement with the theoretical results.

Magnetic resonances without radiofrequency fields.
A different type of resonance can be observed if the multimode laser is used in the absence of any radiofrequency acting on the vapour. The pattern of the longitudinal modes of the laser can be sketched as in Fig. 7, when a narrow band of emission of the dye is selected around the D-line. If the magnetic field axis is slightly tilted to form a small angle with the direction of the light beam (see Fig. 8) the transparency of the vapour decreases and strong fluorescence is observed inside the cell along the path of the beam. In this situation, the path of the beam appears to be crossed by some, apparently black, lines which evidence that the fluorescence is missing at certain discrete values of the magnetic field. An example of this effect is shown in the picture of Fig.9 which was taken along the plane of Fig.8, in a direction orthogonal to the magnetic field.

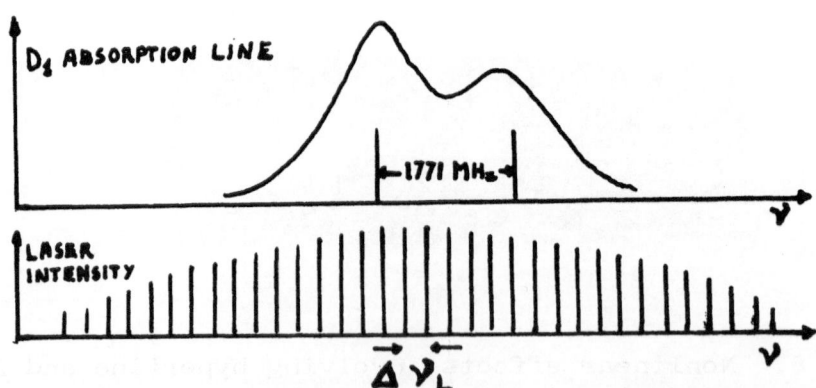

FIG. 7. Multimode laser emission and D_1 absorption profile $\Delta \nu_L$ is the frequency interval between consecutive modes.

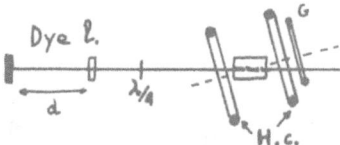

FIG. 8.   Arrangement for the observation of the hyperfine
           coherences.

Due to the residual orientation of the vapour, the r.f.
Zeeman transition (2,-2;2,-1) can be excited and used for
measurements of the magnetic fields where the black lines
appear.   The following rule has been found to hold: the
fluorescence of the vapour is missing for values of the
magnetic field such that an integer number of mode intervals of the dye laser, $n \Delta \nu_L$, matches the frequency
interval $\nu_{h.f.}$ between two hyperfine levels, i.e.

$$\nu_{h.f.} = n \Delta \nu_L$$

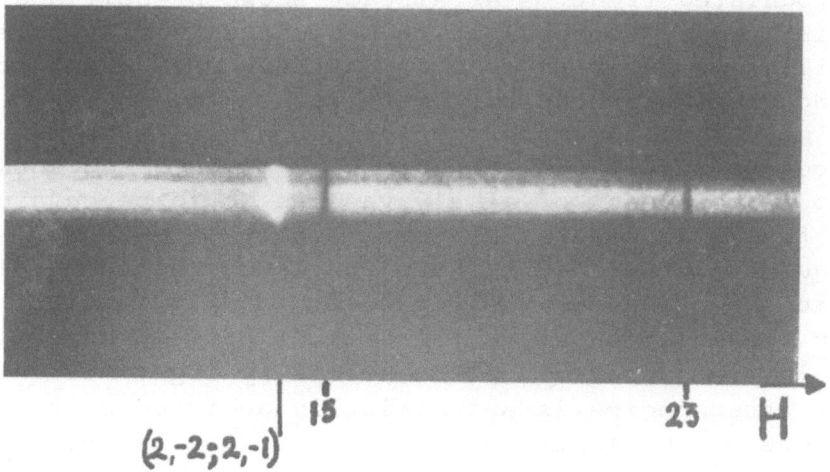

FIG. 9.   The path of the dye beam inside the cell, crossed
           by two black lines.

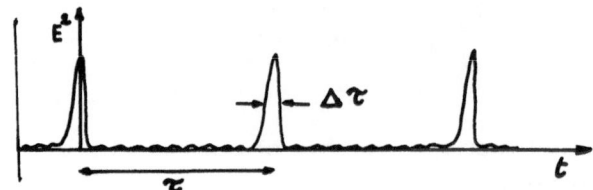

FIG. 10.   Output of a multimode dye laser.

With a dye laser cavity of 52,7 cm, the mode frequency
interval $\Delta \nu_L$ = c/2d is about 290 MHz and 1740 MHz (6x290)
correspond to the frequency difference between (2,-2) and
(1,-1) hyperfine levels in a field of 15 gauss. With a
σ^- light polarization at such a field the most intense
black line is found.   The same frequency,1740 MHz,also
matches the frequency interval between (2,-1) and (1,-1)
levels in a field of 23 gauss.   At this field a second
line is found.   Other lines can be found at higher values
of the magnetic field.
These effects can be explained in terms of optical pumping
in modulated light.   As already shown by Bell and Bloom
in 1961(2) it is possible to produce sizeable, coherently
oscillating polarizations in a transversal magnetic field
by modulating at the Larmor frequency the intensity of the
ordinary lamps used for optical pumping. They named the
effect "optically driven spin precession" and showed
that transparency of alkaly vapours could be obtained,
and transitions between Zeeman sublevels detected, if the
frequency difference between the levels matched the modulation frequency of the light. In our experiment precession frequencies due to transitions between hyperfine
levels are involved and the necessary light modulation at
these frequencies is automatically achieved by the interference of the longitudinal modes of the laser.
As a matter of fact, due to superposition of many longitudinal modes, the output of the laser can be described as

$$E = \sum_{-N}^{+N} {}_{n} E_{o} e^{it(\omega_{o} + n\Delta\omega)}$$

where E is the amplitude of the electric field, ω_{o} is
the central angular frequency of the 2N+1 modes and $\Delta\omega$
is the frequency interval between the modes. The phase
difference between the modes is supposed to be zero at
t=0. By the use of Snellius formula it is easily seen
that the amplitude can also be written as

$$E = E_{o} \frac{\sin(2N+1)\frac{\Delta\omega t}{2}}{\sin\frac{\Delta\omega t}{2}} e^{i\omega_{o}t}$$

Its temporal behaviour is sketched in Fig. 10.
For a cavity length of 50 cm, the output consists of
short pulses of light separated by a time interval
τ = 2d/c $\approx$ 3.10^{-9} sec. The width of the pulses is about
$\Delta\tau$ = 2π/(2N+1)$\Delta\omega$ and by taking into consideration only
10 modes $\Delta\tau$ is of the order of 3 . 10^{-10} sec.
The situation for Na vapour irradiated by the multimode
laser can be pictured as follows. At t=0 the first pulse
of light pumps a small amount of polarization in the direction of the beam (population or depopulation pumping being
equivalent for the process under consideration). This
polarization precesses at hyperfine frequencies and is
reinforced by the polarization produced by the subsequent
pulses of light only if τ is an integer multiple of
the precession period. Optical pumping then takes
place as if the magnetic field were parallel to the light
beam, absorption and fluorescence decrease and the vapour
becomes transparent for that value of the magnetic field
for which nT_{hf} = τ , T_{hf} being the period of the h.f.
precessional motion. By displacing the output mirror
of the dye cavity, the time interval τ can be changed and
a shift of the black lines to different values of the
magnetic field can be observed.
It is worthwhile to remark that a steady sequence of
light pulses requires that if at t=0 all the modes are
in phase, this phase should be maintained for the

subsequent times.   The observed narrowness and stability of the pattern of the black lines test that in the
multimode laser a natural phase locking of the modes is
present.   This observation agrees with the theoretical
views of Lang, Scully and Lamb(3) who have shown that
at sufficiently large intensity of the emitted radiation,
all the modes of a free running multimode laser oscillate with a definite phase ralationship between them.

References
1) A.Di Giacomo and F.Feo : Nuovo Cimento, 30B, 193 (1975).
2) W.E.Bell and L.Bloom : Phys.Rev.Lett. 6, 280, 623 (1961)
3) R.Lang and M.O.Scully, W.E.Lamb : Phys.Rev.A 7, 5, 1788-
   -1799, (1973).

SUPERCONDUCTIVITY AND QUANTUM OPTICS

A. DiRienzo,*† D. Rogovin,* and M. Scully†

Optical Sciences Center
University of Arizona, Tucson, Arizona 85721
and

Science Applications, Inc.
La Jolla, California 92038
and

R. Bonifacio, L. Lugiato, and M. Milani

Istituto di Fisica, Università di Milano
Via Celoria, 16, 20133 Milan, Italy

INTRODUCTION

In these lectures we will be describing the applications of quantum optical techniques to superconducting tunnel junctions. This work has been motivated, in part, by the close analogy between the quantum mechanics of BCS electron pairs in a superconducting tunnel junction and the quantum mechanics of two-level systems.[1] Since two-level systems are a mainstay of quantum optics, it is natural that we begin considering problems such as coherent transients in super-conducting junctions (a classic two-level system) from a quantum optical point of view. Two such problems, supperradiance and Rabi flopping effects in Josephson devices will be discussed. Some of the coherent transient effects we will see are different from those

---

*Supported by the Office of Naval Research.

†This work is supported by the Air Force Office of Scientific Research (AFSC), United States Air Force, and the Army Research Office, United States Army.

observed in the normal two-level atom problem, but there are also many similarities.

We hope to convince you that by applying the kinds of many-body non-equilibrium techniques developed for the quantum theory of the laser,[2] in our analysis of superconducting tunnel junctions, new physics will emerge. In particular, we will be considering transient phenomena as opposed to the steady-state, thermodynamic equilibrium properties which are usually studied by many-body physicists. Since there is considerable interest within the engineering community for using Josephson devices as ultra-fast computer elements, the problems we will be considering are extremely relevant to modern technology.

REVIEW OF SUPERCONDUCTIVITY

Before looking into the problems on Josephson junctions we will give a short review of some pertinent aspects of the BCS theory of superconductivity at zero temperature. At the same time this will give us an opportunity to introduce the quantum optical formalism of two-level systems we intend to use.

The Central Role of Electron Pairs

Electron pairing, discovered by Cooper,[3] is the central concept of the BCS theory of superconductivity.[4] The point is that electrons are attracted to each other to form pairs, the interaction being mediated by phonons. Physically we can visualize this by considering the following (Fig. 1). Let an electron enter a region of the crystal. As it passes through this region the positively charged ion cores are pulled toward it, deforming the lattice. Another electron outside this region sees not just the first electron but also the deformed lattice. Together, the first electron and the deformed lattice have an overall positive charge, to which this second electron is attracted.

Since the potential is attractive and central, we argue that the two electrons go into orbit about each other (just like positronium). This is an intuitive argument, but it can be made much stronger (Fig. 2). We claim, in fact, that the electrons are paired up in an S-wave. Here we are approximating an S-wave, by plane waves (Bloch states). To see this, consider two electrons initially with momenta $\hbar\vec{k}$ and $-\hbar\vec{k}$. An instant later, if we want to mimic an S-wave, these must scatter into other states with momenta $\hbar\vec{k}'$ and $-\hbar\vec{k}'$, respectively. Note that these momenta always occur in pairs. A Cooper pair, then, is a pair of electrons with wave-vector $\pm\vec{k}$. They are said to be in a paired state.

Figure 1.   Illustration of the attractive interaction between two electrons.   (a) An electron enters a region of the crystal.   (b) As it passes through the region, positively charged ion cores are pulled toward it.   A second electron, outside the region, sees an overall positive charge.

Figure 2.   Positronium analogy.   (a) Two electrons initially in opposite momentum states $\pm\hbar\vec{k}$.   An instant later they have scattered a new set of state $\pm\hbar\vec{k}'$.

We now introduce some notation to describe the wave function for the pair, which is useful in a pictorial sense.

$$b^{\dagger}_{\vec{k}} = c^{\dagger}_{\vec{k}} c^{\dagger}_{-\vec{k}} \qquad \text{creates a pair} \qquad (1)$$

$$b_{\vec{k}} = c_{-\vec{k}} c_{\vec{k}} \qquad \text{destroys a pair}$$

$$|full\rangle_{\vec{k}} = |1\rangle_{\vec{k}} = |1_{\vec{k}}, 1_{-\vec{k}}\rangle = b^{\dagger}_{\vec{k}}|0\rangle_{\vec{k}} = \left|\ \vphantom{\bigotimes}\right. \!\!\!\!\bigoplus\!\!\!\! \left.\vphantom{\bigotimes}\ \right\rangle_{\vec{k}} \qquad (2)$$

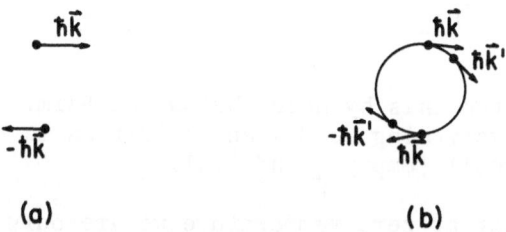

Here the $c^{\dagger}_{\vec{k}}$'s and $c_{\vec{k}}$'s are the usual Fermion creation and annihilation operators with:

$$[c_{\vec{k}}, c^{\dagger}_{\vec{k}'}]_{+} \ = \ \delta_{\vec{k}\,\vec{k}'}$$

$$[c_{\vec{k}}, c_{\vec{k}'}]_{+} \ = \ 0 \ = \ [c^{\dagger}_{\vec{k}}, c^{\dagger}_{\vec{k}'}]_{+}$$

The circle in the ket represents the Fermi sea, while the pair-state, represented by dots, is definitely occupied, i.e., filled. Obviously there is also the possibility, equally important, that the pair-state is not occupied.

$$|\text{empty}\rangle_{\vec{k}} \ = \ |0\rangle_{\vec{k}} \ = \ |0_{\vec{k}}, 0_{-\vec{k}}\rangle \ = \ b_{\vec{k}}|1\rangle_{\vec{k}} \ = \ \left| \begin{array}{c} \end{array} \right\rangle_{\vec{k}} \qquad (3)$$

We have represented this by holes below the Fermi sea which indicate empty states. Everything we do can be written in terms of these states which we call $|\text{empty}\rangle_{\vec{k}}$ and $|\text{full}\rangle_{\vec{k}}$.

Remember that at zero temperature we are only interested in wave functions for a pair. As we have shown, for a given $\vec{k}$ there are only two possibilities, $|\text{empty}\rangle_{\vec{k}}$ and $|\text{full}\rangle_{\vec{k}}$. Therefore, the general state for this $\vec{k}$ is

$$|\psi(\text{pair})\rangle_{\vec{k}} \ = \ U_{\vec{k}}|0\rangle_{\vec{k}} \ + \ V_{\vec{k}}|1\rangle_{\vec{k}} \qquad (4)$$

where $U_{\vec{k}}$ and $V_{\vec{k}}$ are the respective probability amplitudes. Note that we have not mentioned the electron spins. Nothing in our problems really requires them. But to have an S-state as tightly bound as possible the spins must be antiparallel. Consequently, we will take

$$\vec{k} \text{ to mean } \vec{k}\uparrow$$

and

$$-\vec{k} \text{ to mean } -\vec{k}\downarrow$$

but never explicitly write it out.

BCS Ground State, Hamiltonian, and Pseudospin Notation

We note from Eq. (4) that we have essentially two-level quantum mechanics for the wave-function of the $\vec{k}$-th pair. Consequently we write[5]

$$|\psi(pair)\rangle_{\vec{k}} \;=\; U_{\vec{k}} \binom{1}{0}_{\vec{k}} \;+\; V_{\vec{k}} \binom{0}{1}_{\vec{k}} \tag{5}$$

where $\binom{1}{0}_{\vec{k}}$ is the pseudospin state corresponding to $|empty\rangle_{\vec{k}}$ and $\binom{0}{1}_{\vec{k}}$ the pseudospin state corresponding to $|full\rangle_{\vec{k}}$.

There may seem, at first, to be no advantage to this notation. But it becomes a matter of calculational convenience if one transcribes the hamiltonian,[4] which is here given in terms of Fermion operators,

$$H \;=\; \sum_{\vec{k}} \varepsilon_{\vec{k}} \, c^{\dagger}_{\vec{k}} \, c_{\vec{k}} \;-\; V \sum_{\vec{k}} \sum_{\vec{k}'} c^{\dagger}_{\vec{k}} c^{\dagger}_{-\vec{k}} \, c_{-\vec{k}'} c_{\vec{k}'}, \tag{6}$$

$$\varepsilon_{\vec{k}} \;=\; \frac{\hbar k^2}{2m} \qquad\qquad \text{kinetic energy of an electron}$$

$$V \qquad\qquad\qquad \text{the phonon mediated attractive energy between electrons}$$

into pseudospin language. In the process of doing this, we will show that there is a complete analogy between this problem and that encountered in spin-1/2 quantum mechanics.

Consider first the kinetic energy part and rewrite it in terms of energies measured relative to the Fermi sea, $\bar{\varepsilon}_k = \varepsilon_k - \varepsilon_F$. This only means that we are working in the grand canonical ensemble and does not change the problem in any sense.

$$H_{KE} \;=\; \sum_{\vec{k}} \bar{\varepsilon}_{\vec{k}} \, c^{\dagger}_{\vec{k}} \, c_{\vec{k}}$$

Since we will be dealing only with electron pairs we explicitly write H_{KE} in terms of these[6]

$$H_{KE} = \sum_{\vec{k}>0} \bar{\epsilon}_{\vec{k}} (c^{\dagger}_{\vec{k}} c_{\vec{k}} + c^{\dagger}_{-\vec{k}} c_{-\vec{k}})$$

Adding and subtracting a constant $\sum_{\vec{k}>0} \bar{\epsilon}_{\vec{k}}$ we have

$$H_{KE} = -\sum_{\vec{k}>0} \bar{\epsilon}_{\vec{k}} (1 - c^{\dagger}_{\vec{k}} c_{\vec{k}} - c^{\dagger}_{-\vec{k}} c_{-\vec{k}}) + \sum_{\vec{k}>0} \bar{\epsilon}_{\vec{k}} \qquad (7)$$

The second term can now be dropped since it is just a constant.

We can begin to see the correspondence between our problem and spin-1/2 quantum mechanics by noting that $(1 - c^{\dagger}_{\vec{k}} c_{\vec{k}} - c^{\dagger}_{-\vec{k}} c_{-\vec{k}})$ acts like $2\sigma_{\vec{k}z}$ when operating on $|full\rangle_{\vec{k}}$ and $empty\rangle_{\vec{k}}$.

$$(1 - n_{\vec{k}} - n_{-\vec{k}})|1\rangle_{\vec{k}} = -|1\rangle_{\vec{k}} \quad \Longleftrightarrow \quad 2\left[\frac{1}{2}\begin{pmatrix} 1 & 0 \\ 0 & -1 \end{pmatrix}_{\vec{k}}\right]\begin{pmatrix} 0 \\ 1 \end{pmatrix}_{\vec{k}} = -\begin{pmatrix} 0 \\ 1 \end{pmatrix}_{\vec{k}}$$

$$(1 - n_{\vec{k}} - n_{-\vec{k}})|0\rangle_{\vec{k}} = |0\rangle_{\vec{k}} \quad \Longleftrightarrow \quad 2\left[\frac{1}{2}\begin{pmatrix} 1 & 0 \\ 0 & -1 \end{pmatrix}_{\vec{k}}\right]\begin{pmatrix} 1 \\ 0 \end{pmatrix}_{\vec{k}} = \begin{pmatrix} 1 \\ 0 \end{pmatrix}_{\vec{k}}$$

with $n_{\vec{k}} = c^{\dagger}_{\vec{k}} c_{\vec{k}}$ the number operator for electrons of momentum $\vec{k}$.

Consequently, in our pseudospin notation we define

$$\sigma_{\vec{k}z} = \left(\frac{1}{2}\right)(1 - n_{\vec{k}} - n_{-\vec{k}}) = \frac{1}{2}\begin{pmatrix} 1 & 0 \\ 0 & -1 \end{pmatrix}_{\vec{k}} \qquad (8)$$

for our two-dimensional Hilbert space denoted by $\vec{k}$. We can now re-write the kinetic energy term as

$$H_{KE} = -2 \sum_{\vec{k}>0} \bar{\varepsilon}_{\vec{k}} \sigma_{\vec{k}z} \tag{9}$$

The potential energy can, just as easily, be put into this pseudospin notation. From the second quantized form of H

$$H_{PE} = -V \sum_{\vec{k}} \sum_{\vec{k}'} c^{\dagger}_{\vec{k}} c^{\dagger}_{-\vec{k}} c_{-\vec{k}'} c_{\vec{k}'} \tag{10}$$

we see that it first annihilates a pair of momentum $\vec{k}'$ and then creates a pair with momentum $\vec{k}$. Noting that

$$c_{-\vec{k}} c_{\vec{k}} |0_{\vec{k}}, 0_{-\vec{k}}> = 0 \tag{11a}$$

$$c_{-\vec{k}} c_{\vec{k}} |1_{\vec{k}}, 1_{-\vec{k}}> = |0_{\vec{k}}, 0_{-\vec{k}}> \tag{11b}$$

$$c^{\dagger}_{\vec{k}} c^{\dagger}_{-\vec{k}} |1_{\vec{k}}, 1_{-\vec{k}}> = 0 \tag{11c}$$

$$c^{\dagger}_{\vec{k}} c^{\dagger}_{-\vec{k}} |0_{\vec{k}}, 0_{-\vec{k}}> = |1_{\vec{k}}, 1_{-\vec{k}}> \tag{11d}$$

we see the analogous equations in pseudospin language

$$\begin{pmatrix} 0 & 1 \\ 0 & 0 \end{pmatrix}_{\vec{k}} \begin{pmatrix} 1 \\ 0 \end{pmatrix}_{\vec{k}} = 0 \tag{12a}$$

$$\begin{pmatrix} 0 & 1 \\ 0 & 0 \end{pmatrix}_{\vec{k}} \begin{pmatrix} 0 \\ 1 \end{pmatrix}_{\vec{k}} = \begin{pmatrix} 1 \\ 0 \end{pmatrix}_{\vec{k}} \tag{12b}$$

$$\begin{pmatrix} 0 & 0 \\ 1 & 0 \end{pmatrix}_{\vec{k}} \begin{pmatrix} 0 \\ 1 \end{pmatrix}_{\vec{k}} = 0 \tag{12c}$$

$$\begin{pmatrix} 0 & 0 \\ 1 & 0 \end{pmatrix}_{\vec{k}} \begin{pmatrix} 1 \\ 0 \end{pmatrix}_{\vec{k}} = \begin{pmatrix} 0 \\ 1 \end{pmatrix}_{\vec{k}} \tag{12d}$$

Consequently, we identify

$$\sigma^-_{\vec{k}} = c^\dagger_{\vec{k}} c^\dagger_{-\vec{k}} = \begin{pmatrix} 0 & 0 \\ 1 & 0 \end{pmatrix}_{\vec{k}} \tag{13}$$

$$\sigma^+_{\vec{k}} = c_{-\vec{k}} c_{\vec{k}} = \begin{pmatrix} 0 & 1 \\ 0 & 0 \end{pmatrix}_{\vec{k}}$$

and write H_{PE} directly in terms of the spin-1/2 operators.

$$H_{PE} = -V \sum_{\vec{k}} \sum_{\vec{k}'} \sigma^-_{\vec{k}} \sigma^+_{\vec{k}'} \tag{14}$$

Our problem now is completely specified in terms of the spinor notation[7] (see Table I).

We want now an expression for the total pair wave function. Since for the $\vec{k}$-th pair we know that the state is just a coherent superposition of empty and full states, and that we will only be concerned with the physics going on in the pair subspace of the total Hilbert space, we are led to the BCS wave function.

$$|BCS\rangle = \prod_{\vec{k}} |\psi(pair)\rangle_{\vec{k}} = \prod_{\vec{k}} \left[U_{\vec{k}} |0\rangle_{\vec{k}} + V_{\vec{k}} |1\rangle_{\vec{k}} \right]$$

It would be helpful to pause now to consider what the $U_{\vec{k}}$'s and $V_{\vec{k}}$'s look like. For a normal metal we plot them in Fig. 3a. The situation for a superconductor is slightly different. Remember that $V < \hbar\omega_D$ with $\hbar\omega_D \simeq 10^{-2}$ eV, while $\varepsilon_F \simeq 1 - 10$ eV. So we do not expect much to happen far below the Fermi energy, but do expect changes near the Fermi surface (Fig. 3b).

TABLE I

	Second quantized form	Pseudospin form
$\lvert \text{empty}\rangle_{\vec{k}}$	$\lvert 0_{\vec{k}}, 0_{-\vec{k}}\rangle$	$\begin{pmatrix} 1 \\ 0 \end{pmatrix}_{\vec{k}}$
$\lvert \text{full}\rangle_{\vec{k}}$	$\lvert 1_{\vec{k}}, 1_{-\vec{k}}\rangle$	$\begin{pmatrix} 0 \\ 1 \end{pmatrix}_{\vec{k}}$
Creation operator	$b^{\dagger}_{\vec{k}} = c^{\dagger}_{\vec{k}} c^{\dagger}_{-\vec{k}}$	$\sigma^{-}_{\vec{k}} = \begin{pmatrix} 0 & 0 \\ 1 & 0 \end{pmatrix}_{\vec{k}}$
Destruction operator	$b_{\vec{k}} = c_{-\vec{k}} c_{\vec{k}}$	$\sigma^{+}_{\vec{k}} = \begin{pmatrix} 0 & 1 \\ 0 & 0 \end{pmatrix}_{\vec{k}}$
Number operator	$\frac{1}{2}(1 - c^{\dagger}_{\vec{k}} c_{\vec{k}} - c^{\dagger}_{-\vec{k}} c_{-\vec{k}})$	$\sigma_{kz} = \frac{1}{2}\begin{pmatrix} 1 & 0 \\ 0 & -1 \end{pmatrix}_{\vec{k}}$
H_{KE}	$\sum_{\vec{k}} \bar{\varepsilon}_{\vec{k}} c^{\dagger}_{\vec{k}} c_{\vec{k}}$	$-2\sum_{\vec{k}>o} \bar{\varepsilon}_{\vec{k}} \sigma_{kz}$
H_{PE}	$-V \sum_{\vec{k}} \sum_{\vec{k}'} c^{\dagger}_{\vec{k}} c^{\dagger}_{-\vec{k}} c_{-\vec{k}'} c_{\vec{k}'}$	$-V \sum_{\vec{k}} \sum_{\vec{k}'} \sigma^{-}_{\vec{k}} \sigma^{+}_{\vec{k}'}$

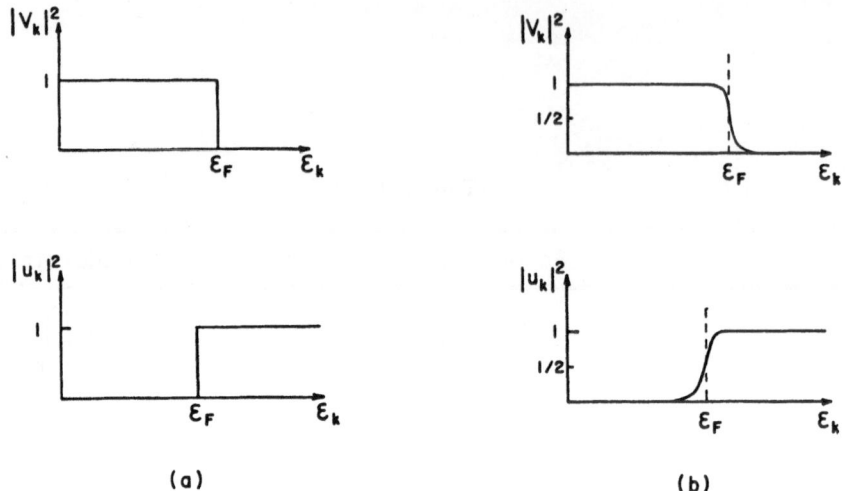

Figure 3. Plots of $\left|U_k\right|^2$ and $\left|V_{\vec{k}}\right|^2$. (a) For a normal metal we see that the electronic states are completely filled up to the Fermi energy and empty thereafter. (b) For a superconductor the probabilities far from the Fermi energy are the same as those for a normal metal. But near the Fermi energy they have been altered due to electron pairing.

We gain some insight into the BCS ground state by returning to the hamiltonian and rewriting it in slightly different form. For each $\vec{k}$-subspace we have already defined $\sigma_{\vec{k}z}$, to this we add $\sigma_{\vec{k}x}$ and $\sigma_{\vec{k}y}$, and identify

$$\sigma_{\vec{k}}^{-} = \sigma_{\vec{k}x} - i\sigma_{\vec{k}y} \tag{15}$$

$$\sigma_{\vec{k}}^{+} = \sigma_{\vec{k}x} + i\sigma_{\vec{k}y}$$

Substituting these into Eq. (14), we get

$$H_{PE} = -V \sum_{\vec{k}} \sum_{\vec{k}'} (\sigma_{\vec{k}x} \sigma_{\vec{k}'x} + \sigma_{\vec{k}y} \sigma_{\vec{k}'y})$$

and the total hamiltonian becomes

$$H = -\sum_{\vec{k}>0} (2\bar{\epsilon}_{\vec{k}} \sigma_{\vec{k}z}) -V \sum_{\vec{k}} \sum_{\vec{k}'} (\sigma_{\vec{k}x} \sigma_{\vec{k}'x} + \sigma_{\vec{k}y} \sigma_{\vec{k}'y}) \qquad (16)$$

Factoring out $\vec{\sigma}_{\vec{k}}$ from the right hand side puts the hamiltonian into a form reminiscent of a magnetism problem

$$H = -\sum_{\vec{k}} \left[2\bar{\epsilon}_{\vec{k}} \hat{z} + V \sum_{\vec{k}'} (\sigma_{\vec{k}'x} \hat{x} + \sigma_{\vec{k}'y} \hat{y}) \right] \cdot \vec{\sigma}_{\vec{k}}$$

$$H = -\sum_{\vec{k}} \vec{H}_{eff.}^{\vec{k}} \cdot \vec{\sigma}_{\vec{k}}$$

This is similar to the hamiltonian of a ferromagnetic problem, which we know how to solve in at least a mean field approximation. There the coupling between the i-th and j-th spins in a ferromagnet is measured by the exchange coefficient J_{ij}.

$$H = \sum_{i} \sum_{j>i} J_{ij} \vec{\sigma}_i \cdot \vec{\sigma}_j$$

So, by analogy, we have basically the same type of problem in a superconductor. Consequently, if we can physically interpret the ground state of the ferromagnet problem in some simple way, then we should be able to get some insight into our problem.

In the ferromagnet problem, if all the spins (considered in a plane, Fig. 4a) point in the same direction, then we are in the lowest energy state. But if we are in a situation such that the spins begin to twist as we come out of the plane then we are not in the lowest energy state (see Fig. 4b). This twisting is associated with the term in the hamiltonian coming from the y- component of the $\vec{\sigma}$'s. Consequently, by choosing a given direction and keeping the spins in that plane we will get the lowest energy piece without spin excitations.

Similarly, in our problem we consider all the spins to be in the xz-plane. The hamiltonian is then

$$H = -\sum_{\vec{k}} 2\bar{\epsilon}_{\vec{k}} \sigma_{\vec{k}z} -V \sum_{\vec{k}} \sum_{\vec{k}'} \sigma_{\vec{k}x} \sigma_{\vec{k}'x}$$

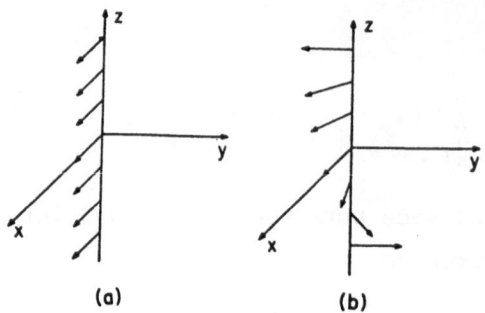

Figure 4. (a) The lowest energy state, with all spins in the xz-plane. (b) Here the spins have twisted out of the xz-plane and have obtained a y-component.

If we want the added energy associated with the y-twist then we will have to add the y-components we dropped. But for now, this is all we need.

Continuing our examination of the hamiltonian we make one last approximation. Assume we are in a strong-coupled superconductor; that is, a superconductor in which the potential energy per pair is large, compared to the kinetic energy per pair $\bar{\epsilon}_k$, for $k \simeq k_F^8$ (Fig.5). Remembering that we are only interested in pairs near the Fermi energy, we now approximate our hamiltonian by dropping the kinetic energy piece.

$$H = -V \sum_{\vec{k}} \sum_{\vec{k}'} \sigma_{\vec{k}x} \sigma_{\vec{k}'x}$$

For our purposes we can get the insight we need by looking at just this strong-coupled conductor.

The eigenvalue problem for this hamiltonian can be solved trivially.

$$H|\psi>_{sc} = E|\psi>_{sc}$$

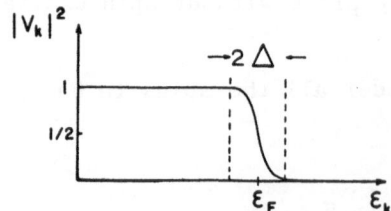

Figure 5. $|V_k|^2$ is the occupation distribution for the BCS ground state. Note that $\Delta \sim 10^{-3}$ eV while $\epsilon_F \simeq 1\text{-}10$ eV.

Obviously,

$$-V \sum_{\vec{k}} \sum_{\vec{k}'} \sigma_{\vec{k}x} \sigma_{\vec{k}'x} \left[\prod_{\vec{k}''} \frac{1}{\sqrt{2}} \begin{pmatrix} 1 \\ 1 \end{pmatrix}_{\vec{k}''} \right] = E \left[\prod_{\vec{k}''} \frac{1}{\sqrt{2}} \begin{pmatrix} 1 \\ 1 \end{pmatrix}_{\vec{k}''} \right]$$

The final representation of the state for our problem then is,

$$|\psi\rangle_{sc} = \prod_{\vec{k}} \left[\frac{1}{\sqrt{2}} \begin{pmatrix} 1 \\ 1 \end{pmatrix}_{\vec{k}} \right]$$

$$|\psi\rangle_{sc} = \prod_{\vec{k}} \left[\frac{1}{\sqrt{2}} \begin{pmatrix} 1 \\ 0 \end{pmatrix}_{\vec{k}} + \frac{1}{\sqrt{2}} \begin{pmatrix} 0 \\ 1 \end{pmatrix}_{\vec{k}} \right] \qquad (17)$$

where we now have equal probabilities for the full and empty states (Fig. 6). Remember though, that this strong-coupled superconductor state is an approximation. In general, we have

$$|BCS\rangle = \prod_{\vec{k}} \left[U_{\vec{k}} \begin{pmatrix} 1 \\ 0 \end{pmatrix}_{\vec{k}} + V_{\vec{k}} \begin{pmatrix} 0 \\ 1 \end{pmatrix}_{\vec{k}} \right]$$

with the distributions of Fig. 3b. But the essential physics is represented by the fact that we are in a coherent superposition of empty and full states.

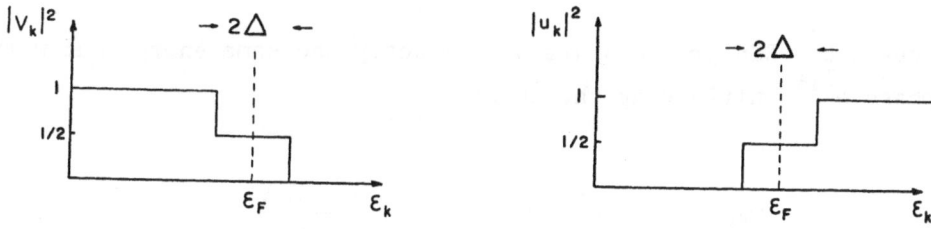

Figure 6. $|V_k|^2$ and $|U_k|^2$ for a strong-coupled superconductor. Note that we are only interested in those states which have energies within the region $\pm\Delta$ of the Fermi energy, and m is the number of pairs in this region.

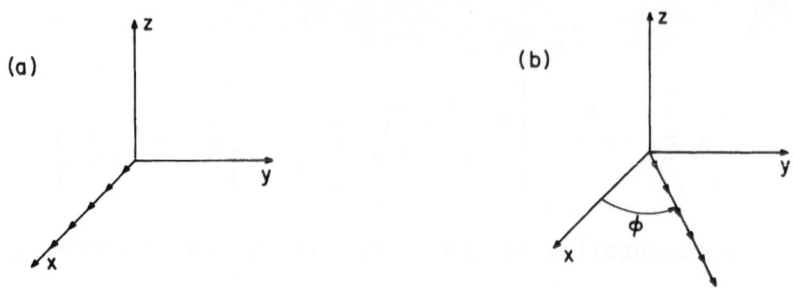

Figure 7. Phase rotation about the z-axis. (a) Initially each spin is along the x-axis. (b) After rotation each spin has been rotated through the same angle ϕ.

Finally, we consider the ground state of a superconductor under a phase rotation about the z-axis of our pseudospin space. This will be important later for understanding the Josephson effects. First, remember that the spins for the strong-coupled superconductor are all pointing in the x-direction (Fig. 7). If we then define

$$U_\phi \;=\; \exp\left(i\phi \sum_{\vec{k}} \sigma_{\vec{k}z}\right)$$

and let it act on our state $|\psi\rangle_{sc}$

$$|\psi(\phi)\rangle \;=\; U_\phi \prod_{\vec{k}}\left[\frac{1}{\sqrt{2}}\begin{pmatrix}1\\0\end{pmatrix}_{\vec{k}} + \frac{1}{\sqrt{2}}\begin{pmatrix}0\\1\end{pmatrix}_{\vec{k}}\right]$$

we get a new BCS ground state with exactly the same energy but with a phase $e^{-i\phi}$ multiplying the spinor $\begin{smallmatrix}0\\1\end{smallmatrix}_{\vec{k}}$

$$|\psi(\phi)\rangle \;=\; \prod_{\vec{k}}\left[\frac{1}{\sqrt{2}}\begin{pmatrix}1\\0\end{pmatrix}_{\vec{k}} + e^{-i\phi}\frac{1}{\sqrt{2}}\begin{pmatrix}0\\1\end{pmatrix}_{\vec{k}}\right] \tag{18}$$

It is easy to see this if we look at how $e^{i\phi\sigma_{\vec{k}z}}$ acts on the $\vec{k}$-th subspace of $|\psi\rangle_{sc}$. The important thing to note is that each $\vec{k}$ pseudospin has the same phase.

BCS, SUPERRADIANT AND COHERENT STATES: ANALOGIES

Consider now the superradiant state.[9] That is, a state where all the atoms in the superradiant problem are in a coherent superposition. This is directly analogous to the present superconducting state which has a coherent superposition of empty and filled states. Superradiance:

$$|\psi> \;=\; \prod_i \; U_i \left[\begin{pmatrix} 1 \\ 0 \end{pmatrix}_i \;+\; V_i \begin{pmatrix} 1 \\ 0 \end{pmatrix}_i \right]$$

- a coherent superposition of atomic states:

$$\underset{\rule{1cm}{0.4pt}}{\overset{\bullet\rule{0.8cm}{0.4pt}}{}} \;=>\; \begin{pmatrix} 1 \\ 0 \end{pmatrix} \qquad\qquad \underset{\bullet\rule{0.8cm}{0.4pt}}{\overset{\rule{1cm}{0.4pt}}{}} \;=>\; \begin{pmatrix} 0 \\ 1 \end{pmatrix}$$

Superconductivity:

$$|BCS> \;=\; \prod_{\vec{k}} \left[U_{\vec{k}} \begin{pmatrix} 1 \\ 0 \end{pmatrix}_{\vec{k}} \;+\; V_{\vec{k}} \begin{pmatrix} 0 \\ 1 \end{pmatrix}_{\vec{k}} \right]$$

- a coherent superposition of empty and filled electron pair states.

This analogy can be pushed much further when we look at the dynamics of a Josephson junction in the presence of external radiation. Some details of this comparison will be given later (also see Ref. 10).

We can also draw a close correspondence between the BCS ground state and the Glauber coherent photon state.[11] To see this we write the BCS ground state in second quantized notation using the operators of Eq. (1).

$$|BCS> \;=\; \prod_{\vec{k}} (U_{\vec{k}} \;+\; V_{\vec{k}} b_{\vec{k}}^\dagger) |0>$$

By factoring out a U from each term of the product we get

$$|BCS> \;=\; \eta \prod_{\vec{k}} \left[1 + \left(\frac{V_{\vec{k}}}{U_{\vec{k}}} \right) b_{\vec{k}}^\dagger \right] |0>$$

with $\eta = \prod_{\vec{k}} U_{\vec{k}}$. Recalling that $b_{\vec{k}}^{\dagger} = c_{\vec{k}}^{\dagger} c_{-\vec{k}}^{\dagger}$, and therefore,

$$(b_{\vec{k}}^{\dagger})^n |0> = 0$$

for n>1 we can write

$$|BCS> = \eta \prod_{\vec{k}} \left[1 + \left(\frac{V_{\vec{k}}}{U_{\vec{k}}}\right) b_{\vec{k}}^{\dagger} + \frac{1}{2}\left(\frac{V_{\vec{k}}}{U_{\vec{k}}}\right)^2 \left(b_{\vec{k}}^{\dagger}\right)^2 + \cdots \right] |0>$$

which yields

$$|BCS> = \eta \prod_{\vec{k}} \exp\left[\left(\frac{V_{\vec{k}}}{U_{\vec{k}}}\right) b_{\vec{k}}^{\dagger} \right] |0>$$

and, finally,

$$|BCS> = \eta \exp\left[\sum_{\vec{k}} \left(\frac{V_{\vec{k}}}{U_{\vec{k}}}\right) b_{\vec{k}}^{\dagger} \right] |0> \qquad (19)$$

We see immediately the close correspondence between this and the Glauber multimode coherent state.

$$|\alpha> = \eta' \exp\left[\sum_k \alpha_k a_k^{\dagger} \right] |0>$$

where a_k is the k-mode photon creation operator.

It is also clear that the same analysis which brought the BCS state into the form of Eq. (19) can be used on the Dicke superradiant state. Consequently, the states representing the three systems can all be written as

$$|\psi> = \eta^{*} \exp\left[\sum_i \gamma_i b_i^{\dagger} \right] |0>$$

Table II gives a comparison of these states.[1]

TABLE II

	γ_i	$b_i^\dagger$	$|0\rangle$
Photon	α_k defined by $a_k|\alpha_k\rangle = \alpha_k|\alpha_k\rangle$	$a_k^\dagger$ photon creation operator	photon vacuum
Atom	ratio of prob. amp. for being in ground and excited states	$\alpha_k^\dagger$ raising operator	all atoms in ground state
Electron pair	$V_{\vec{k}}/U_{\vec{k}}$	pair creative operator $b_{\vec{k}}^\dagger = c_{\vec{k}}^\dagger c_{-\vec{k}}^\dagger$	electron vacuum

QUANTUM MECHANICAL ANALYSIS OF THE TUNNEL JUNCTION

We are now in a position to use our BCS ground state in studying the tunnel junction. Consider the problem of tunneling a pair from the left to the right hand side of the junction (Fig. 8). This proces is described by two operators.[12] One destroys a pair on the left, whi the other creates a pair on the right.

$$S_L^- = \sum_{\vec{k}} \sigma_{\vec{k}}^+ \quad - \text{ destroys a pair on the left;} \tag{20}$$

$$S_R^+ = \sum_{\vec{q}} \sigma_{\vec{q}}^- \quad - \text{ creates a pair on the right.}$$

The sums of σ operators are required because a given pair is not in a single moment state, but rather, a superposition of these states. For later use, we also define the operators

$$S_{ZL} = -\sum_{\vec{k}} \sigma_{\vec{k}z} \tag{21}$$

$$S_{ZR} = -\sum_{\vec{q}} \sigma_{\vec{q}z}$$

which describe the number of electron pairs, in excess of the value required to maintain charge neutrality, on the respective side. Also, we use the notation the $\vec{k}$ refers to the left and $\vec{q}$ to the right.

The tunneling hamiltonian[13] involves obviously, a matrix element T measuring the tunneling strength between the two superconductors, and operators describing pair annihilation on the left and creation on the right.

$$H_T = T \sum_{\vec{k}} \sum_{\vec{q}} (\sigma_{\vec{q}}^- \sigma_{\vec{k}}^+ + \text{ adjoint}) \tag{22}$$

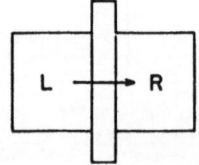

Figure 8. Two superconductors separated by a thin oxide layer with coherent tunneling of electron pairs.

DC Josephson Effect

Classically the current J, across an isolated junction, is related to the time derivative of the number of electron pairs on one side.

$$J = -2e\dot{N}_L = \frac{-2ei}{\hbar}\left[H_T, N_L\right]$$

The operator corresponding to N_L in our pseudospin notation is

$$N_L = \sum_{\vec{k}} \begin{pmatrix} 0 & 0 \\ 0 & 1 \end{pmatrix}_{\vec{k}} = \sum_{\vec{k}}\left(\frac{1}{2} - \sigma_{\vec{k}z}\right) \tag{23}$$

And, we can easily see that

$$J = \frac{-2ei}{\hbar} T \sum_{\vec{k}} \sum_{\vec{q}} (\sigma^+_{\vec{k}}\sigma^-_{\vec{q}} - \sigma^+_{\vec{q}}\sigma^-_{\vec{k}})$$

To calculate the matrix element of this operator, we take a direct product of two BCS ground states as the state describing the junction.

$$|\psi> = |BCS>_L \otimes |BCS>_R$$

$$|\psi> = |\psi(\phi_L)> \otimes |\psi(\phi_R)> \tag{24}$$

$$|\psi> = \prod_{\vec{k}}\left[U_{\vec{k}}\begin{pmatrix}1\\0\end{pmatrix}_{\vec{k}} + V_{\vec{k}}\begin{pmatrix}0\\1\end{pmatrix}_{\vec{k}} e^{i\phi_L}\right] \otimes$$

$$\prod_{\vec{q}}\left[U_{\vec{q}}\begin{pmatrix}1\\0\end{pmatrix}_{\vec{q}} + V_{\vec{q}}\begin{pmatrix}0\\1\end{pmatrix}_{\vec{q}} e^{i\phi_R}\right] |0>$$

Using this to form the matrix element we get

$$J_{DC} = <\psi|J|\psi>$$

$$J_{DC} = \frac{-2eTi}{\hbar} \sum_{\vec{k}} \sum_{\vec{q}} \left[U_{\vec{k}} V_{\vec{k}} U_{\vec{q}} V_{\vec{q}} \; e^{i(\phi_L - \phi_R)} - c.c. \right]$$

$$J_{DC} = J_o \sin^- (\phi_L - \phi_R)$$

which is the famous DC Josephson relation.[14]

AC Josephson Effect

Consider now a voltage V_o applied directly across our junction (Fig. 9). Before calculating the current we must take into consideration the changes this battery brings about in our hamiltonian and state vector. The battery causes the junction to act as a capacitor, building up a charge difference between the left and the right hand sides, $N_L - N_R$. In our pseudospin formalism this becomes

$$N_L - N_R = - \sum_{\vec{k}} \sigma_{\vec{k}z} + \sum_{\vec{q}} \sigma_{\vec{q}z}$$

Consequently, a term should be added to the hamiltonian to represent this capacitive energy.

$$H_C = -eV_o \left(\sum_{\vec{k}} \sigma_{\vec{k}z} - \sum_{\vec{q}} \sigma_{\vec{q}z} \right)$$

The state vector which describes our junction in the presence of this voltage is

$$|\psi>_{V_o} = e^{-(ieV_o/\hbar) (\sum_{\vec{k}} \sigma_{\vec{k}z} - \sum_{\vec{q}} \sigma_{\vec{q}z})t} |\psi>$$

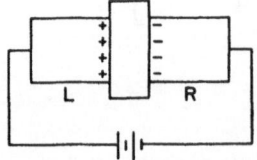

Figure 9. A voltage V_o is placed across the junction causing it to act as a capacitor.

We obtained this by going into the Heisenberg picture and letting $e^{iH_C t/\hbar}$ act on our direct product BCS ground state $|\psi\rangle$. But as we know from the previous section, an operator like this just introduces an additional phase factor. These phases are

$$\phi_L(t) = \phi_L(o) - \frac{eV_o t}{\hbar}$$

$$\phi_R(t) = \phi_R(o) + \frac{eV_o t}{\hbar}$$

Note that the phases are now time-dependent and the second terms differ by a sign. Using $|\psi\rangle_{V_o}$ to calculate the current $J = \langle J\rangle$ we obtain

$$J = J_o \sin\left(\phi_L(t) - \phi_R(t)\right)$$

$$J = J_o \sin\left(\phi_L(o) - \phi_R(o) - \frac{2eV_o t}{\hbar}\right)$$

There is now a term in the argument corresponding to a frequency $\omega = 2eV_o/\hbar$. Since the current is now oscillating, we expect radiation of frequency ω. This is the AC Josephson relation.

JOSEPHSON JUNCTION DYNAMICS

Up to this point we have rewritten certain aspects of the BCS theory in a form more identifiable to the world of quantum optics. In the process, it has been shown that this language is completely adequate to describe Josephson junction phenomena at $T = 0^\circ K$. We now wish to look at the types of problems which lend themselves to solution when viewed with the insight brought from typical quantum optical studies (see references 10 and 15).

Equations of Motion

We will be looking at the transient dynamics, in particular the current, as a junction evolves from some initial state in the presence of external radiation. Consider the situation where initially tunnel junction is in an external circuit with Josephson currents flowing from one side to the other (Fig. 10). Remember that there is now a charge imbalance between the two sides. Quantum optically we would say that this corresponds to a population inversion.

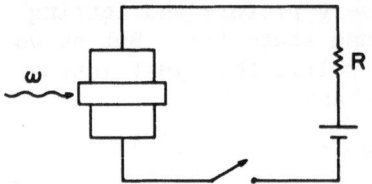

Figure 10. A Josephson junction in a circuit while being emersed in an external radiation field of frequency

$$\omega = \omega_0 = \frac{2eV_0}{\hbar}$$

The hamiltonian for such a problem is very simple. Namely, we are talking about tunneling from one side to the other and a charge imbalance between the two sides. The charge imbalance is measured by $N_L - N_R$. This is now used to define a new operator S_Z.

$$S_Z = \frac{1}{2} (N_L - N_R) \tag{25}$$

$$= \frac{1}{2} \left(-\sum_{\vec{k}} \sigma_{\vec{k}z} + \sum_{\vec{q}} \sigma_{\vec{q}z} \right)$$

$$= \frac{1}{2} (S_{ZL} - S_{ZR})$$

Remember we must sum because the electrons are not really in a single $\vec{k}$- or $\vec{q}$-state. There are three types of energies that must be considered in this problem. First, the capacitive energy due to the pair imbalance. This we write as

$$H_C = \frac{(2e)^2 \, S_Z^2}{2C} \tag{26}$$

Second, the free field energy

$$H_{field} = \hbar \sum_{k} \Omega_k a_k^+ a_k \tag{27}$$

and, finally, the energy describing the interaction of the currents with the radiation field.

$$H_I = i\hbar g \sum_{k} (S^- a_k^+ - S^+ a_k) \tag{28}$$

Here

$$S^- = \frac{1}{\sqrt{2}m} (S_L^- S_R^+) \tag{29}$$

$$S^+ = \frac{1}{\sqrt{2}m} (S_R^- S_L^+)$$

with m equal to the number of Cooper pairs taking part in the inter-action (Fig. 6). As we have seen, these operators describe the tunneling from one side to the other and are therefore related to the current.

The total hamiltonian then is

$$H = \frac{(2e)^2 S_z^2}{2C} + i\hbar g \sum_k (S^- a_k^\dagger - S^+ a_k) + \hbar \sum_k \Omega_k a_k^\dagger a_k$$

We will treat problems semiclassically so that the field incident on the function is classical. This then just gives us the capacitive and interaction terms in the hamiltonian.

$$H = \frac{(2e)^2 S_z^2}{2C} + i\hbar g (A^+ S^- - A^- S^+) \qquad (30)$$

where $A^\pm = \frac{1}{2} A_o e^{\pm i\omega_o t}$

It is useful to notice that this problem is similar to the one treated by Dicke in atomic physics.[9] There

$$H_{atomic} = \hbar\omega_o S_z + \hbar g' (A^+ S^- + A^- S^+)$$

The only difference being that this hamiltonian is linear in S_z whereas in our problem it is quadratic.

In order to calculate the equations of motion for S_z and $S^\pm$ we must know their commutation rules. To find these we note from Eqs. (20) and (21) that $\vec{S}_L$ and $\vec{S}_R$ are both angular momentum operators, and therefore have the commutation relations

$$[S_{z\alpha}, S_\alpha^+] = S_\alpha^+$$

$$[S_{z\alpha}, S_\alpha^-] = -S_\alpha^- \qquad\qquad \alpha = L, R \qquad (31)$$

$$[S_\alpha^+, S_\alpha^-] = 2S_{z\alpha}$$

Using these along with the definitions of S_z, $S^\pm$ (Eqs. 25 and 29) it is easy to show

$$[S_z, S^+] = S^+ \qquad\qquad (32a)$$

$$[S_z, S^-] = -S^- \tag{32b}$$

$$[S^+, S^-] = \frac{1}{m^2} \left[S_{ZL} \left(S_R^- S_R^+ \right) - S_{ZR} \left(S_L^- S_L^+ \right) \right] \tag{33}$$

Conservation of angular momentum allows us to write

$$S_\alpha^- S_\alpha^+ = \vec{S}_\alpha^2 - S_{z\alpha}^2 - S_{z\alpha}$$

with $\vec{S}_\alpha^2 = j_\alpha (j_\alpha + 1)$. Then by defining

$$D = S_{ZL} + S_{ZR}$$

$$j_{LR} = j_L(j_L + 1) - j_R(j_R + 1)$$

we can show

$$S_L^- S_L^+ = S_R^- S_R^+ - 2(1 + D) S_z + j_{LR}$$

$$S_R^- S_R^+ = j_R(j_R + 1) - \left(\frac{D}{2} - S_z \right) \left(\frac{D}{2} - S_z + 1 \right)$$

Substituting these into our commutator we get, finally

$$[S^+, S^-] = \frac{1}{m^2} \left[2j_R(j_R + 1) S_z - 2S_z^3 - j_{LR} \left(\frac{D}{2} - S_z \right) \right.$$

$$\left. + \frac{D^2}{2} S_z \right] \tag{34}$$

We can approximate this result in the following way. Charge conservation allows us to set $D = 0$, while the physical interpretation of $\vec{S}_L$ and $\vec{S}_R$ gives us $j_L \simeq j_R \simeq m$. Consequently, Eq. (33) reduces to

$$[S^+, S^-] \simeq \left(1 - \frac{S_z^2}{m^2} \right) (2S_z)$$

Now by considering the maximum charge the capacitor can sustain we can show that $(S_z/m)^2 \approx 0(10^{-6})$. Therefore, our commutator becomes

$$[S^+, S^-] \simeq 2S_z \tag{35}$$

By using the hamiltonian H (Eq. 30) and the commutation rules (Eqs. 32a, 32b, 35) we can calculate the equations of motion for S_z and $S^\pm$. These are

$$\dot{S}_z = -g(S^+A^- + S^-A^+) \tag{36}$$

$$\dot{S}^+ = i \frac{(2e)^2}{\hbar c} S_z S^+ + 2gA^+S_z \tag{37}$$

Eqs. 36 and 37 determine the transient response of the system in the presence of external radiation.

Rabi Flopping in a Josephson Junction

We consider now a problem which is analogous to the Rabi flopping problem in quantum optics.[10] The situation we will describe has the added advantage of allowing a comparison of our result with that typically found in textbooks on superconductivity.

The situation is that shown in Fig. 11. A Josephson junction, initially in an external circuit, is subject to an incident radiation field. At t = 0, the junction is cut free from the circuit, and we try to describe how the junction evolves in time.

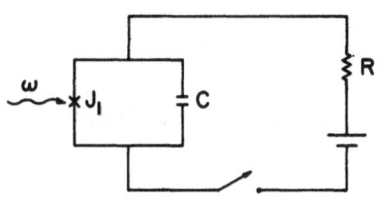

Figure 11. A Josephson junction with parameters J_1 and C initially in a circuit while being emersed in an external radiation field of frequency $\omega = \omega_o = \frac{2eV_0}{\hbar}$. V_B is the voltage of the battery.

First let us consider the approach described in the typical super-conductivity texts.[16] There are three equations describing the junction.

$$J = -J_1 \sin\phi - C\frac{dV}{dt} \qquad (38a)$$

$$V = V_B + RJ \qquad (38b)$$

$$\frac{d\phi}{dt} = \frac{2eV}{\hbar} \qquad (39)$$

Here V is the total voltage across the junction. Solving for $\ddot{\phi}$ by eliminating V and J, we obtain

$$\ddot{\phi} = \frac{-2eJ_1}{\hbar C}\sin\phi + \frac{2eV_B}{\hbar RC} - \frac{1}{RC}\dot{\phi}$$

Setting the junction free from the circuit can be reflected in this equation by letting $R \to \infty$, giving us

$$\ddot{\phi} = \frac{-2eJ_1}{\hbar C}\sin\phi \qquad (40)$$

This is just a pendulum equation describing the time rate of change of ϕ, similar in form to that seen in other quantum optical problems.

Let us consider now how this phase ϕ varies as a function of all the different voltages of the problem.

$$V = V_0 + \tilde{V}(t) - \int \vec{E} \cdot d\vec{\ell}$$

Here:

V_0 = initial dc-voltage measured across the junction

$-\int \vec{E} \cdot d\vec{\ell}$ = voltage due to the electric field across the junction

$\tilde{V}(t)$ = the time rate of change of the voltage relative to V_0. This voltage is non-zero as we cut the junction free from the circuit.

We can get an expression for ϕ as a function of time directly from Eq. (41) by using

$$\phi = \frac{2e}{\hbar} \int V dt \tag{42}$$

In order to facilitate the comparison to our work we define

$$\Phi = \phi(0) + \frac{2e}{\hbar} \int^t \tilde{V}(t') dt' \tag{43}$$

where Φ measures the difference between the phase the junction has at any instant and the phase it would have had if we had left the switch closed.

Eq. (42) can then be substituted into Eq. (40) to give

$$\ddot{\Phi} = \frac{-2eJ_1}{\hbar C} \left(\frac{e\ell A_o}{\hbar} \right) \cos \Phi \tag{44}$$

Here we have used the rotating wave approximation, the details of which are given in reference 10. Also, A_o is the amplitude of the applied field, while ℓ is the barrier thickness. Eq. (44) is then basically the expression taken from the textbooks plus three or four intermediate steps.

We now apply the angular momentum formalism we developed earlier to this problem. Note first that the hamiltonian we are using conserves total angular momentum.

$$S_z + \frac{(S^+S^- + S^-S^+)}{2} = \text{constant}$$

Consequently we define

$$S_z = m \cos \theta \tag{45}$$

$$S^\pm = m \sin \theta \, e^{\pm i\phi} \tag{46}$$

The equations of motion for θ, the Bloch angle, and ϕ, the phase difference, can be obtained by inserting Eqs. (45) and (46) into Eqs. (36) and (37). This gives

$$\dot{\theta} = \Omega_R \cos \Phi \tag{47}$$

$$\dot{\phi} = \frac{(2e)^2}{\hbar C} \ m \cos \theta - \Omega_R \frac{\cos \theta}{\sin \theta} \sin \Phi \qquad (48)$$

where $\Phi = \phi - \omega_o t$ and $\Omega_R = gA_o$ is the Rabi flopping frequency.

Taking a derivative of Eq. (48) and using Eq. (47), we obtain, finally

$$\ddot{\phi} = -\frac{(2e)^2}{\hbar C} \ m \sin \theta \ \dot{\theta} - \frac{d}{dt} \left[\Omega_R \frac{\cos \theta}{\sin \theta} \sin \Phi \right]$$

The first term is of order ω while the second is of order Ω_R. Dropping the second term for the sake of comparison, we get

$$\ddot{\phi} \simeq - \left[\Omega_R \frac{(2e)^2}{\hbar C} \ m \sin \theta \right] \cos \Phi \qquad (49)$$

We can now compare this result to that of Eq. (44). As is shown in Ref. 10, the two coefficients are the same. The major difference between these two approaches is that our equation for $\ddot{\phi}$ has the $\cos \Phi$ term modulated by $\sin \theta$.

In conclusion, then, we see that the angular momentum approach of quantum optics leads to an equation of motion for Φ with considerably more structure than the equation we would obtain using the standard phenomenological approach of the textbooks. Further development of these observations will be published elsewhere.

REFERENCES

1. D. Rogovin and M. Scully, Phys. Rept., 25C, 175 (1976).

2. M. Sargent, M. Scully, and W. Lamb, Laser Physics, Addison-Wesley, 1974; H. Haken, "Laser Theory", Encyclopedia of Physics, Vol.XXV/Zc, ed. by S. Flügge, Springer-Verlag, Berlin (1970); M. Lax, Brandeis University Summer Institute in Theoretical Physics, (1966).

3. L. N. Cooper, Phys. Rev., 104, 1189 (1956).

4. J. Bardeen, L. Cooper, and J. Schrieffer, Phys. Rev., 108, 1175 (1957).

5. P. Anderson, Phys. Rev., $\underline{112}$, 1900 (1958).

6. $\vec{k} > 0$ means the sum is to include only those $\vec{k}$'s with $k_z > 0$.

7. With this representation we can do simple quantum mechanics of a superconductor. Everything we will deal with is at $T = 0^{\circ}$. But to understand the physics of the Josephson problem, the (zero temperature) spinor representation is all we need.

8. We can see this by writing

$$H = \sum_{\vec{k}} \left[\bar{\varepsilon}_{\vec{k}} \, \hat{Z} + \sum_{\vec{k}} V \sigma_{\vec{k}'x} \, \hat{X} \right] \cdot \vec{\sigma}_{\vec{k}}$$

and noting that whereas the expectation value of the kinetic energy for the $\vec{k}$-th pair is $\bar{\varepsilon}_{\vec{k}}$ the expectation value of its potential energy is $\langle V \sum_{\vec{k}} \sigma_{k'x} \rangle \simeq Vm$ (m is defined in Fig. 6).

9. R. Dicke, Phys. Rev., $\underline{93}$, 99 (1954).

10. A. DiRienzo, D. Rogovin, and M.O. Scully, Proc. of Fourth Rochester Conference on Coherence and Quantum Optics, Rochester, New York, June 1977.

11. R. Glauber, Phys. Rev., $\underline{130}$, 2529 (1963); Phys. Rev., $\underline{131}$, 2766 (1963).

12. P. Lee and M. Scully, Phys. Rev., $\underline{B3}$, 769 (1971).

13. P. Wallace and M. Stavn, Can. J. Phys., $\underline{43}$, 411 (1965).

14. B. Josephson, Phys. Letters, $\underline{1}$, 251 (1962).

15. R. Bonifacio, D. Rogovin, and M. Scully, Opt. Comm., $\underline{21}$, 293, (1977).

16. W. A. Harrison, <u>Solid State Theory</u>, McGraw-Hill, New York (1977).

SURFACE PLASMON EFFECTS ON MOLECULAR DECAY PROCESSES

NEAR METALLIC INTERFACES

H. Morawitz

IBM Research Laboratory

San Jose, California 95193, U.S.A.

I.   INTRODUCTION

This article deals with the consequences of a metal surface on the decay and energy shift of excited molecules, both for the isolated molecular case and for a many molecule system.   The simplest type of theory (image theory) already makes the single molecule-metal surface problem analogous to a model system with cooperative properties (Dicke superradiant enhancement of the linewidth and a cooperative Lamb shift).

For a treatment of the metallic surface, which goes beyond the model of a perfect conductor, the surface plasmon spectrum of the metal surface dominates the de-excitation of the molecule and leads to large near-resonant shifts of the excited state energy near $\omega_p/\sqrt{2}$.

Finally, it is shown that for the decay of a many molecule system near a metal surface a phenomenon analogous to cooperative spontaneous decay – superfluorescence – may occur, in which the energy originally stored in a highly excited molecular monolayer appears as a soliton-like pulse of surface plasma oscillations travelling along the metal-dielectric interface.

II. a.) Experimental Background

The radiative decay of excited molecules at distances less than the fluorescence wavelength of the optical transition were systematically studied some 12 years ago [1].   The crucial step in

achieving well-controlled distances, at which the fluorescing
molecule (a Eu-chelate complex) was deposited, was an ingenious
adaptation of the Langmuir-Blodgett technique utilizing fatty acid
chains of variable chain length as spacers. By varying the number
of Cd-arachidate monolayers of average length of 26.8Å deposited
onto a variety of metal films (Au, Ag, Al, Cu) a passive spacer
layer of thickness between 26.8Å to several thousand Angstrom
could be built up by careful preparation. The Eu-chelate complex
was attached chemically to the topmost fatty acid chain before
deposition in low concentration. The preparation of this metal
- dielectric - radiator assembly in its excited state was performed
by exciting the chelate ligands by a u.v. pulse of microsecond
duration. The excitation energy deposited in the complex transfers
by intramolecular energy transfer to the Eu^{3+} ion, dissipating
the excess energy in vibrational motion. The central Eu ion has
a $^7F_2 \rightarrow {}^5D_0$ transition at $\lambda = 6120$Å with a free space radiative
lifetime of about 1 millisecond. The experiments of Drexhage,
Schaefer, Kuhn, and co-workers [1-4] revealed an oscillatory
dependence of the radiative decay rate on the distance D from the
metal surface. For distances considerably less than the
fluorescence wavelength, the lifetime of the excited Eu^{+++} ion is
much shorter than the free space lifetime. The qualitative
behavior of this effect was found to be similar for the different
metal surfaces employed, although the detailed distance dependence
varied from metal to metal. These experimental results were
qualitatively explained [1-4] by a semiclassical calculation
equating the energy radiated into the half space above the metal
surface to the damping of the molecule - the oscillatory behavior
arising from the interference of the directly emitted
electromagnetic field with the reflected field.

II. b.)   Theoretical Description

   Although at first sight surprising, the modification of an
atomic property such as an excited state lifetime due to a metallic
surface several hundred Angstroms distant becomes easily
understandable if one recalls the constituents of a golden rule
calculation of a decay rate. The relevant parameters are the
square of a suitable matrix element for the transition and the
available phase space density of the electromagnetic radiation
field at the transition frequency of the atom. While the former
quantity is hardly going to be affected by a modification of the
environment, the latter senses through the change in the mode
density of the electromagnetic field due to the boundary conditions
holding at the metal-dielectric spacer interface the presence of
the metallic surface. It is further plausible that such effects
are determined by distances of the order of the wavelength of the
transition as the electromagnetic field mode structure is one of

the standing-wave type due to the vanishing transverse electric field at the metal.

The simplest theoretical approach to deal with the decay of an excited atom near a conducting surface [5,6] is to assume the metal to be perfectly reflecting. In that case, it is possible to substitute an image molecule at a distance of 2D from the excited molecule for the metal surface. Such systems of two radiators have been discussed in the literature with the intent to understand cooperative effects in the radiative decay of many atom systems [7,8]. The approximation of treating the metal surface as a perfect conductor assumes that the conduction electrons can screen electric fields of arbitrarily high frequencies. This assumption clearly breaks down for frequencies of the order of the bulk plasma frequency (ω_p~10eV) of ordinary metals.

In order to describe the atomic response, we assume a single atom scattering amplitude

$$t(\omega) = \frac{3}{4k} \frac{\gamma_0}{\left(\omega-\omega_0 + \frac{i\gamma_0}{2}\right)} \vec{e}^{(1)} \cdot \vec{e}^{(2)*} \tag{1}$$

where $k=\omega_0/c$ and γ_0 is the free space linewidth of a two-level atom with an excitation energy of $E=\hbar\omega_0$; $\vec{e}^{(1)}$, $\vec{e}^{(2)*}$ are the polarization vectors of the ingoing and outgoing electromagnetic fields.

In order to find the connection between an external field and the induced polarization field, we require the polarizability tensor $\varepsilon_{\alpha\beta}$, which for a single atom is simply

$$\varepsilon_{\alpha\beta} = p\delta_{\alpha\beta} \tag{2}$$

The dipole moment induced by an external radiation field of polarization $\varepsilon_\beta^{(1)}$ is given by

$$p_\alpha = p\delta_{\alpha\beta} e_\beta^{(1)} = p e_\alpha^{(1)} \tag{2a}$$

The scattering amplitude is given by the contraction of the induced dipole moment with the scattered radiation field of polarization vector $\vec{e}^{(2)*}$

$$T(\omega) = \vec{p} \cdot \vec{e}^{(2)*} \tag{3}$$

As asserted earlier, the problem of an excited molecule at a distance D from a metal surface can be replaced by a two-molecule

system at a distance 2D, with one of the molecules in the excited
state, the other (image molecule) in the ground state [5]. For
a two-level model of each molecule one therefore finds oneself
with an example of Dicke's original two-spin model for
superradiance [6] - as the correct decay of this system has to be
described in terms of symmetrized (anti-symmetrized) states of
the two molecules leading to super- and subradiant states of the
emitted radiation field.

The image approximation, which may be assumed to be reasonable
for optical frequencies much less than the bulk plasma frequency
of the metal and for distances D larger than $\lambda_F/10$, allows a
reduction of the excited molecule-metal surface problem to a
soluble two-body scattering problem.

II. c.)   Analogy to Dicke's Problem

We need to consider two molecules a distance 2D apart in an
incident electromagnetic field, which we will ultimately replace
by the transition dipole of the excited molecule. For the two
molecule problem, the symmetry is reduced from the spherical
symmetry of the polarizability tensor of a single atom to
cylindrical symmetry around the axis through the positions of the
two molecules.

Consider the electric field of an oscillating dipole at
location $\vec{R}_1$, at the position of a second dipole $\vec{R}_2$ and vice versa

$$\vec{E}_{1,2} = e^{\mp i\vec{k}\cdot\vec{R}}[\vec{P}_{1,2}k^2 + \vec{\nabla}(\vec{P}_{1,2}\cdot\vec{\nabla})]\,\frac{e^{ikR}}{R} \qquad (4)$$
$$\vec{R} = \vec{R}_1 - \vec{R}_2$$

The most general polarizability tensor that can be constructed is

$$\varepsilon_{\alpha\beta}^{(1,2)} = \alpha^{(1,2)}\delta_{\alpha\beta} + \beta^{(1,2)}n_\alpha n_\beta \qquad (5)$$

with the unit vector $\vec{n}=\vec{R}/|R|$. The transition dipole of molecules
1 or 2 in the field of the other molecule can be written

$$\vec{P}_{1,2} = a_{1,2}\left(\vec{e}^{(1)} + \vec{E}_{2,1}\right) \qquad (6)$$

with

$$a_{1,2} = \pm a\,\frac{3}{4k}\,\frac{\gamma_0}{\omega_0 - \omega - \frac{1}{2}i\gamma_0} \qquad (7)$$

The coefficients $a_{1,2}$ are the resonant scattering amplitudes for the molecules and in addition, we have introduced a phase convention for the image molecule. As can easily be checked by using the standard expression for the image change $q_I = q(1-\varepsilon)(1+\varepsilon)^{-1}$ in terms of the dielectric function of the metal ε (for a perfect conductor $\varepsilon \to -\infty$, hence $q_I = -q$) a dipole perpendicular to the metal surface produces an image dipole parallel to itself. For the other possibility - a dipole parallel to the surface - the image dipole is pointed in the opposite direction, explaining the $\pm$ signs in Eq. 7 for transition dipole orientations parallel and perpendicular to the metal surface. We stress again that the dipole oscillator frequencies have to be substantially less than the limiting frequency of the electron gas, the bulk plasma frequency, which limits the response of the metallic electrons to form an image dipole. It is now possible utilizing Eqs. 4, 5 and 6 to reduce the multiple scattering integral equation for the scattering amplitude of the two-molecule (molecule-image molecule) system to the two-coupled equations

$$A_{1,2} = a_{1,2}\left[1 + \frac{A_{2,1}e^{i\kappa}}{R}\left(1 + \frac{i}{\kappa} - \frac{1}{\kappa^2}\right)e^{\pm i\vec{k}\cdot\vec{R}}\right]$$

$$B_{1,2} = -\frac{a_{1,2}e^{i\kappa \pm i\vec{k}\cdot\vec{R}}}{R}\left[A_{2,1}\left(1 + \frac{3i}{\kappa} + \frac{3}{\kappa^2}\right) - 2\frac{B_{2,1}}{\kappa}\left(\frac{1}{\kappa} - i\right)\right] \quad (8)$$

$$A_i = k^2\alpha^{(i)} \; ; \quad B_i = k^2 B^{(i)} \quad i = 1,2 \; ; \quad \kappa = k|R|$$

The deceptive simplicity of these equations stems from neglecting the recoil of the scattering molecules, which couples the incident photon frequencies to a continuum of scattered frequencies, necessitating integration over frequencies. In the impulse approximation, which we have made, the standard coupled T-matrix integral equations reduce to the algebraic Equations (8) with solutions

$$A_{1,2} = \frac{a\left[1 + a\frac{e^{i\kappa}}{R}\left(1 + \frac{i}{\kappa} - \frac{1}{\kappa^2}\right)e^{\pm i\vec{k}\cdot\vec{R}}\right]}{\left[1 - \frac{a}{R}\left(1 + \frac{i}{\kappa} - \frac{1}{\kappa^2}\right)e^{i\kappa}\right]\left[1 + \frac{a}{R}\left(1 + \frac{i}{\kappa} - \frac{1}{\kappa^2}\right)e^{i\kappa}\right]}$$

$$(8a)$$

$$B_{1,2} = -\frac{ae^{i\kappa}}{R}\frac{\left[\left(1 + \frac{3i}{\kappa} - \frac{3}{\kappa^2}\right)A_{2,1}e^{\pm i\vec{k}\cdot\vec{R}} + \frac{2a}{\kappa R}A_{1,2}\left(1 + \frac{1}{\kappa}\right)e^{2i\kappa}\right]}{\left[1 + \frac{2a}{\kappa R}\left(\frac{1}{\kappa} - i\right)e^{i\kappa}\right]\left[1 - \frac{2a}{\kappa R}\left(\frac{1}{\kappa} - i\right)e^{i\kappa}\right]}$$

These solutions are <u>exact</u> within the impulse approximation [9], containing <u>all</u> orders of multiple scattering. The impulse approximation is very well fulfilled for this case as the molecular recoil is very small compared to photon momenta.

We are now able to construct the total scattering amplitude of the two-molecules or equivalently the excited molecule-perfect metal surface system, so we form

$$T(\omega) = k^2 \epsilon_{\alpha\beta} e_\alpha^{(2)*} e_\beta^{(1)} = \left[A_1 \left(\vec{e}^{(1)} \cdot \vec{e}^{(2)*} \right) + B_1 \left(\vec{e}^{(1)} \cdot \vec{n} \right) \left(\vec{e}^{(2)*} \cdot \vec{n} \right) \right]$$

$$e^{-\vec{q} \cdot \vec{R}_1} + \left[A_2 \left(\vec{e}^{(1)} \cdot \vec{e}^{(2)*} \right) + B_2 \left(\vec{e}^{(1)} \cdot \vec{n} \right) \left(\vec{e}^{(2)*} \cdot \vec{n} \right) \right] e^{-i\vec{q} \cdot \vec{R}_2}$$

where $\quad \vec{q} = \vec{k}' - \vec{k}$.   (9)

The dipole orientations relative to the metal surface are then given by choosing $\vec{e}_1$ (the polarization vector of the incident field, which is chosen to be given by the direction of the transition dipole moment of the excited molecule) parallel or perpendicular to the vector $\vec{R} = \vec{R}_1 - \vec{R}_2$ or equivalently the normal to the metal surface. For the former can $\vec{e}_1 = \vec{n}$ and the two-molecule scattering amplitude (Eq. 9) reduces to

$$T(\omega) = A_1 + A_2 + B_1 + B_2 = 2a \left[\frac{\sin^2(\frac{1}{2}\vec{k} \cdot \vec{R})}{1 + \frac{2a}{\kappa R}(\frac{1}{\kappa} - i)e^{i\kappa}} + \frac{\cos^2(\frac{1}{2}\vec{k} \cdot \vec{R})}{1 - \frac{2a}{\kappa R}(\frac{1}{\kappa} - i)e^{i\kappa}} \right]$$

(9a)

To find the effect of the image molecule which simulates the metal surface, we examine the poles of the scattering amplitude $T(\omega)$. We find in the forward direction $\vec{k} || \vec{R}$ after writing

$$G \equiv 1 - \frac{2a}{\kappa R} \left(\frac{1}{\kappa} - i \right) e^{i\kappa} ; \quad a = \frac{3}{4k} \frac{\gamma_0}{\omega_0 - \omega - i\frac{\gamma_0}{2}}$$

$$T(\omega) = (aG)^{-1} = (\omega_1 - \omega - i\frac{\gamma_1}{2})^{-1}$$

(10)

$$\omega_1 = \omega_0 - \frac{3}{2} \gamma_0 \left(\frac{\cos\kappa}{\kappa^3} + \frac{\sin\kappa}{\kappa^2} \right)$$

(10a)

$$\gamma_1 = \gamma_0 + 3 \gamma_0 \left(\frac{\sin\kappa}{\kappa^3} - \frac{\cos\kappa}{\kappa^2} \right)$$

where R=2D, κ=2kD. The interaction with the perfectly conducting
metal surface has renormalized the energy and width of the excited
molecule [5]. Note that in the limit D→0, the frequency shift
contribution diverges as D^{-3} indicating the image force would
ionize the molecule. Obviously, the treatment used, which
described only radiative effects between nonoverlapping electronic
systems, breaks down. The width renormalization, however, remains
finite and asymptotically as D→0 attains the value $\gamma_1(0)=2\gamma_0$,
i.e., the molecule would decay twice as fast for 0 separation from
the metal surface. This is the behavior characteristic of a
two-atom Dicke super-radiant state with in-phase transition
dipoles.

For the second orientation choice $\vec{e}_1 \cdot \vec{n}=0$, i.e., the transition
dipole of the excited dipole oriented parallel to the surface we
find for the scattering amplitude, choosing $a_2=-a$,

$$T(\omega) = A_1 + A_2 = 2a \left[\frac{\sin^2(\frac{1}{2}\vec{k}\cdot\vec{R})}{1 + \frac{a}{R}(1 + \frac{i}{\kappa} - \frac{1}{\kappa^2})e^{i\kappa}} + \frac{\cos^2(\frac{1}{2}\vec{k}\cdot\vec{R})}{1 - \frac{a}{R}(1 + \frac{i}{\kappa} - \frac{1}{\kappa^2})e^{i\kappa}} \right]$$

(11)

Proceeding as before, we find new renormalized energy
eigenvalues and widths of the molecule-metal surface given by

$$\omega_2 = \omega_0 - \frac{3}{4}\gamma_0 \left[(\frac{1}{\kappa^3} - \frac{1}{\kappa}) \cos\kappa + \frac{1}{\kappa^2} \sin\kappa \right]$$

(11a)

$$\gamma_2 = \gamma_0 + \frac{3}{2}\gamma_0 \left[(\frac{1}{\kappa^3} - \frac{1}{\kappa}) \sin\kappa - \frac{1}{\kappa^2} \cos\kappa \right]$$

The frequency shift $\omega_2-\omega_0$ again diverges for D→0 as D^{-3}, but the
renormalized width of the excited molecule γ_2→0 as D→0, which
corresponds to the subradiant phasing of eigenstates of the
two-molecule system in Dicke terminology [5]. It suggests that
from a purely radiative point of view, an excited dipole transition
with this polarization very close to a metal surface would tend
to have an infinite lifetime or avail itself of higher order
multipolar decay channels. (For the more realistic consequences
of the surface plasmon spectrum of the metal surface, see
Section III.) The functional form for both the renormalized
frequency and width as a function of the distance from the metal
surface is shown in the figures 1 and 2 and for both the "super"
and "subradiant" spatial arrangements.

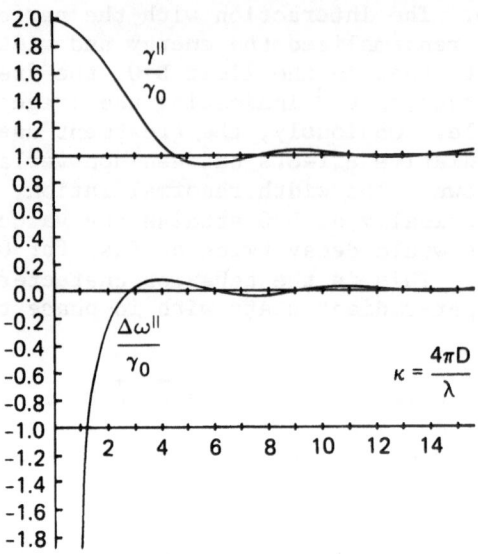

Figure 1. Width and Frequency Shift for Case I (superradiant combination) as a function of distance.

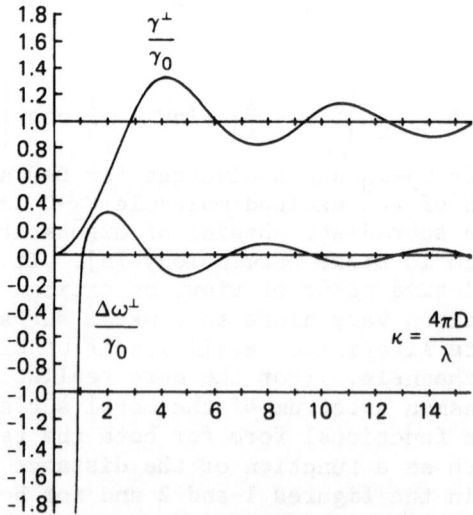

Figure 2. Width and frequency shift for Case II (subradiant combination) as a function of distance.

A reasonably good fit with the experimental data of Drexhage et al. [1-4] down to distances of 1000Å can be obtained by weighting the in-phase (superradiant) and out-of-phase (subradiant) image dipole results by a ratio of 1:2, corresponding to a <u>random</u> distribution of the molecular transition moment, i.e.

$$\gamma_{OBS}(D) = \frac{1}{3} \left[\gamma_1(D) + 2\gamma_2(D) \right] . \tag{12}$$

The treatment of the molecule-perfectly conducting metal surface system so far has relied on a purely classical description of the polarization tensor in the source dipole field of an excited molecule represented as a damped dipole oscillator.

In the next section we show that a <u>semiclassical</u> treatment of the same problem leads to identical results. The only difference is in the treatment of the source term as a transition dipole of the excited two-level molecule described quantum mechanically. We still appeal to image theory, that is we are treating the metallic surface as a perfect conductor and consequently have a fictitious image molecule at distance 2D to be considered quantum mechanically as well [5].

II. c.) Level Shift and Damping Effects

We consider a two-level molecule described by a nondegenerate ground state wavefunction $\psi_0(\vec{R}_{1,2})$ and triply degenerate excited state $\phi_m(\vec{R}_{1,2})$ (m=0,±1) with an allowed transition dipole matrix element $p_1^m = e\langle\phi_m r_i \psi_0\rangle$, coupling to the radiation field still treated classically. The correct eigenstates to describe the decay of a two-molecule system at a distance $2D<\lambda_F$ are symmetric and antisymmetric product states

$$\psi_m^{(\pm)}(1,2) = \frac{1}{\sqrt{2}} \left[\phi_m(1)\psi_0(2) \pm \phi_m(2)\psi_0(1) \right] . \tag{13}$$

The semiclassical interaction Hamiltonian in the dipole approximation, which induces the coupling and decay of the molecule-image molecule system is

$$H_I = - \vec{p}_2 \cdot \vec{E}^{(1)} - \vec{p}_1 \cdot \vec{E}^{(2)} = - k^2 \left[2(-\frac{i}{\kappa^2} + \frac{1}{\kappa^3}) \, p_1^{||} p_2^{||} + (1 + \frac{i}{\kappa} - \frac{1}{\kappa^2}) \right.$$
$$\left. \times p_1^{\perp} p_2^{\perp} \right] \frac{e^{i\kappa}}{\kappa} \tag{14}$$

where $p_{1,2}^{||}$ and $p_{1,2}^{\perp}$ are the projections of the transition dipole vector on the unit vector $\vec{n}=\vec{R}_1-\vec{R}_0/|\vec{R}_1-\vec{R}_2|$ and a vector perpendicular to $\vec{n}$, respectively. The electromagnetic self-energy

$\Sigma E_m^{\pm}$ due to the resonant coupling of the molecules in the symmetric and antisymmetric states of system is given by

$$\Sigma E_m^{\pm} = \langle \psi_m^{(\pm)} H_I \psi_m^{(\pm)} \rangle \tag{14a}$$

If we use the $\vec{n}$ vector as the quantization axis for the system we obtain

$$\Sigma E_0^{\pm} = \pm \frac{3}{4} \gamma_0 \left(\frac{i}{\kappa} - \frac{1}{\kappa^2} \right) \frac{e^{i\kappa}}{\kappa}$$

$$\tag{15}$$

$$\Sigma E_{m=\pm1}^{\pm} = \pm \frac{3}{4} \gamma_0 \left(-1 - \frac{i}{\kappa} + \frac{1}{\kappa^2} \right) \frac{e^{i\kappa}}{\kappa}$$

Separating real and imaginary parts of Eq. (15), we recover the earlier renormalized energy shift and width equations obtained from the damped harmonic oscillator description of the excited molecule. To put it very simply, the excited molecule in front of a metallic surface acts very similarly to an antenna in front of a metallic reflector [10].

III. SURFACE PLASMONS AND THEIR EFFECT ON MOLECULAR DECAY

III. a) Introduction to Jellium

In this section we will show that a quantitatively correct theoretical description of the decay of an excited molecule near a metal surface has to go beyond image theory. In particular, it is clear that the detailed modification of the decay process varies from one metal to another and consequently the role of the metal surface in the decay has to be incorporated in a form containing parameters specific to each metal, such as the plasma frequency ω_p electron-electron scattering time τ, optical constants or most generally the frequency and wavevector dependent dielectric function of the metal [11].

We begin by giving a very simple picture of a metal, representing the conduction electrons by free electrons of density n and replacing the ionic lattice by a uniform positive background of the same density. The electron fluid can respond to external perturbations and rearrange itself to screen out local charges and fields.

The motion of the electrons within the metal and the effect of electromagnetic fields is related by Maxwells' equation, and a constitutive relation between the induced current and the electric field

$$j_\mu = \sigma_{\mu\nu} E_\nu \qquad (16)$$

as magnetic field effects are reduced by a factor $\beta = v_F/c$ (v_F Fermi velocity of the electron gas, c the speed of light), we will set $\mu = \mu o$ and neglect them. We assume a single Fourier component of the electric field at angular frequency ω and expect a longitudinal field $\vec{E}_\ell$ to result in the electron gas in response to external fields. The total electric field in the metal is decomposed into transverse and longitudinal components

$$\vec{E} = \vec{E}_\ell + \vec{E}_t \;\; ; \;\; \vec{E}_\ell \cdot \vec{E}_t = 0 \; .$$

Maxwells' equations can be written in terms of $\vec{E}_\ell$, $\vec{E}_T$ and the electron current $\vec{j}$

$$\nabla x \vec{E}_\ell = 0 \;\; ; \;\; \nabla \cdot \vec{E}_\ell = \frac{4\pi i}{\omega} \nabla \cdot \vec{j}$$

$$\nabla x \vec{E}_t = -i \frac{\omega}{c} \vec{H} \;\; ; \;\; \nabla \cdot \vec{E}_t = 0 \qquad (17)$$

$$\nabla x \vec{H} = i \frac{\omega}{c} \vec{E} + \frac{4\pi}{c} \vec{j} \;\; ; \;\; \vec{j} = \partial_\ell \vec{E}_\ell + \partial_t \vec{E}_t \; .$$

For a general linear constitutive relationship the system of Eqs. (17) is closed, although $\partial_{\ell,t}$ are, in general, integral operators.

There are two cases in which longitudinal and transverse fields decouple and one can find purely longitudinal (Case I) or purely transverse (Case II) solutions for the electric field.

Case I: $\frac{i\omega}{c} \vec{E}_\ell + \frac{4\pi}{c} \partial_\ell \vec{E}_\ell = 0$ everywhere and $\vec{E}_\ell$ not

identically zero.

Case II: $\nabla \cdot (\partial_t \vec{E}_t) \equiv 0$ everywhere and $\vec{E}_t$ not

identically zero.

For an infinite system, the Eqs. (17) simplify for a single plane wave Fourier component of the field

$$\nabla x \vec{E}_\ell = 0 \;\; => \;\; k x \vec{E}_\ell (\vec{k}, \omega) = 0$$

$$\nabla \cdot \vec{E}_\ell = \frac{4\pi i}{\omega} \nabla \cdot \vec{j} \;\; => \;\; \hat{\epsilon}_\ell (\vec{k}, \omega) \; \hat{k} \cdot \vec{E}_\ell (\vec{k}, \omega) = 0 \; . \qquad (17a)$$

We can, therefore, define the generalized longitudinal dielectric function for a simple metal metal $\hat{\epsilon}_\ell (\vec{k}, \omega)$

$$\hat{\varepsilon}_\ell(\vec{k},\omega) = 1 - \frac{4\pi i}{\omega}\,\hat{\sigma}_\ell(\vec{k},\omega) \tag{18}$$

The nontrivial electric field solution of Eq. (17c) can then be given by the zeroes of $\varepsilon_\ell(\vec{k},\omega)=0$, which are the <u>bulk</u> plasmons of the infinite metal [12], corresponding to collective longitudinal modes of the electron gas. The simplest (jellium) form of the longitudinal dielectric function with <u>no</u> dispersion can be written in the form

$$\varepsilon_\ell(0,\omega) \equiv \varepsilon(\omega) = 1 - \frac{\omega_p^2}{\omega^2}\frac{1}{1 - i\gamma\omega} \tag{19}$$

where $\omega_p^2=4\pi Ne^2/m$ is the square of the plasma frequency of the infinite metal, N the electron density, and m the effective mass of the electron; γ^{-1} is an electron-electron scattering time.

III. b.) Metal-Vacuum Interface – Surface Plasmons

The infinite metal is, however, not the physical system we are interested in describing microscopically. We, therefore, next consider the modifications introduced by a metal-vacuum interface, separating the half-space z<0 with $\varepsilon_I(\omega)=1-\omega_p^2/\omega^2$ (metal) and z>0 $\varepsilon_{II}=\varepsilon_0$ (vacuum). The existence of the interface requires application of the appropriate boundary conditions, namely, continuity of the transverse magnetic and electric field components at z=0, to the solution of Maxwells' equation in both half-spaces. We assume an isotropic metal (no y dependence in the fields) and set $H_x=H_z=E_y=0$. We will only consider solutions corresponding to p polarized light (e.g., polarization vector in the plane of incidence), which corresponds to TM modes in waveguide terminology and omit TE modes (polarization vector perpendicular to the plane of incidence (s polarized light). We will see that the former electromagnetic modes couple strongly to the mixed polarization-electromagnetic excitations (surface plasmons) of the metal vacuum interface [13]. Defining the x axis along the direction of the incident field and the z axis perpendicular to the metal surface, we find that Maxwells' equations reduce to ordinary differential equations

$$E_x(z) = -\frac{i}{k}\frac{dE_z}{dx}$$

$$H_y(z) = \left(\frac{\omega\varepsilon_{I,II}}{ck}\right)E_z \tag{20}$$

$$\frac{d^2 E_z}{dz^2} - \kappa_{I,II}^2\,E_z = 0$$

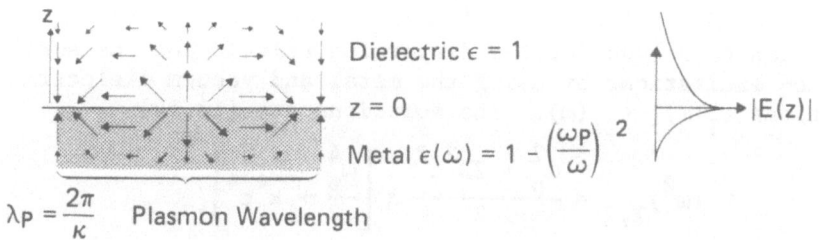

Figure 3. Electric field distribution of a SP on a metal-dielectric interface.

where the propagation constants are given by $\kappa_{I,II}^2 = k^2 - \epsilon_{I,II} \, \omega^2/c^2$ and the magnetic permeability in both half-spaces have been taken equal. Physically allowed solutions of (20) remaining finite as $z \rightarrow \pm \infty$ are

$$E_z^{II} = A_{II} \, e^{\kappa_{II} z} \qquad z < 0$$

$$(20a)$$

$$E_z^{I} = A_{I} \, e^{-\kappa_{I} z} \qquad z > 0$$

and the positive square root has been chosen for $\kappa_{I,II}$. The continuity equations for E_x and H_y at the metal-vacuum interface $z=0$ lead to the equation

$$R = \frac{\kappa_{II} \epsilon_{II}^{-1}}{\kappa_{I} \epsilon_{I}^{-1}} = -1 \qquad (21)$$

which can be solved for the dispersion relation of the surface plasmon excitations by using the metal and vacuum dielectric functions $\epsilon_I(\omega)$, $\epsilon_{II}(\omega)$. The solutions are [13,14]

$$(\omega^2)_{1,2} = \frac{\omega_p^2 + 2k^2 c^2}{2} \pm \left[\frac{\omega_p^4}{2} + k^4 c^4 \right]^{1/2} \qquad (21a)$$

and are plotted in Figure 4. Note that the lower lying branch starts with a slope (group velocity) equal to the speed of light and bends over for large k to a flat branch asymptotically approaching $\omega_s(k \rightarrow \infty) = \omega_p/\sqrt{2}$. This is the value of the surface plasma frequency obtained by electrostatic considerations, neglecting propagation effects (i.e., the finite velocity of light).

The electric field distribution corresponding to a surface plasmon is plotted in Figure 5 and corresponds to an electric field distribution decreasing over the distance of a screening length ($\sim 1-2$Å) into the metal while falling off slowly into the vacuum. It is the spatial extent of the surface plasmon field into the vacuum or dielectric region which is responsible for the strong coupling effects of molecular decay processes to these surface excitations. In addition, the electric field distribution is of mixed transverse-longitudinal polarization and consequently similar to the dipole near field and evanescent optical field components [14].

III. c.) Radiating Dipole near a Conducting Surface

We shall now discuss the problem of the decay of an excited molecule which we model as a radiating dipole, near a metal

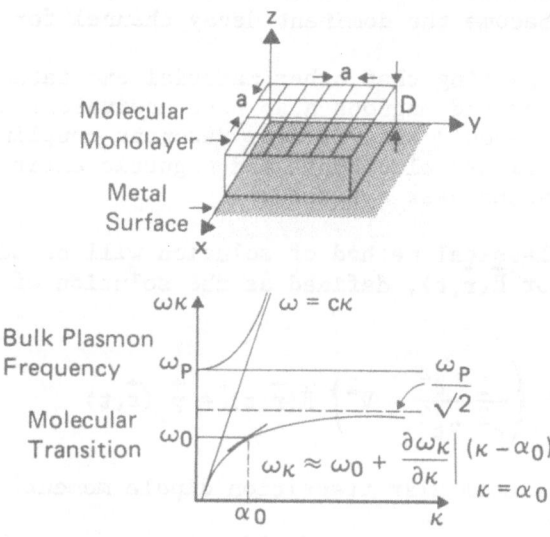

Figure 4. Mixed photon-plasmon dispersion relation – the wave-vector κ is two-dimensional.

surface, using the information obtained in the previous section. The excited molecule is represented by its transition dipole moment $\vec{p}_i(\vec{r},t)$ oscillating at an angular frequency $\omega_0 = (E_e - E_g)/h$, where E_e^i, E_g are the excited and ground state energies of the molecule, which provides the source term in the solution of the coupled Maxwell equations.

The loss of translational symmetry perpendicular to the metal surface has been shown in the previous section to lead to a new brand of mixed electromagnetic field-charge oscillation excitations, which are called surface plasmons. We will demonstrate in this section how they crucially affect the decay of excited molecules in the vicinity (D<500Å) of metal surfaces and, in fact, become the dominant decay channel for $\omega_0 \lesssim \omega_p/\sqrt{2}$ [15].

We note in passing that other material excitations such as phonons, excitons and magnons also have equivalent surface branches, which should show similar resonant coupling effects for molecular vibrations, electronic and magnetic excitations of molecules deposited near surfaces.

A purely classical method of solution will be used relying on the Hertz vector $\vec{\Pi}(\vec{r},t)$, defined as the solution of the wave equation

$$\left(\frac{1}{c^2} \frac{\partial^2}{\partial t^2} - \nabla^2 \right) \vec{\Pi}(\vec{r},t) = \vec{p}_1(\vec{r},t) \tag{22}$$

where $\vec{p}_i$ is the molecular transition dipole moment.

The electric and magnetic field components can be derived by differentiation of $\vec{\Pi}(\vec{r},t)$, namely

$$\vec{E}(\vec{r},t) = k_1^2 \, \vec{\Pi}(\vec{r},t) + \nabla(\nabla \cdot \vec{\Pi}(\vec{r},t))$$

$$\vec{H}(\vec{r},t) = \frac{k_1^2}{i\omega_0 \mu_0} \nabla \times \vec{\Pi}(\vec{r},t) \tag{23}$$

The necessity for discussing the two possible orientations separately was indicated in Section 1.

Case I - transition dipole perpendicular to the metal surface. The cylindrical symmetry for an isotropic metal surface is obvious and we express the Hertz vector in the vacuum

$$\vec{\Pi}^I(\vec{r},t) = \Pi^I(\rho=(x^2 + y^2)^{1/2},z;t)\hat{z} \tag{24}$$

The wave equation (22) for $\vec{\Pi}(\rho,z;t)$ can be solved assuming a periodic time-dependence $\sim e^{-i\omega_0 t}$ by utilizing the retarded Greens function to give

$$G(r,r') = \frac{\exp(ik|r-r'|)}{|r-r'|}$$

$$\vec{\Pi}^I(\rho',z') = \frac{e^{ikR}}{R}\vec{P}_1 \tag{24a}$$

$$R = \left[\rho'^2 + (z'-D)^2\right]^{1/2}$$

Where D is the z coordinate of the source dipole and ρ',z' are the field points in medium I.

In the metal it is necessary to take into account the polarization of the free electrons for z<0 and its contribution to the total field. We have, therefore, to find the contribution to the Hertz vector from the response of the conduction electrons in the metal, which attempt to screen the external field. We make use of the relation $D^{II}(\rho,z)=\varepsilon_{II}(\omega)E(\rho,z)$ and the existence of incident, transmitted (strongly attenuated) and reflected electromagnetic field components, which satisfy the boundary conditions at z=0.

We will consider optical transitions in molecules such that $\omega 0<\omega p$, and consequently our model metal dielectric function is negative, the index of refraction of the metal $n(\omega)=(\varepsilon(\omega))^{1/2}$ is therefore pure imaginary. Note, that for an even more realistic description of the metal, the knowledge of the optical constants of the metal for all frequencies $\omega\le\omega_0$ are required. In general, however, $\kappa(\omega)>>n(\omega)$, the imaginary part $\kappa(\omega)$ of the metal is considerably larger than the real part $n(\omega)$ in the frequency range of interest $[\varepsilon(\omega)=(n(\omega)+i\kappa(\omega))^2]$. The nonradiative components of the transition dipole, which dominate in the near field region $r\le\lambda F$, couple strongly to the surface plasmon field, especially for $\omega 0\approx\omega p/\sqrt{2}$. Using a cylindrical wave expansion [10] we have for the Hertz vector

$$\vec{\Pi}^I_{Dielectric}(\rho,z) = \vec{P}_1\left\{\frac{e^{ik_1 R}}{R} + \frac{e^{ik_1 R'}}{R'}\right.$$

$$
- 2k_1^2 \int_0^\infty \frac{J_0(\lambda\rho) e^{-(\lambda^2 - k_1^2)^{1/2}(z+D)} \lambda(\lambda^2 - k_2^2) d\lambda}{(\lambda^2 - k_1^2)^{1/2} N} \Bigg\}
$$

$$
z \geq 0 \qquad\qquad (25)
$$

$$
\vec{\Pi}_{\text{Metal}}^{I}(\rho, z) = \vec{P}_1 \, 2k_1^2 \, I_1
$$

$$
I_1 = \int_0^\infty \frac{J_0(\lambda\rho) \exp\left\{ (\lambda^2 - k_2^2)^{1/2} z - (\lambda^2 - k^2)^{1/2} D \right\}}{N} \lambda d\lambda
$$

where $\quad k_2^2 = \varepsilon(\omega) k_1^2, \quad R^1 = [\rho'^2 + (z+d)^2]^{1/2} \quad$ and

$$
N = k_2^2 (\lambda^2 - k_1^2)^{1/2} + k_1^2 (\lambda^2 - k_2^2)^{1/2}
$$

The integral terms represent corrections to the perfect metal result (image theory) and have the following properties:

(I)   The reflectivity $|R| \neq 1$ and the phaseshift $\delta \neq \Pi$, i.e., the Fresnel coefficients for reflection are contained in the integral terms via the dielectric function $\varepsilon(\omega)$ of the metal.

(II)   Singularities of the integral term corresponding to zeroes of N (see above) correspond to the excitation of surface polaritons (surface plasmons).

We can demonstrate point (I) by performing a saddle point integration and combine the image and integral terms of Eq. 25 to the form

$$
\pm \rho(k, \sin\Theta) \frac{e^{ik_1 R}}{R},
$$

when the coefficient

$$
\rho(k, \sin\Theta) = \frac{(\xi^2 - \sin^2\Theta)^{1/2} - \xi^2 \cos\Theta}{(\xi^2 - \sin^2\Theta)^{1/2} + \xi^2 \cos\Theta} \quad (\xi^2 = \varepsilon(\omega))
$$

Taking the angle of reflection $\Theta=0$ as the molecule lies along the z-axis reduces to

$$\rho(k,0) = \frac{1-\xi}{1+\xi} = \frac{1-n(\omega)}{1+n(\omega)} = |R|e^{i\delta} .$$

One can, therefore, view this approximation as a means of deriving a modified image theory [6], which includes the non-perfect optical response of the metal via the Fresnel equations.

The second consequence of the integral term in Eqs. (25) is more profound, however, and leads to qualitatively new effects, which dominate the molecular decay for short distances [15,16,17]. Solving for the roots of $N(\lambda)\equiv0$, we find

$$\lambda_{1,2} = \pm k_1 \sqrt{\frac{\xi^2}{1+\xi^2}} = \pm k_1 \left[\frac{\omega^2-\omega_p^2}{2\omega^2-\omega_p^2}\right]^{1/2} \tag{26}$$

which lead to poles in the integrands of Eqs. (25). These poles correspond to complex angles of incidence or equivalently to waves propagating along the metal-vacuum interface. The surface wave pole moves off to infinity for the asymptotic limit of the surface plasmon spectrum $\omega=\omega p/\sqrt{2}$ (see figure) corresponding to the electrostatic (nonretarded) condition for a surface resonance [13] $1+\epsilon(\omega)=0$.

Case II - Transition dipole oriented parallel to the metal surface is somewhat more complicated and we have to admit nonvanishing x and z components of the Hertz vector. We find by a procedure analagous to the one used in Case I.

$$\Pi_x(\rho,z) = \vec{P}_1\left(\frac{e^{ik_1R}}{R} - \frac{e^{ik_1R'}}{R'} + \right.$$

$$\left. + 2\int_0^\infty \frac{J_0(\lambda\rho)\exp\left\{-(\lambda^2-k_1^2)^{1/2}(z+D)\right\}}{(\lambda^2-k_1^2)^{1/2}+(\lambda^2-k_2^2)^{1/2}}\lambda d\lambda\right)$$

$$\Pi_z(\rho,\Phi,z) = -2p_1\cos\Phi\, I_2$$

$$I_2 = \int_0^\infty \frac{J_1(\lambda\rho)\exp\left\{-(\lambda^2-k^2)^{1/2}(z+d)\right\}\left[(\lambda^2-k^2)^{1/2}-(\lambda^2-k_2^2)^{1/2}\right]\lambda^2 d\lambda}{N}$$

$$\cos\Phi = \frac{x}{\rho}$$

Having expressed the Hertz vector due to the transition dipole moment of the excited molecule and the metal induced response in terms of its dielectric function, we are now able to calculate the time-evolution of the coupled molecule-metal surface configuration. Physically speaking, it is the modification of the total electromagnetic field, which contains contributions of the reflected and phase-shifted transverse electromagnetic field (photon) and the field due to excitation of the surface plasmon spectrum of the metal with its exponentially decaying amplitude, which result in drastically different values of the frequency and damping of the molecule.

We calculate the molecular decay process by assuming at t=0 the molecule to be in the excited state. The molecular transition dipole $p_1(\vec{r},t)$ (t≥0) radiates for a time $t_{RET}=2D/c$ as if in free space, as causality requires this minimum time-delay before the reflected field begins to influence the molecular decay [20].

In contrast to image theory, in which the perfectly conducting surface forces a standing wave description with nodes at z=0 on the electromagnetic field, in the present more realistic and physical description of the reflection process, the surface charge density waves arrange themselves to screen the external field and set-up a reflected field, which acts back on the decaying molecule.

Modelling the driven motion of the excited molecule by a damped harmonic oscillator [5,6] in the presence of the total field amplitude E_{ref}, we have to solve

$$\ddot{p} + \gamma_0 \dot{p} + \omega_0^2 \, p = \frac{e^2}{m} \, E_{ref}(D) \tag{28}$$

for the transition dipole-moment p(t), given the field E_{ref}, which can be obtained from Eqs. 25 and 27. We find, separating real and imaginary parts of Eq. (28)

$$\Omega^R = \omega_0 - \gamma_0 \, Re\{E_{ref}(\rho=0, z=D)\}$$

$$\tag{29}$$

$$\gamma^R = \gamma_0\{1 + Im[E_{ref}(\rho=0, z=D)]\}$$

The image terms $\pm e^{ikR'}/R'$ lead to results obtained in Section I for the perfect conductor, but we find additional contributions due to surface plasmons, which have dramatic consequences on the frequency shift $\Omega^R - \omega_0$, and effective linewidth γ^R of the excited molecule.

Again differentiating between the two possible orientations of the transition dipole $\vec{p}$ with respect to the metal surface we obtain for Case I ($\vec{p}||\vec{n}$), the renormalized width

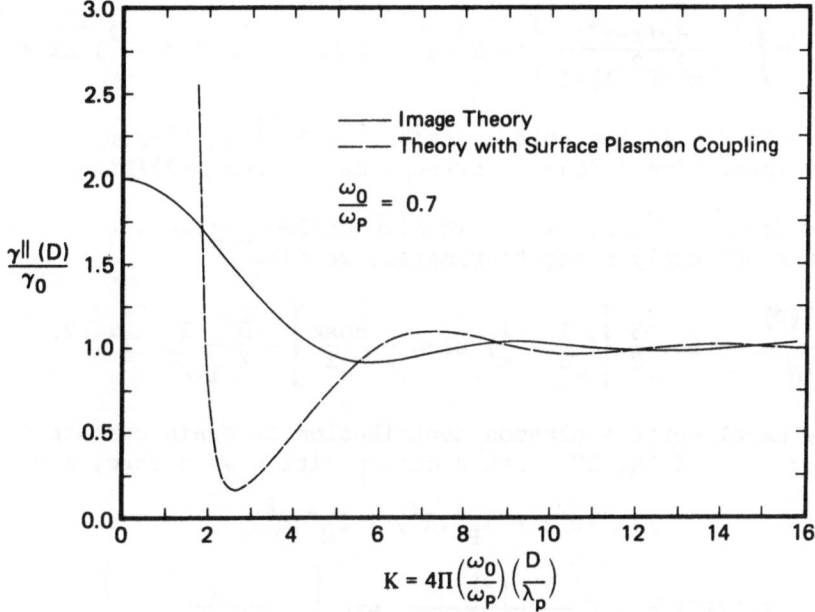

Figure 5. Renormalized width with and without surface plasmon contribution for Case I.

$$\frac{\gamma_I(D)}{\gamma_0} = 1 + 3\left(\frac{\sin\kappa}{\kappa^3} - \frac{\cos\kappa}{\kappa^2}\right) - \frac{3}{1-\xi^2} I^{||}(\xi^2) \tag{30}$$

where the surface plasmon contribution $I^{||}(\xi^2)$ can be decomposed into a pole term $I_P^{||}(\xi^2)$ and a principal value integral $I_N^{||}(\xi^2)$ with definition

$$I_P^{||}(\xi^2) = -\Pi \frac{|\xi|^6}{(|\xi|^2-1)^{5/2}} \exp\left\{-\frac{\kappa}{(|\xi|^2-1)^{1/2}}\right\}$$

$$\tag{31}$$

$$I_N^{||}(\xi^2) = \int_0^1 \frac{dx(1-x^2)}{x^2(\xi^2+1)-1}\left[(1-x^2-\xi^2)\cos\kappa x - \xi^2 x(1-x^2-\xi^2)^{1/2}x\sin\kappa x\right]$$

The singularity in the pole contribution $I_P^{||}$ at $\xi^2=\varepsilon(\omega)=-1$ is seen by inspection ($\varepsilon(\omega)=-1$ corresponds to $\omega_s=\omega_p/\sqrt{2}$)(Figure 5).

For Case II ($\vec{p}_1 \perp \vec{n}$), which we also called <u>subradiant</u> for the case of a perfectly conducting metal, we find

$$\frac{\gamma_2^\perp(D)}{\gamma_0} = 1 + \frac{3}{2}\left[(\frac{1}{\kappa^3} - \frac{1}{\kappa})\sin\kappa - \frac{\cos\kappa}{\kappa^2}\right] - \frac{3}{2}\frac{1}{1-\xi^2} I^\perp(\xi^2) \tag{32}$$

The metal surface plasmon contribution is again contained in the last term of Eq. 32, with a decomposition as before, i.e.

$$I^\perp(\xi^2) = I_P^\perp(\xi^2) + I_N^\perp(\xi^2)$$

$$I_P^\perp(\xi^2) = -\Pi \frac{|\xi|^4}{(|\xi|^2-1)^{5/2}} \exp\left\{-\frac{\kappa}{(|\xi|^2-1)^{1/2}}\right\}$$

$$I_N^\perp(\xi^2) = \int_0^1 \frac{dx}{x^2(\xi^2+1)-1}\left\{x^2[2(1-x^2)-\xi^2(1+x^2)]\cos\kappa x\right.$$

$$\tag{33}$$

$$\left. + (x^2+2x^2\xi^2-1)(1-x^2-\xi^2)\sin\kappa x\right\}$$

In Figure 6 we have plotted the behavior of $\gamma_I(D)/\gamma_0$ as a function of distance in dimensionless variables for $\omega_0/\omega_p=0.7$, i.e., for an optical transition frequency very close to the surface plasmon frequency $\omega_s=\omega_p/\sqrt{2}$.

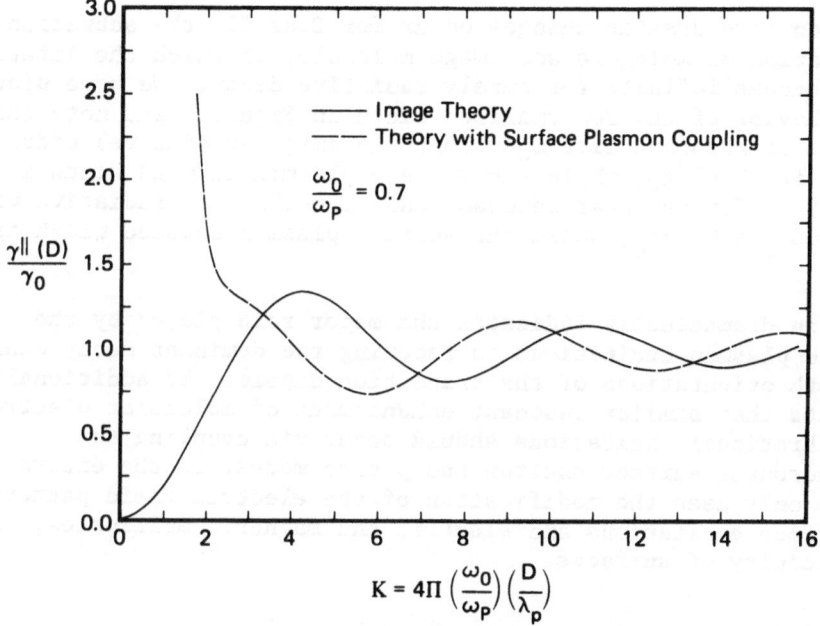

Figure 6. Renormalized width with and without surface plasmon contribution for Case II.

Comparing the contributive of the radiation channel and the surface plasmon channel to the width of the excited molecule, we find that the former has almost reached the asymptotic value of $2\gamma_0$ (superradiant case) for $D=500\text{Å}$, while the surface plasmon contribution is $0.87\gamma_0$ even for an optical transition frequency $\omega_0=0.16\omega_p$, i.e. far from resonance with the surface plasmon branch. At the same distance of 500Å for the near resonant value of $\omega_0=0.7\omega_p$, the radiative channel gives a width of $1.92\gamma_0$, while the surface plasmon induced width is $198\gamma_0$, i.e., the decay into surface plasmons has become the dominant decay channel, shortening the lifetime by several orders of magnitude.

Even more drastic changes occur for Case II, the subradiant combination of molecule and image molecule, in which the lifetime would become infinite for purely radiative decay. We have plotted the behavior of the renormalized width in Figure 7 and note that for the nonresonant case $\omega_0=0.16\omega_p$ the image (radiative) width at $D=500\text{Å}$ is $8\times10^{-3}\gamma_0$, while the surface plasmon induced width is $9\times10^{-2}\gamma_0$. For the near resonant case $\omega_0=0.7\omega_p$ the radiative width at $D=500\text{Å}$ is $0.15\gamma_0$, while the surface plasmon induced width is $95\gamma_0$.

This dramatically indicates the major role played by the surface plasmon excitations in becoming the dominant decay channel for both orientations of the transition dipole. It additionally suggests that similar resonant enhancement of molecular electronic and vibrational excitations should occur via coupling to near-resonant surface exciton and phonon modes, as the entire theory only uses the modification of the electric field patterns of surface excitations and electric and magnetic multipoles, in the vicinity of surfaces.

III. a.) Van der Waals Effects in Excited Molecules

Most of the experimental work, which underlies the subject of this article, has been based on measurements of the changes in lifetime of excited molecules near metal surfaces [1-4]. A recent experiment [21] has demonstrated directly the conversion of excited electronic energy into surface plasmons, which are detected by coupling out through a prism. In addition, experiments in the infrared region by Burstein and co-workers [22], have demonstrated macroscopic propagation distances of surface polaritons.

These effects measure only one of the properties calculated in Section C, namely the renormalized width of the molecule induced by its coupling to the surface plasmon spectrum. As shown in the preceding section, this effect is dramatic and, in fact, has been shown to lead to quantitative agreement with experiment in the nonresonant region [16,17].

In this section we draw attention to a second effect, which arises from the change in the energy of the excited state [23]. In earlier work [5,24] it was pointed out, that the renormalization of the the shift might be observable as a frequency shift (cooperation Lamb shift) in the emitted fluorescence. Due to the preferred emission into the surface plasmon channel and the resulting large increase in the width of the excited state, it seems very difficult to observe this shift in an optical experiment.

The effects of the surface plasmon spectrum on the energy shift were shown to be very large and a new contribution to the Van der Waals force on an excited molecule was predicted recently [23].

The starting point for the calculation is the first of Eqs. 29, utilizing the full expression for the electric field amplitude at the position of the radiating molecule $\vec{R}=(\vec{\rho}=0,z=D)$. The shift of the excited state energy from its free space value $E_e=\hbar\omega_0$ (E_g is taken to be zero) is given for Case I by

$$\frac{\Delta E^I}{\gamma_0} = \frac{\hbar(\Omega_I^R-\omega_0)}{\gamma_0} = -\frac{3}{2}\left(\frac{\cos\kappa}{\kappa^3} + \frac{\sin\kappa}{\kappa^2}\right) + \frac{3}{1-\xi^2}\, J^{||}(\xi^2) \tag{34}$$

The first term again is the image contribution for the perfect conductor obtained in Section I, while the last term contains the modifications due to the nonrealistic treatment of the metal surface accounting for $|R|\neq1$, phaseshift $\delta\neq\Pi$ and the presence of the surface plasmon excitation spectrums a new decay channel. $J^{||}(\xi^2)$ can be calculated from the knowledge of the dielectric function of the metal

$$J^{||}(\xi^2) = I_3 + I_4$$

$$I_3 = -\int_0^1 \frac{dx(1-x^2)}{1-x^2(\xi^2+1)}\left[x\xi^2(1-x^2-\xi^2)\cos\kappa x + (1-x^2-\xi^2)\sin\kappa x\right] \tag{35}$$

$$I_4 = -P\int_0^\infty \frac{(x^2+1)(x^2+1-\xi^2)^{1/2}\exp\{-\kappa x\}dx}{x^2(\xi^2+1)+1}\left[(x^2+1-\xi^2) - \xi^2 x(x^2+1-\xi^3)^{1/2}\right]$$

For Case II we obtain for the level shift of the molecule in the excited state

$$\frac{\Delta E^{II}}{\gamma_0} = \frac{\hbar(\Omega_{II}^R-\omega_0)}{\gamma_0} = -\frac{3}{4}\left[(-\frac{1}{\kappa^3} - \frac{1}{\kappa})\cos\kappa + \frac{1}{\kappa^2}\sin\kappa\right] + \frac{3}{1-\xi^2}\, J^\perp(\xi^2) \tag{36}$$

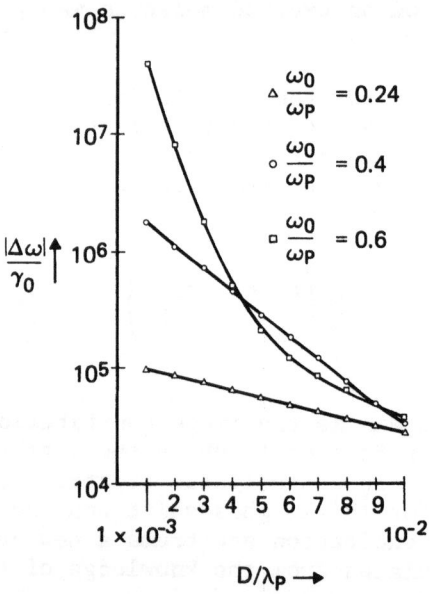

Figure 7.   Surface plasmon induced shift.

with a similar definition of $J^{\perp}(\xi^2)$. The behavior of the line shift $\Delta E/\gamma_0$ has been plotted in Figure 7 for $\omega_0/\omega_p = 0.24$, 0.4 and 0.6 as a function of D/λ_p.

The enhancement of the level shift and width for $\omega_0 \gtrless \omega_p/\sqrt{2}$ is shown in Figure 8. Note that the shift vanishes at the asymptotic surface plasma frequency $\omega_p/\sqrt{2}$ and peaks near it with opposite sign within some characteristic frequency width Γ determined by the amount of damping in $\varepsilon(\omega)$. This feature may be understood from general principles on resonant decay and the Kramers-Kronig relations connecting real and imaginary parts of the self-energy of the molecular excited state. As the decay of an unstable system such as an excited molecule presumes energy conservation between the initial state (the excited molecule) and the final state (the ground state molecule and one quantum of either the electromagnetic field or the surface plasmon field with energy $\hbar\omega_0$), the contribution to the imaginary part of the self-energy of the excited molecule contains only resonant (energy-conserving) states [damping of the surface plasmons due to their coupling to electron-hole states, phonons, etc. will relax this condition]. In contrast, the real part of the self-energy of the excited molecule is given by a principal value integral over all energies of the resulting final state and is strictly zero on resonance. The general slope of the excited energy shift is of the characteristic dispersion form. It is necessary to introduce damping into the metal dielectric function to obtain a finite result for the width. The physical origin of damping in $\varepsilon(\omega)$ are electron-electron scattering processes and its phenomenological inclusion leads to the correct zero-frequency response guaranteeing stability of the model jellium metal.

The experimental consequences of the large enhancement of the radiative (image-theory) shift result are most intriguing: The standard theory of molecule-metal surface interaction [25] (physisorption) derives a D^{-3} dependence of the interaction potential on the distance D. The proportionality constant is given by the polarizability of the molecule. The standard polarizability expression

$$\alpha = \sum_n \frac{\langle 0|ex_\mu|n\rangle\langle n|ex_\mu|0\rangle}{E_n - E_0} \qquad (37)$$

contains the relevant atomic parameter for the radiation decay constant $\gamma_0 - |p|^2 = |\langle 0|ex_\mu|n\rangle|^2$ for a two-level system [26]. The additional quantity determining γ_0 is the phase space for the radiation field at $\omega_0 [\rho_k = \int d^3 k \delta(c|k| - \hbar\omega_0)]$. Note, however, that in contrast to the ground state Van der Waals interaction between molecule and metal surface, in which the zero point fluctuations of the bound electronic system induce a polarization in the metal

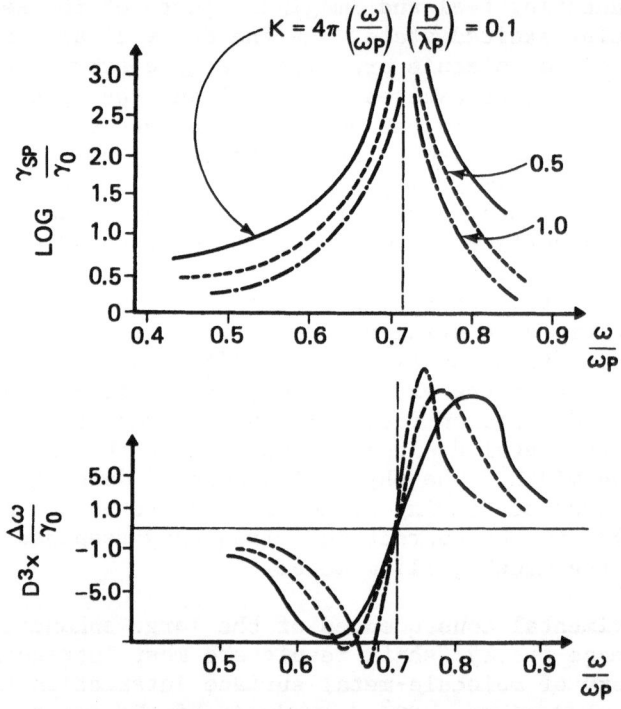

Figure 8. Schematic plot of width and shift effects for three different molecular energies.

(image dipole), which in turn reacts back on the molecule via the dipole-induced dipole interaction and results in a D^{-3} potential, for the case of an <u>excited</u> molecule and a metal surface the situation is different: One can think of the interaction between an excited molecule and the surface plasmon quasiparticle system along the lines of the resonant interaction between two molecules, one excited, the other in its ground state [27]. As in the simplest model for the calculation of dispersion forces the molecules or atoms are represented by a set of oscillators corresponding to the various excited electronic states, the only difference between the Van der Waals interaction between two identical or similar molecules, one excited, the other in its ground state, and the excited molecule-unexcited surface plasmon system is the different origin of the set of oscillator frequencies and their different oscillator strengths.

The idea of viewing the surface plasmon quasiparticle system as a <u>single</u> oscillator at the frequency $\omega_\xi = \omega_p/\sqrt{2}$ has also been used by others [28].

The notion of near-resonant exchange between a localized electronic excitation and the delocalized set of surface plasmon oscillators is a new and exciting possibility for a better understanding of surface reactivity [28].

IV. COOPERATIVE EFFECTS - TWO-DIMENSIONAL SUPERFLUORESCENCE

IV. a.) Molecular Monolayer - Metal Surface Hamiltonian

In the first three sections of this article we have established profound modifications in the decay and energy of an excited molecule in the vicinity of a metal surface, particularly for near degeneracy of molecular transition frequency ω_0 and limiting surface plasmon frequency $\omega_p/\sqrt{2}$.

In this final section we discuss a physically more complex system, which is, at least theoretically, capable of developing cooperative behavior in its decay [29]. We consider the application of the principal result obtained in Section III for a single molecule to a many molecule system, specifically a plane, regular array of molecules such as molecular monolayer near a metallic surface. Such systems can be prepared by techniques similar to the ones described in References 1-4.

We propose that a phenomenon similar to superradiance [7] should occur in a monomolecular layer deposited at some distance D much less than the wavelength of fluorescence or phosphorescence of its lowest excited state [29].

The generalization made by us is to propose that cooperative behavior between different molecules in the same monolayer can occur starting from a phase-incoherent initial state taken, for simplicity, to be a fully inverted molecular monolayer. In contrast to superfluorescence [30], however, in which cooperative behavior between different atoms, which have been prepared incoherently by an intense pulse of radiation, arises due to the interaction with a common radiation field mode, the end-fire mode of the pencil geometry, for example, the role of the coherence-inducing mode is played in the present case by the surface plasmon mode resonant with the molecular transition and propagating along the metal surface in the most favorable direction (parallel to the long axis of the molecular monolayer).

It becomes possible then to envisage the generation of a highly nonlinear propagating state of surface plasmon excitation, which resembles the soliton-like solution of the standard superfluorescence problem [30]. The crucial difference, however, lies in the spatial behavior of this pulse, which is bound to the metal-dielectric interface and propagates in the plane of the interface.

We will use a Hamiltonian description of the interacting molecular monolayer-surface plasmon-photon system, in which the vector potential $\vec{A}(\vec{r},t)$ is expanded in terms of both radiative modes (s and p polarized transverse photons) and surface plasmons [31]. The molecules are modelled by two-level Pauli spin matrices $\sigma_3^{(i)}$, $\sigma_{+,-}^{(i)}$ (i=1,..N) and the molecule-field interaction is taken to result from electric dipole coupling.

The molecular monolayer is assumed to be arranged on a rectangular lattice of dimension Lx=Sa, Ly=Ta (S>>T, a molecule-molecule spacing).

The vector potential $\vec{A}(\vec{r},t)$ can be expanded in the various field modes

$$\vec{A}(\vec{r},t) = \vec{A}^s(\vec{r},t) + \vec{A}^p + \vec{A}_{SP}(\vec{r},t)$$

(38)

$$\vec{A}_{SP} = \sum_\kappa \left(\frac{4\Pi\hbar c}{\vec{A}p_\kappa}\right)^{1/2} \left(i\hat{\kappa} + \frac{\kappa}{v_0}\hat{z}\right) \left[a_\kappa e^{-i\omega_\kappa t} + a_{-\kappa}^+ e^{i\omega_\kappa t}\right]\exp\{-v_0 z + i\vec{\kappa}\cdot\vec{\rho}\}$$

The distances D we will be considering will be such that the coupling to the surface plasmon decay channel dominates the radiative channels. A molecular transition frequency ω_0 is chosen near $\omega_p/\sqrt{2}$, which may can be attained by choosing metal surface (e.g., Ag) and molecular (e.g., substituted benzenes) judiciously.

We use cylindrical coordinates $\vec{r}=(\vec{\rho},z)$ in the half-space $z \geq 0$, which contains the molecular monolayer; κ in Eq. 39 is the two-dimensional wavevector in the reciprocal space of the metal surface of area A. The surface plasmon mode frequencies ω_κ and coupling parameters ν_0 and p_κ are given by [14,31]

$$\omega_\kappa^2 = \frac{1}{2}\,\omega_p^2 + c^2\kappa^2 - \left(\frac{\omega_p^4}{4} + c^4\kappa^4\right)^{1/2} \;;$$

$$\nu_0^2 = \kappa^2 - \frac{\omega_\kappa^2}{c^2} \;;\quad p_\kappa = \frac{\varepsilon^4(\omega_\kappa)-1}{[-\varepsilon(\omega_\kappa)-1]^{1/2}}\,\frac{1}{\varepsilon^2(\omega_\kappa)}$$

(39)

For the metal we again use an idealized jellium dielectric function $\varepsilon(\omega)=1-\omega_p^2/\omega^2$, although for given optical constants of a metal the calculation can be easily refined.

The total Hamiltonian for the molecular monolayer-surface plasmon system can be written

$$H = H_0^m + H_0^{SP} + H_I^{m-SP}; \quad H_0^m = \hbar \sum_{i=1}^{N} \omega_0 \sigma_3^{(i)} \;;\quad H_0^{SP} = \hbar \sum_\kappa \omega_\kappa a_\kappa^\dagger a_\kappa$$

(40)

$$H_I^{m-SP} = -\hbar \sum_{i=1}^{N} \vec{d}^{(i)} \cdot \vec{E}(\vec{r}_i,t) = -\sum_{\kappa,i} \left(\frac{4\pi\omega_\kappa^2}{\hbar A c p_\kappa}\right)^{1/2} [\hat{\kappa}\cdot\hat{p}_i + i\,\frac{\kappa}{\nu_0}\,\hat{z}\cdot\hat{p}_i]$$

$$\times\ (\sigma_+^{(i)} + \sigma_-^{(i)})\ \exp[-\nu_0 D + i\hat{\kappa}\cdot\hat{\rho}_i]$$

The creation and destruction operators a_κ^+, a_κ describe surface plasmon emission and absorption in the molecular transition from excited to ground state, the transition dipole operator for the ith molecule is given by

$$\vec{P}_i = -\,e\,\langle \Psi_e^{(i)}|\vec{x}|\Psi_g^{(i)}\rangle \;;\quad \vec{d}^{(i)} = \vec{P}_i(\sigma_+^{(i)} + \sigma_-^{(i)})$$

(41)

IV. b.) Superfluorescence Treatment for Quasiparticles

The previous section has served to establish a formal correspondence between the excited molecular monolayer-surface plasmon system and the well-established model of superradiance [7,30,32] consisting of two-level atoms in a certain spatial geometry and the resonant end-fire electromagnetic field mode of this system, which <u>preferentially</u> receives the energy initially stored in the fully inverted atomic system.

We now analyze the possibility of observing a similar effect in the decay of a highly excited molecular monolayer system, with the crucial distribution: The final state of the system is characteristic of the metal-dielectric interface and consists of a soliton-like single or multiple pulse of surface charge density oscillation traveling along the interface [29]. In Table I, we give a comparison between the various features and parameters of traditional superfluorescence and the proposed two-dimensional surface plasmon surperfluorescence.

The theoretical treatment follows recently developed techniques utilizing a generalized master equation [30].

As a first step, the individual localized transition operators $\sigma_{\pm}^{(i)}$ are transformed to reciprocal space form suitable for the description of the collective behavior of the highly excited molecular ensemble

$$R_{\pm}(\vec{\alpha}) = \sum_{j=1}^{N} \exp\{\pm i\vec{\alpha}\cdot\vec{\rho}_j\}\sigma_{\pm}^{(j)} \tag{42}$$

The two-dimensional reciprocal-lattice vector $\vec{\alpha}=(\alpha_x^s, \alpha_m^t)$ are defined with $\alpha_x^s=2\Pi/L_x$ s, $\alpha_m^t=2\Pi/L_m$ t where $-1/2$ S$\leq$s$\leq$1/2S, $-1/2$T$\leq$t$\leq$1/2T and the total number of molecules is given by N=ST.

It follows from the commutative relations of the local Pauli matrices $\sigma_{\pm}^{(i)}$, $\sigma_3^{(i)}$, that the collective operators $R_{\pm}(\vec{\alpha})$ obey

$$[R_+(\vec{\alpha}), R_-(\vec{\alpha'})] = \sum_{j,\ell} \exp\{i[\vec{\alpha}\cdot\vec{\rho}_j-\vec{\alpha'}\cdot\rho_\ell]\}[\sigma_+^{(j)}, \sigma_-^{(\ell)'}] = 2R_3(\vec{\alpha}-\vec{\alpha'}) \tag{43}$$

The total Hamiltonian $\hat{H}$ in the interaction representation may be expressed in terms of the collective operators $R_{\pm}(\vec{\alpha})$ in the form

$$\hat{H} = \hbar\sum_{\kappa}\delta_\kappa a_\kappa^+ a_\kappa + \frac{h}{A^{1/2}} \sum_{\kappa,\alpha} g_\kappa[a_\kappa^+(R_A(\alpha)f(\vec{\kappa}-\vec{\alpha},t)-h.c.] \tag{44}$$

The diffraction function $f(\vec{\kappa}-\vec{\alpha},t)$ is defined as

$$f(\vec{\kappa}-\vec{\alpha},t) = \frac{1}{N} \sum_{j=1}^{N} \exp\{i(\vec{\kappa}-\vec{\alpha})\cdot\vec{\rho}_j\} \exp\{i\delta_j t\} \tag{45}$$

TABLE I

	Optical Superfluorescence	Surface Plasmon Superfluorescence
Dimensions:	3D	2D
Radiating Ensemble:	Pencil-shaped Volume Excited Atoms	Monolayer of Excited Molecules
Resonant Field Mode:	<u>Transverse</u> Axial Mode (Photon)	Mixed Transverse – Longitudinal Surface Mode (Surface Plasmon)
Dispersion of Field Mode:	$\omega_{\vec{k}} = c\lvert\vec{k}\rvert$	$\omega_{\kappa}^2 = \dfrac{\omega_p^2}{2} + \kappa^2 c^2$ $-\left[\dfrac{\omega_p^4}{4} + c^4\kappa^4\right]^{1/2}$
Final State of Total System – Atoms & Field:	Atoms in Ground State, Field in Special Pulse Form – Sech	Molecules in Ground State, Field in Surface Plasmon Propagating in Surface – Sech
Dissipative Process:	Leakage into Free Space	Leakage into Free Surface and Plasmon Damping
Coupling Constant:	$g_0 = \left(\dfrac{\omega_0 \mu^2}{2h}\right)^{1/2}$	$g_{\alpha_0} = \left(\dfrac{4\pi\omega_\kappa^2 \mu^2}{hcp_\kappa}\right)^{1/2}$
Phase-Space:	$d^3\vec{k}$	$d^2\vec{\kappa}$

We have made the assumption that the molecules are arranged
on a square lattice with lattice constant a and that the molecular
transition frequency ω_0 is homogeneously broadened.

We assume separability of the spatial and temporal variation
of $f(\vec{\kappa}-\vec{\alpha},t)$ to decompose it into a product

$$f(\vec{\kappa}-\vec{\alpha},t) = F(\vec{\kappa}-\vec{\alpha})g(t)$$

(46)

$$g(t) = \int_{-\infty}^{\infty} G(\varepsilon)e^{i\varepsilon t}dt$$

with

$$F(\vec{\kappa}-\vec{\alpha}) = \frac{1}{N} \sum_{j=1}^{N} \exp\{i(\vec{\kappa}-\vec{\alpha})\cdot\vec{x}_j\}$$

(46a)

$$= \sum_{j=x,y} \frac{\sin\frac{1}{2}(\kappa_j-\alpha_j)L_j}{N_j \sin[\frac{1}{2}N_j^{-1}(\kappa_j-\alpha_j)L_j]}$$

and $G(\varepsilon)$ is a line-shape function. For no inhomogeneous broadening
$G(\varepsilon)=\delta(\varepsilon)$ and $g(t)=1$.

The interaction between collective molecular operators $R_\pm(\vec{\alpha})$
and the surface plasmon field modes a_κ, a_κ^+ is through the
diffraction functions $F(\vec{\kappa}-\vec{\alpha})$, which is strongly peaked about $\vec{\kappa}=\vec{\alpha}$
and has the periodicity of the reciprocal lattice: We next define
surface plasmon quasimodes $A(\alpha)$ by

$$A(\vec{\alpha}) = (L_xL_y/A)^{1/2} \sum_{\kappa} a_\kappa F(\vec{\kappa}-\vec{\alpha})$$

(47)

and note that the $A(\vec{\alpha})$, $A^+(\vec{\alpha})$ operators have Bose commutation
relations as a result of the orthonormality of the diffraction
functions $F(\vec{\kappa})$

$$\frac{1}{(2\pi)^2} L_xL_y \int d^2\vec{\kappa}F(\vec{\kappa}-\vec{\alpha})F(\vec{\kappa}-\vec{\alpha}') = \delta_{\vec{\alpha}\vec{\alpha}'},$$

(48)

As the width of the diffraction function $F(\vec{\alpha})$ is of the order of
the mode spacing $\sim 1/S$ in reciprocal space about $\vec{\alpha}=0$, we may neglect
the variation of the coupling constant g_κ in the interaction
Hamiltonian and replace it by $g_{\vec{\alpha}}$; we have therefore

$$H_I^{m-SP} = \frac{ih}{(L_xL_y)^{1/2}} \sum_{\vec{\alpha}} g_{\vec{\alpha}} (+) [A^+(\vec{\alpha})R_-(\vec{\alpha}) - h.c.]$$

(49)

The surface plasmon quasimodes are damped and the Hamiltonian is no longer <u>unitary</u> in this representation. The time evolution of the density operator ρ in the Hilbert space of quasimodes obeys the standard master equation [30] with a dissipation term $\Lambda_F \rho$ for the irreversible aspects of its motion

$$\frac{d\rho}{df} = \frac{1}{i\hbar} [H_0 + H_I, \rho] + \Lambda_F \rho \;;\; \Lambda_F \rho = \sum_\alpha \left([\kappa(\vec{\alpha}) + j] \left\{ A(\vec{\alpha}), \rho \right. \right.$$
$$\left. \left. A^+(\vec{\alpha})] + h.c. \right\} \right)$$

$$H_0 + H_I = \hbar \sum_{\vec{\alpha}} (\omega_{\vec{\alpha}} - \omega_0) A^+(\vec{\alpha}) A(\vec{\alpha}) + \frac{i\hbar}{(L_x L_y)^{1/2}}$$

$$\sum_{\vec{\alpha}} g_{\vec{\alpha}}^{(+)} [A^+(\vec{\alpha}) R_-(\vec{\alpha}) - h.c.] \qquad (50)$$

The quantity $\kappa(\vec{\alpha})$ is the inverse surface plasmon escape time from the radiating monolayer region is defined as

$$\kappa(\vec{\alpha}) = \frac{v_{SP}}{2L_x} \;,\; \text{with } v_{SP} = \left(\frac{\partial \omega_\kappa}{\partial \kappa} \right)_{\vec{\kappa} = |\alpha|} \;,$$

the group velocity of the resonant surface plasmon mode. A phenomonological intrinsic damping term γ has been added in the anti-Hermitian part of Eq. 50 to account fo the decay of surface plasmons into electron-hole pairs, phonons, etc.

The presence of the intrinsic damping of the surface plasmon modes is an additional complication as it leads to a finite lifetime of the surface plasmon. In optical superfluorescence [30], the damping of the <u>electromagnetic</u> quasimode is due to its escape from the radiating region [in a time of order $L/c = 1/\kappa$, i.e., the transit time of a pulse across a sample length L, and we have to satisfy the set of inequalities $L/c << (g_{\alpha 0}\sqrt{\rho})^{-1} << \gamma_{inc}^{-1}$, where $(g_{\alpha 0}\sqrt{\rho})^{-1}$ is the cooperation time and γ_{inc}^{-1} the dephasing time] – in the case studied here the field mode is itself unstable. From the analogy with the optical case the leakage term $\kappa(\vec{\alpha})$ will cause formation of highly directional surface plasmon pulse analogous to the end-fire modes of Dicke's model [7], but only if $\gamma \leq \kappa(\vec{\alpha})$.

It seems possible to meet this condition in the knee region of the surface plasmon dispersion relation (Figure 4) when the surface plasmon damping is small [33] and $v_{SP} \sim c/5$.

Using the master-equation (50) we may directly calculate the time-evolution of $\langle R_3(\vec{\alpha})\rangle$, the excitation-density ensemble average. By definition

$$\frac{d\langle R_3(\alpha)\rangle}{dt} = \frac{1}{i\hbar} \, tr_{SP} \, \{[H_0 + H_I^{m-SP}, \, R_3(\vec{\alpha})] + \sum_{\alpha'} [\kappa(\alpha') + \gamma]$$

(51)

$$\times \, [A(\vec{\alpha}'), \, R_3(\alpha)A^+(\alpha')] + h.c.]\rho$$

The density matrix ρ for the coupled molecular monolayer-surface plasmon system has been decomposed $\rho = \rho_{SP} \times \rho_m$ and the plasmon degrees of freedom have been traced over in Eq. (52).

We further specify to a __single__ resonant ($\omega_0 = \omega_{\kappa = \alpha_0}$) mode $R_\pm(\vec{\alpha}_0)$, which has the largest gain and couples to the corresponding quasimode of the surface plasmon field

$$A_{\alpha_0}, \, A_{\alpha_0}^+.$$

The collective molecular operators define a constant of the motion

$$\langle R_+(\vec{\alpha}_0) \, R_-(\vec{\alpha}_0) + R_3^2(\vec{\alpha}_0) - R_3(\vec{\alpha}_0)\rangle = \frac{1}{2} \, N \, (\frac{1}{2} \, N+1)$$

(52)

The equation of motion for $\langle R_3(\vec{\alpha}_0)\rangle$ (Eq. 51) expresses balance between the total internal energy

$$\langle A_{\alpha_0}^+ A_{\alpha_0}\rangle + \langle R_3(\alpha_0)\rangle$$

and the loss term

$$[2\kappa(\alpha_0) + 2\gamma] \times \langle A_{\alpha_0}^+ A_{\alpha_0}\rangle$$

in the form

$$\frac{d}{dt} \, [\langle A_{\alpha_0}^+ A_{\alpha_0}\rangle + \langle R_3(\alpha_0)\rangle] = - \, 2[\kappa(\alpha_0) + \gamma]\langle A_{\alpha_0}^+ A_{\alpha_0}\rangle$$

(53)

The corresponding equation of motion for $\langle R_3(\vec{\alpha}_0)\rangle$ can be directly obtained from Eq. 51 utilizing the commutation relations for $R_3(\alpha)$, i.e.

$$[R_3(\vec{\alpha}), R_{\pm}(\vec{\alpha}')] = \pm 2R_{\pm} \delta_{\alpha\alpha'} \ .$$

Decoupling higher order products of operators in the self-consistent field approximation and using a modified Bloch angle $\psi(t)$ as the dynamical variable Eq. 51 reduces to the damped pendulum equation

$$\ddot{\psi} + [\kappa(\alpha_0) + \gamma]\dot{\psi} - \frac{g\alpha_0^2}{L_x L_y} (N+1) \sin\psi = 0 \qquad (54)$$

where we used the definition

$$\langle R_3 \rangle - \frac{1}{2} = \frac{1}{2} (N+1) \cos\psi(t) \ . \qquad (55)$$

IV. C.) Physical Considerations for Experimental Observation

In the last section, the cooperation emission of a model molecular monolayer into a cooperative surface plasmon mode has been reduced to the solution of a damped Sine-Gordon equation [34]. It is well established that particular solutions exist for this equation with highly unusual properties (solutions) [34]. The most spectacular feature of Eq. (54), which is valid for $t_2^* \to \infty$ (no inhomogeneous broadening), can be seen by inspection on neglecting the inertial term $(\sim\ddot{\psi})$ in the pendulum equation. The overdamped pendulum solution for the decay of a fully inverted system is already presented in Dicke's classic paper [7] and reappears in other nonlinear effects such as self-induced transparency [35], photon echoes [36]. The inertial term leading to oscillatory superfluorescence was derived in the recent work of Bonifacio Lugiato and Banfi [30].

For the occurrence of surface plasmon superfluorescence as a single soliton-like pulse, we require $\tau_c/\tau_d \gg 1$. Here the cooperation time

$$\tau_c = (g_{\alpha_0} \sqrt{\rho})^{-1}$$

arises from collective spontaneous decay and $\tau_d = [\kappa(\alpha_0)+\gamma]^{-1}$ is the damping time accounting for escape of the excitation pulse from the active region and intrinsic damping due to conversion into electron-hole pairs, phonons and finally heat. For the following set of parameters $\nu_{SP} \sim c/5$, $\gamma \sim 10^{12} \sec^{-1}$, $L_x = 3 \times 10^{-3}$cm, we obtain $\tau_d \sim 0.5 \times 10^{-12}$sec. Taking a molecular monolayer of density $\rho \sim 10^{12}$ molecules/cm^2 and a typical phosphorescence lifetime $\tau_0 \sim 10^{-3}$sec, we obtain as an estimate for the cooperation time

$$\tau_c = (g_{\alpha_0} \sqrt{\rho})^{-1} = 2 \times 10^{-11} \text{sec} ,$$

meeting the requirement for superfluorescence $\tau_c \gg \tau_d$. We have implicitly assumed that $T_2^* \to \infty$ in our treatment so far, which is equivalent to a system with a dephasing time of 10^{-9}-10^{-10} sec. It is very likely that a low temperature experiment will be necessary to attain such long dephasing times.

The most suitable surface for such an experiment is silver, which has the asymptotic surface plasmon frequency occur at $\hbar\omega_{SP}=3.6$eV, which may be further reduced by a dielectric overlayer of large dielectric constant. Organic molecules, such as benzene have their lowest triplet transition in the same energy range and it is possible to modify the lifetime by heavy atom substitution, which increases the spin-orbit coupling responsible for spin-changing triplet-singlet transition.

The detection of the propagating surface plasmon pulse could utilize a prism-airgap geometry (inverse attenuated - total reflection technique [36] or surface Raman scattering in the time-domain to look for an anti-Stokes component due to the intense surface plasmon pulse with its hyperbolic secant shape).

We note, parenthetically, that the most convenient spectral region may not lie in the optical domain, but in a lower frequency region. The effects we have been describing are of a general nature and also occur if molecular vibrations are considered as the local oscillators, and surface phonons or excitons as the surface polariton field.

IV. d.) Summary and Conclusions

This article has considered single molecule spontaneous decay and level shift effects near a metallic surface. Very large resonant and quasi-resonant effects occur for near-degeneracy of the molecular transition frequency and the surface plasmon limiting frequency $\omega_p/\sqrt{2}$. Image theory leads to good agreement with experiment to distances of $\lambda/4$, but only redistributes the intensity between the suitably phased linear conbinations of molecule-image molecule. The existence of surface modes of mixed longitudinal and transverse polarization allow very strong mixing via the near field components of the molecular dipole field and leads to a new decay channel for short distances, which dominates the decay [15].

Finally a new cooperative phonomenon, surface plasmon superfluorescence is predicted, which suggests a way of generating very intense pulses of surface charge oscillation, by a generalization of optical superfluorescence to a molecular

monolayer in a highly excited state emitting a soliton-like pulse
of surface plasmons into a metal-dielectric interface.

REFERENCES

1. K. H. Drexhage, Habilitationsschrift, University of Marburg,
   Germany (1966).

2. K. H. Drexhage, M. Fleck, H. Kuhn, F. P. Schäfer and
   W. Sperling, Ber. Bunsenges. Phys. Chem. 70, 1179 (1966).

3. H. Kuhn, Spectroscopy of Monolayer Assemblies, in: Physical
   Methods of Chemistry, Part III B, eds. A. Weissberger and
   B. W. Rossiter (Wiley-Interscience, New York 1972).

4. K. H. Drexhage in "Progress in Optics XII," ed. E. Wolf,
   North-Holland (1974).

5. H. Morawitz, Phys. Rev. 187, 1792 (1969).

6. H. Kuhn, J. Chem. Phys. 53, 101 (1970).

7. R. H. Dicke, Phys. Rev. 93, 99 (1954), and in "Quantum
   Electronics," Proceedings of the Third International Congress,
   Paris, 1964, eds. P. Grivet and N. Bloemberger (Columbia
   University Press, New York, 1964).

8. M. J. Stephen, J. Chem. Phys. 40, 669 (1963); J. H. Eberly
   and N. E. Rehler, Phys. Rev. A2, 2038 (1970).

9. M. L. Goldberger and K. M. Watson, Collision Theory (John Wiley
   and Sons, Inc., New York, 1964), pg. 762.

10. A. Sommerfeld, Ann. Phys. 28, 665 (1908); also in A. Sommerfeld,
    Partial Differential Equations in Physics, Academic Press,
    New York 1949, Chap. 4.

11. D. Pines, Revs. of Mod. Physics 28, 184 (1956).

12. H. E. Bennett and J. M. Bennett, in Optical Properties and
    Electronic Stucture of Metals and Alloys, ed. F. Abelès
    (North-Holand, Amsterdam, 1966).

13. R. H. Ritchie, Phys. Rev. 106, 874 (1957); R. A. Ferrell,
    Phys. Rev. 111, 1214 (1958).

14. E. N. Economou, Phys. Rev. 182, 539 (1969).

15. H. Morawitz and M. R. Philpott, Phys. Rev. 810, 4863 (1974).

16. R. Chance, A. Prock and R. Silbey, J. Chem. Phys. 62, 2245
    (1975); ibid. 60, 2184, 2744 (1974).

17. K. Tews, Ann. Phys. (Leipz.) 29, 97 (1973).

18. M. Babiker and G. Barton, J. Phys. A9, 129 (1976).

19. G. S. Agarwal, Phys. Rev. 11A, 253 (1975).

20. J. Hamilton, Proc. Phys. Soc. (London) A62, 12 (1949).

21. Y. M. Gerbshtein, I. A. Merkulov and D. N. Mirlin, JETP Lett.
    22, 35 (1975).

22. J. Schoenwald, E. Burstein and J. M. Elson, Sol. St. Comm.
    12, 185 (1973).

23. H. Morawitz, Bull. Am. Phys. Soc. 21, 241 (1975).

24. H. Morawitz, Phys. Rev. A7, 1148 (1973); H. Morawitz, Bull.
    Am. Phys. Soc. 20, 1148 (1975).

25. C. Mavroyannis, Mol. Phys. 6, 593 (1963).

26. G. Barton, J. Phys. B7, 2134 (1974).

27. J. O. Hirschfelder, C. F. Curtis and R. B. Bird, "Molecular
    Theory of Gases and Liquids," Chap. 13, John Wiley and Sons,
    Inc., New York.

28. J. I. Gerstein and N. Tzoar, Phys. Rev. B9, 4038 (1974).

29. R. Bonifacio and H. Morawitz, Phys. Rev. Lett. 36, 1559 (1976);
    H. Morawitz (to be published).

30. G. Banfi and R. Bonifacio, Phys. Rev. Lett. 33, 1259 (1974);
    R. Bonifacio and L. Lugiato, Phys. Rev. A11, 1507 (1975).

31. J. M. Elson and R. H. Ritchie, Phys. Rev. B4, 4129 (1971).

32. V. Ernst and D. Stehle, Phys. Rev. 176, 1456 (1968).

33. J. G. Endriz and W. Spicer, Phys. Rev. B4, 4144 (1971).

34. A. Scott, F. Chu and D. McLaughlin, Proc. IEEE 61, 1443 (1973).

35. S. L. McCall and E. Hahn, Phys. Rev. 183, 4579 (1969).

36. A. Otto, Z. f. Physik 216, 398 (1968).

INTRODUCTION TO PICOSECOND SPECTROSCOPY

Bruno Bosacchi*

Optical Sciences Center

University of Arizona, Tucson, Arizona 85721

The aim of this short course was to provide an introductory survey of the present status of picosecond spectroscopy. The first lecture covered the methods of generating, handling, and measuring picosecond pulses. The second one introduced the basic techniques and ideas of picosecond spectroscopy for the study of ultrashort processes in matter, with examples taken from biology and chemistry. These examples had been chosen for their pedagogical and illustrative value and did not represent necessarily the most recent and updated results. Finally, the third lecture dealt with the application of picosecond spectroscopy to selected problems of solid state physics. Given the amount of material, which could not have been covered in reasonable detail in a manuscript of limited length, we have chosen to give here a detailed table of contents of the first two lectures, together with a corresponding set of references. This has allowed us to deal somewhat more extensively with the subject of picosecond spectroscopy in solid state physics.[1] Whereas several review papers exist that cover exhaustively the topics dealt with in the first two lectures,[2-19] only two contributions have recently appeared in which topics discussed in the third lecture have been partially covered, and from a somewhat different viewpoint.[20,21] A monograph containing an expanded version of the entire course will be published elsewhere.[22]

---

*On leave from the University of Parma, Parma, Italy.

TABLE OF CONTENTS

3. Dynamical processes in liquids[7,10-13,21]
4. Relaxation of coherent vibrational modes[7,21]
Active picosecond physics. Prospects in laser induced
chemistry

VI. Applications of Picosecond Pulses. II. Picosecond Spectroscopy and Solid State Physics[1,20,21]

REFERENCES

1. B. Bosacchi, this volume, p.
2. A. J. DeMaria, in Progress in Optics, Vol. 9, ed. by E. Wolf
    (North-Holland, 1971), p. 31.
3. R. C. Greenhow and A. J. Schmidt, in Advances in Quantum
    Electronics, Vol. 2, ed. by D. W. Goodwin (Academic Press,
    1974), p. 157.
4. P. W. Smith, M. A. Duguay, and E. P. Ippen, in Progress in
    Quantum Electronics, Vol. 3, ed. by J. H. Sanders and S.
    Stenholm (Pergamon Press, 1975), p. 107.
5. M. M. Malley, in Creation and Detection of the Excited
    State, Vol. 2, ed. by W. R. Ware (Dekker, 1974), p. 99.
6. D. von der Linde, Appl. Phys. 2, 281 (1973).
7. A. Laubereau and W. Kaiser, Opto-Electron. 6, 1 (1974).
8. D. J. Bradley and G. H. C. New, Proc. IEEE 62, 313 (1974).
9. D. von der Linde, IEEE J. Quantum Electronics QE-8, 328 (1972).
10. K. B. Eisenthal, Acc. Chem. Res. 8, 118 (1975).
11. K. J. Kaufmann and P. M. Rentzepis, Acc. Chem. Res. 8, 407
    (1975).
12. A. Laubereau and W. Kaiser, Annu. Rev. Phys. Chem. 26, 83
    (1975).
13. G. E. Busch and P. M. Rentzepis, Science, Oct. 1976, p. 276.
14. S. L. Shapiro, ed., Ultrashort Light Pulses (Springer Verlag,
    1977).
15. S. L. Shapiro, ibid, p. 1.
16. D. J. Bradley, ibid, p. 18.
17. E. P. Ippen and C. V. Shank, ibid, p. 83.
18. K. B. Eisenthal, ibid, p. 275.
19. A. J. Campillo and S. L. Shapiro, ibid, p. 317.
20. D. H. Auston, ibid, p. 123.
21. D. Von der Linde, ibid, p. 204.
22. B. Bosacchi and M. O. Scully, Introduction to Picosecond
    Lasers and Spectroscopy, forthcoming.

PICOSECOND SPECTROSCOPY AND SOLID STATE PHYSICS

Bruno Bosacchi*

Optical Sciences Center

University of Arizona, Tucson, Arizona 85721

I. INTRODUCTION

In the previous lectures it was shown how the unique features of picosecond pulses, i.e., their ultrashort duration and ultra-high power, have made possible the development of picosecond spectroscopy as a new and promising technique for the investigation of ultrafast and/or nonlinear processes in matter.[1] The study of the decay processes from the excited states of chemical substances served as an example to illustrate some basic ideas and applications of picosecond spectroscopy. In this lecture, the obvious extension of the technique to solid state physics will be considered. Here the situation is somewhat more complicated than in chemistry, particularly when one comes to the interpretation of the experimental results, because of the much more complex pattern of interactions taking place in a solid. On the other hand, the prospect that we will learn something about these interactions through the use of picosecond techniques looks rather good. Let us consider, for example, the case of electron-phonon interaction in solids.

The many-body theory of the solid state shows, in first approximation, that the electron-phonon (EP) interaction can be included, in many cases, within a one-electron frame, provided that some parameters of the one-electron theory, like the effective mass and the dielectric constant, are suitably renormalized. Experimentally, this implies a lack of qualitative new effects caused by the EP-interaction. It follows that information about this interaction can in general be obtained only by comparing the

---

*On leave from the University of Parma, Parma, Italy.

experimental results, which are "renormalized" by the interaction,
with the predictions of the one-electron theory without EP interaction. The reliability of this procedure rests on the reliability
of the one-electron results, which are not directly susceptible
to experimental verification. Here is where picosecond techniques
might show their effectiveness. Since the times associated with
the EP interaction are $\sim 10^{-13}$ to 10^{-10} sec, it is expected that
the study of phenomena on a picosecond, or subpicosecond, time
scale, might reveal effects where EP interaction does not have
the time to come in. The comparison with the same phenomena on a
larger time scale, where the EP interaction is effective, might
therefore provide direct experimental information on the interaction itself. Though the above considerations are somewhat spec-
ulative, they illustrate, on a very general ground, the potentialities of picosecond spectroscopy in solid state physics.

    In the following sections we will discuss a number of cases
in which picosecond techniques have been applied to study problems
of solid state physics. Specifically, we will consider

(1) The optical properties of the electron-hole plasma in
     Ge (Section II).
(2) The picosecond spectroscopy of the excitonic phases in
     CdSe (Section III).
(3) The two photon absorption of GaAs measured with pico-
     second pulses (Section IV).
(4) The question of self-induced transparency in semicon-
     ductors (Section V).

    The above selection does not claim any completeness, and
might be somewhat arbitrary, being mostly dictated by personal
experience and interest. Yet, we believe that it provides a good
illustration of the use of picosecond spectroscopy in solid state
physics and can be considered at present as a meaningful representation of the entire field.

II. THE OPTICAL PROPERTIES OF HIGHLY EXCITED GE

    In this section we will attempt to provide a critical over-
view of the phenomena associated with the excitation of a Ge
sample by picosecond pulses from a Nd:glass laser (λ = 1.06 μm).
The pioneering work in this field, which is still actively investigated, has been done by M. O. Scully and his group,[2,3] at the
University of Arizona, and by Auston and Shank at Bell Labs.[4-6]
We first recall some background information on the optical
properties of Ge at moderate light intensities, as a framework
for the discussion that follows. The band structure of Ge in
the proximity of the Γ point is shown in Fig. 1. The elementary
process of absorption for a photon of energy 1.17 eV, corre-

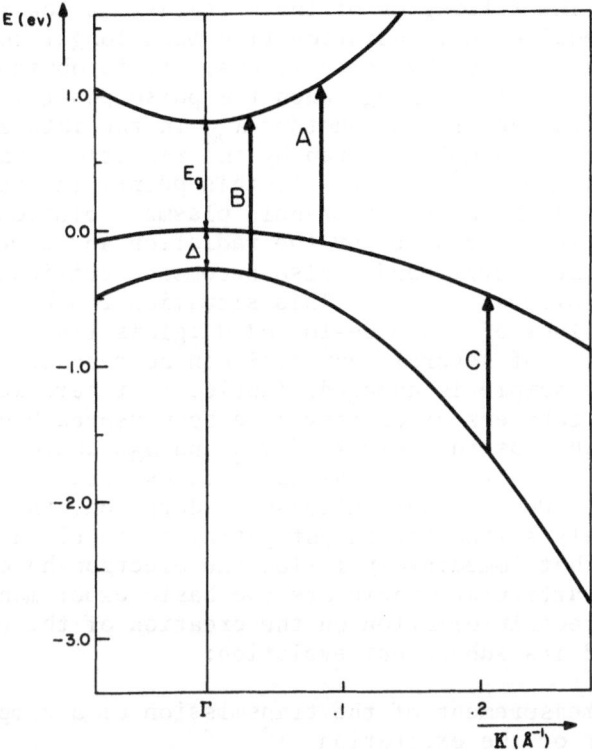

Fig. 1. Schematic energy band structure of Ge near the center Γ
of the Brillouin zone.

sponding to $\lambda = 1.06$ μm, is the creation of an electron-hole pair
by the direct interband transitions A and B. A consideration of
the magnitudes involved shows indeed that both transitions from
the spin orbit split branches of the valence band are effective.
Other absorption processes, like the indirect transitions, require
the intervention of phonons in order to conserve momentum, and
are negligible in comparison to A and B. The above absorption
process is characterized by a coefficient $\alpha \simeq 1.2 \cdot 10^4$ cm^{-1}. The
subsequent evolution of the electrons and holes so created will
ultimately lead to their recombination, radiative or not. This
is, however, a rather slow process ($\sim 10^{-9}$ sec^{-1}) that is preceded
by others (intervalley scattering, phonon-assisted relaxation,
electron-hole binding, etc.), which will be discussed later.
Whatever process takes place after the absorption of a photon
has no influence on the absorption process itself, since the density N_d of the electrons states resonant with (i.e., coupled by)
the radiation is much higher than the density N_{ph} of the photons
absorbed per second. To give an idea of the orders of magnitude,
one can estimate that the densities of states associated with
the transitions A and B are $N_{dA} \simeq 5.4 \cdot 10^{17}$ cm^{-3} and $N_{dB} = 3 \cdot 10^{16}$ cm^{-3}

respectively. The filling up of these states, at low radiation power, would require an irradiation time much longer than that spent in the same states by the carriers. It is obvious, however, that the situation might change when the pulse power increases to the point that the density of photons N_{ph} in the interaction volume (the portion of the sample excited by the radiation) can no longer be neglected with respect to N_d. At this point, the rate of the various processes in the electron-hole plasma evolution, which act to deplete the levels coupled by the radiation and spread the carriers over the energy bands, also becomes a critical factor in the absorption process itself. This situation can be realized with the powerful pulses of the mode-locked Nd:glass laser; indeed, carrier densities of several 10^{20} cm^{-3} can be reached, up to the point where the sample is damaged. (Notice that here we refer to the density of carriers after they have been "spread" over the energy bands, whereas the values of N_{dA} and N_{dB} quoted above refer to the density of carriers in the band states directly coupled by the radiation.) Due to their ultrashort duration, these pulses are also uniquely suited for investigating on an ultrafast scale the processes that immediately follow the electron-hole plasma creation. In particular, there are two basic experiments that provide rather direct information on the creation of the electron-hole plasma and its subsequent evolution:

(1) The measurement of the transmission of a sample vs the power of the excitation.[2,3]
(2) The measurement of the transmission of a weak pulse (probe) vs its time delay from a strong excitation pulse (excite-probe experiment).[2,3,5,6]

Data from a transmission experiment on a sample ~ 5.0 μm thick at room temperature and at $\sim 105°$K are reported in Fig. 2.[3] We will limit the discussion to the data at $\sim 105°$K, where the trends are more evident, and will distinguish three regions. Region I (Beer's law region) is associated with interband transitions at low power and is characterized by an absorption coefficient $\alpha \simeq 1.2 \cdot 10^4$ cm^{-1}. Region II shows an increase in transmission, from $\sim 10^{-3}$ to $\sim 8 \cdot 10^{-3}$, when the incident energy ranges from $\sim 7 \cdot 10^{12}$ to $\sim 10^{14}$ photons; correspondingly, α decreases from $\sim 1.2 \cdot 10^4$ to $\sim 7.7 \cdot 10^3$ cm^{-1}. Finally, in region III, the increase in transmission slows down, and a "plateau" is reached at $\sim 10^{-2}$, corresponding to $\alpha \simeq 7.2 \cdot 10^{-3}$ cm^{-1}, before the sample is damaged. The results of two excite-probe experiments (with the probe wavelength at 1.06 μm) are shown in Fig. 3.[3,5] The main features here are a sharp spike in the transmission of the probe, close to zero delay (Fig. 3a), followed by a slower rise and an even slower decrease (Figs. 3a and 3b).

Concerning the transmission experiment, the enhancement of the transmission in region II points clearly to a filling of the states available for the transitions. Quantitatively, however,

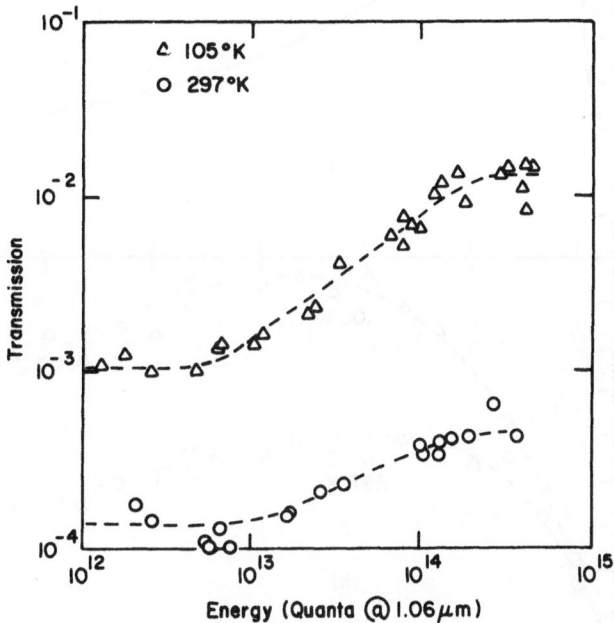

Fig. 2.   Transmission of a single pulse, as a function of its energy, at two different temperatures (from ref. 3).

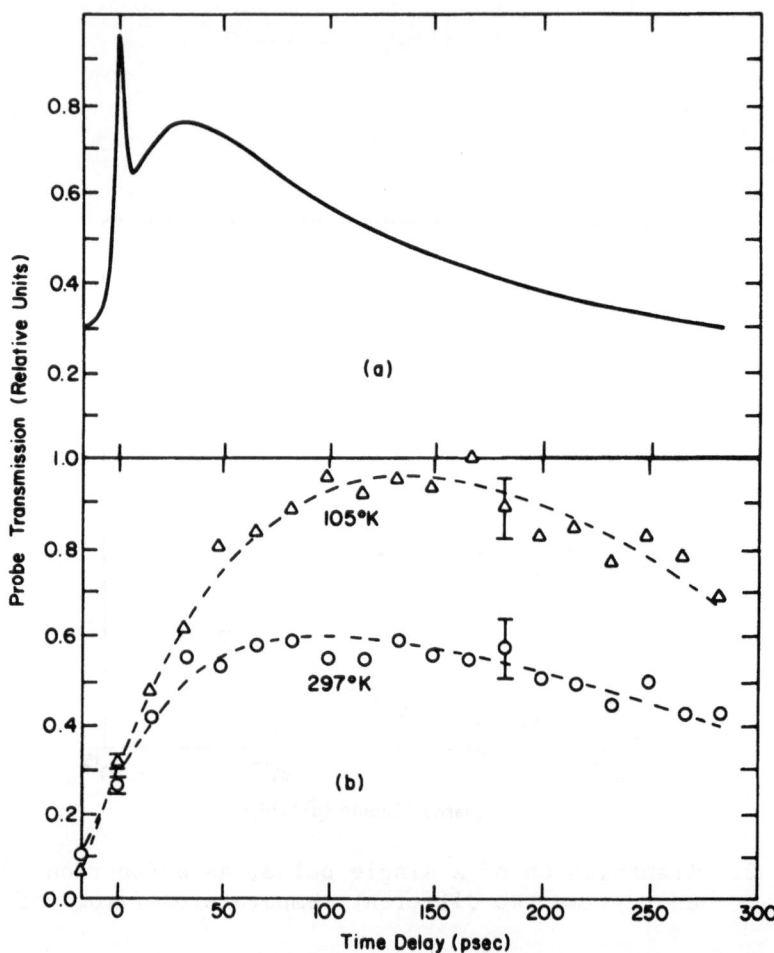

Fig. 3.   (a) Relative transmission of a probe pulse at 1.06 μm
          as a function of its delay from an intense excitation
          pulse (from ref. 5).   (b) Same as (a) with details in
          the spike region at 0 delay suppressed (from ref. 3).

there are some points to be noted:

(1)　The Beer's law region extends at considerably large pulse energies ($\sim 2 \cdot 10^{12}$ photons) where, with an inter- action volume of $\sim 5 \cdot 10^{-7}$ cm^3, one would already expect saturation effects to be present.

(2)　At $\sim 10^{14}$ photons, where $N_{ph} \gg N_d$, the sample would have to be completely transparent, i.e., the transmission, assuming the reflectivity R to be constant all over the excitation range, would have to jump up to $\sim 48\%$, whereas it remains instead ~ 50 times lower (Region III).

It is appropriate, at this point, to distinguish two possible cases of saturation:

(1)　If the excitation rate is large compared with the inverse lifetime of the carriers in the optically coupled states, a saturation results directly from the filling of these states.

(2)　If instead the excitation rate is less than, or compa- rable with, the rate of the processes that empty the coupled states, the saturation is a slower effect, since it will result from the filling of the bands up to the level of the coupled states. (This can be seen as a dynamic analog of the Burstein-Moss shift in degen- erate semiconductors.)

The late onset of the increase in transmission pointed out at point (1) apparently excludes a direct saturation of the coupled states. In the meantime, the leveling of the transmission at a rather low value (Region III) also points to the relevance of processes that oppose the saturation, by removing carriers from the coupled states in a time that is fast, or comparable, to the pulse duration. These processes are likely to have an efficiency that increases with the power of the excitation, so that a dynamic equilibrium is finally established with the process of direct absorption. There is no problem in finding such mechanisms. On the contrary, the real problem is that there are too many candi- dates, and it is difficult to discriminate among them. Since a discussion of the mechanism of carrier removal from the coupled states is useful also for interpreting the electron-hole plasma evolution, and for the discussion of Sections III through V, let us list and consider some of them.

Strictly, only processes with a rate comparable to the direct interband absorption rate ($\approx 10^{14}$ sec^{-1}) affect directly the ele- mentary act of absorption. Yet, slower processes can influence the absorption of the pulse, which is what we measure experi- mentally, provided their inverse rate is comparable to the dura- tion of the pulse. Therefore, processes with rates down to

$\sim 10^{12}$ sec^{-1} may be relevant in discussing the transmission experiment.

(1) <u>Electron-hole spontaneous recombination</u>. This process, which usually occurs between the extremes of the bands, across direct or indirect gaps, is too slow ($\gtrsim 10^9$ sec^{-1}) to require any further consideration in the present context.

(2) <u>Stimulated emission</u>. Recombination of carriers directly between the coupled states might take place under the action of the radiation itself, when a substantial population of carriers is built up. However, other processes of removal [see (3), (4), and (5) below] are expected to compete with this process, so that its relevance is rather doubtful at present. This problem will be further considered in Section V.

(3) <u>Intervalley scattering</u>. Figure 1 shows the band structure of Ge in a limited region near the center Γ of the Brillouin zone. However, over the entire zone, there are four additional minima at L, and six near X in the Γ-X direction. An electron that has just made a transition to the central valley can go to one of these side valleys by emission or absorption of a phonon. According to Elci et al.,[7] this process is extremely fast ($\approx 10^{14}$ sec^{-1}). Therefore, the electrons, as they arrive in the central valley through the direct process of absorption, are instantaneously transferred to the side valleys, which act thus as sinks against the saturation of the coupled states. As a consequence of this process, the actual density of the coupled states is considerably increased, and the late onset of the enhancement of the transmission can be simply explained.[7]

(4) <u>Phonon-assisted relaxation</u>. This is the process in which carriers relax from the coupled states to lower states within the same band. The mechanism requires the intervention of phonons to conserve the momentum. If optical phonons are involved, phonon-assisted relaxation might be fast enough to affect the absorption of the pulse, both through the removal of the carriers and the gradual filling of the bands.

(5) <u>Coulomb thermalization</u>. This is the process by which the carriers, instead of being concentrated in the coupled states, are distributed (possibly according to Fermi law) within the valley. This process, caused by carrier-carrier interaction, is extremely fast and effective ($\gtrsim 10^{14}$ sec^{-1}). However, due to screening effects, its effectiveness might decrease above a certain density of carriers. As a consequence of Coulomb thermalization, one can envisage Fermi-like distributions of carriers obtained simultaneously with the absorption process. Clearly, this mechanism competes effectively not only against the direct saturation of the coupled states, but also against stimulated

emission.

(6) Plasmon-assisted recombination. This is the process in
which an electron in the Γ valley will recombine with a hole by
means of spontaneous emission of a plasmon.[8] This, in turn, increases the temperature of the carrier distribution. This mechanism, whose rate is $\sim 10^{14}$ sec^{-1} would become effective when the
density of carriers is sufficiently high and the plasmon frequency
ω_p becomes comparable to the forbidden gap energy E_g. Actually,
this never happens in the conditions of our experiment. However,
according to Elci et al.,[7] the plasma resonance is sufficiently
broadened to make this mechanism effective.

(7) Auger recombination. This is the process in which an
electron recombines with a hole across the gap, and transfers its
energy to a second electron, whose energy in the conduction band
increases correspondingly. Auger recombination is similar to
plasma assisted recombination but differs in strength and density
dependence. According to Elci et al.,[7] in the range of densities
of interest to us, plasmon-assisted recombination is stronger and
faster than Auger recombination, whose rate is $\lesssim 10^{10}$ sec^{-1}.

(8) Free carrier absorption. This is the process in which
a carrier absorbs a photon and makes a transition to a state
higher in energy within the same valley. In order to conserve
momentum, an additional interaction is required (for example,
emission or absorption of phonons). This process would contribute
to the absorption of the 1.06 μm radiation not only be removing
carriers from the coupled states (as the processes we have discussed so far), but also directly. It also contributes to heating
the carrier distributions. However, according to the estimates
of Elci et al.,[7] the rate of this process (10^{10} to 10^{13} sec^{-1})
remains consistently lower than the direct interband transition
rate, up to carrier densities of 10^{20} cm^{-3}. It must be noted,
however, that the above estimates consider only the interaction
with phonons. At high carrier densities it is expected that
carrier-carrier interaction will act to conserve momentum and
make this mechanism more effective. Therefore, the effectiveness
of this mechanism in shaping the curve of Fig. 2 cannot be ruled
out very easily.

(9) Direct free hole transitions. This process, corresponding to the transition C of Fig. 1, is actually a free carrier
absorption process with a higher rate, due to the direct transition feature. It requires Coulomb thermalization as a prerequisite, and is strongly dependent on the temperature of the hole
distribution. It is not a mechanism of carrier removal from the
coupled states, but provides directly a contribution to the absorption. Experimental evidence shows that, in heavily doped Ge at
$\sim 2 \cdot 10^{20}$ cm^{-3}, this contribution can be as large as 10^4 cm^{-1}.[9]

Values as high are also obtained from preliminary estimates, at distribution temperatures that are easily reached in actual experiments.[10]

(10) <u>Indirect absorption</u>. Like the previous process, this is not a mechanism of removal, but it contributes an additional term to the absorption. It requires an additional interaction to conserve momentum, and if this has to be provided by phonons, the process, as it is well known, is very inefficient. At high densities of carriers, however, carrier-carrier interaction might come in to make its contribution relevant. Experimental evidence in heavily doped Ge indicates that, at $2 \cdot 10^{20}$ cm^{-3}, this process might even overcome the direct absorption.[11]

(11) <u>Excitonic interaction</u>. This process is mentioned just for completeness and for its relevance to the discussion of Part III, though it is very likely that it plays no relevant role in the temperature regime we are considering. This process is the binding of electrons and holes into excitonic pairs, brought about by Coulomb attraction. A multitude of interesting effects, related to interaction among excitons when their density is in an appropriate range, might take place at very low temperatures (excitonic phases). This point will be discussed in Section III.

Let us now come back to the interpretation of the transmission and excite-probe experiments. We note at the outset that the present understanding is rather unsatisfactory, since an unambiguous identification of the various processes has not yet been provided. However, a comprehensive model has been put forward by Elci et al.[7] (ESSM model). Its successes and limitations will serve as a convenient basis for our discussion. The basic interpretation of the ESSM model is as follows.

When the pulse hits the Ge sample, its transmission initially rises due to the partial filling of the available optically coupled states resulting from direct absorption and intervalley scattering. Further increase in the transmission at intensities larger than 10^{14} photons is hindered as the electrons are heated and removed from these states by free carrier absorption and plasmon-assisted recombination. At these excitation intensities, electrons in the Γ valley fall back to the hole pockets in the valence band by means of plasmon emission, at a rate comparable to direct absorption, thus leveling off the transmission. After the passage of the pulse at time $\bar{t}$, the interaction region contains a large number of hot carriers (it is assumed that electron and hole have the same Fermi distribution temperature), and the plasmon recombination is turned off. As time progresses, the distribution temperature T is reduced by phonon-assisted relaxation, and it is essentially the time evolution of this distribution that is monitored by the probe transmission in the excite-probe experiment (with probe at 1.06 µm), since this is a sensitive measurement of whether the coupled states are available for absorption or are occupied. In this perspective, the general interpretation of the excite-probe experiment is rather straightforward, as seen in the pictorial representation of Fig. 4. The transmission of the probe reflects directly the population of the coupled states as the temperature of the distribution cools down. The times involved in both the rise and decay are mainly related to the phonon relaxation process.

The spike at 0 delay (Fig. 3a) is attributed to a process of parameteric scattering, as originally suggested by Shank and Auston,[5] and does not have any deep "solid state" implication.

Alternative explanations have been put forward by the Bell Labs group. In a first tentative interpretation,[5] they attributed the rising part of the excite-probe experiment to band filling and the subsequent decay to diffusion (for which they have strong evidence from ellipsometric measurements),[4] and Auger recombination (used to interprete the excite-probe results with probe at 1.55 µm[6]). More recently, they have suggested that direct free hole transitions, and carrier-carrier assisted indirect transition are effective in the transmission experiment, whereas the excite-probe behavior could be explained by a combination of the above processes and a monotonic decrease in carrier density caused by Auger recombination.[12]

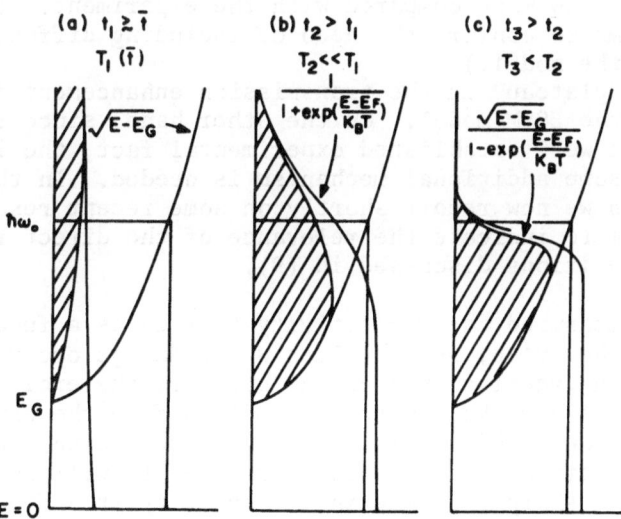

Fig. 4.   Schematic diagram for the temporal evolution of the free carrier distribution created by the excitation pulse (from ref. 7).

Let us now discuss the ESSM model. There are further experimental results that have been satisfactorily explained in its framework. Pressure measurements for samples subject to hydrostatic pressure seem to substantiate the role of plasmon emission recombination, as the ratio between ω_p and E_g is varied.[13] Photoluminescence spectra at high excitation levels also get along well with the carrier distribution assumption of the model, though they probe the system only after the temperature has cooled down.[14] However, there are also some limitations to the ESSM model. First, only a limited number of the possible processes are included. Second, there are some drastic assumptions in it, like the neglect of the spatial variation of the parameters that characterize the electron-hole plasma through the interaction volume.[15] Consequently, the nonhomogeneous distribution of the excitation, and the role of longitudinal diffusion of the carriers in the excite-probe experiment, are also neglected. But diffusion seems to be important in the light of the previously quoted ellipsometric experiment of Auston and Shank.[4] Moreover, the preliminary results of transmission and excite-probe experiments as a function of thickness, also seem to require the inclusion of spatial dependent effects in the ESSM model.[16] Further problems are the following:

(1) A definite prediction of the model, a pulsewidth dependence in the transmission experiment results, has not been found[17] (Fig. 5). One must note, however, that a readjustment of plasmon and phonon coupling constants can void the above prediction.[18]

(2) The decay part of the theoretical excite-probe curve is too slow when compared with the experiment. (This seems to confirm the need of including diffusion effects in the model.)

(3) No "plateau" in the transmission enhancement is produced by the ESSM model. On the other hand, since this plateau is a well established experimental fact, the inclusion of some additional mechanism is needed. In this connection we now report shortly on some recent results that seem to indicate the relevance of the direct free-hole transitions discussed in (9).

We have measured the reflectivity R of Ge as a function of the power of the pulse, up to values where damage occurs. R is found to remain practically constant over all the excitation range, but shows a marked increase ($\sim 30\%$) before the damage threshold is reached.[19] It is tempting to relate this enhancement to the concentration N of free carriers. As is well known, the condition for a sharp increase in reflectivity at a frequency ω is the occurrence of a plasma resonance at that frequency, i.e., $\omega \simeq \omega_p$. If we analyze the problem in the free electron model, however, we find that the occurrence of ω_p at ~ 1.17 eV in Ge

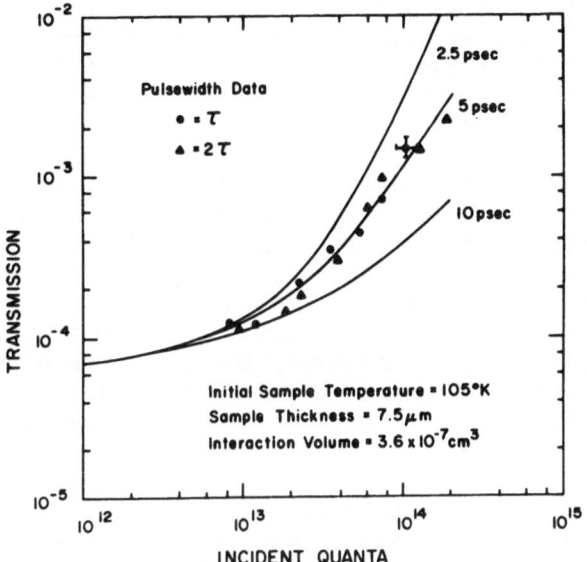

Fig. 5. Pulsewidth experiment data, compared with predictions
from the ESSM model (from ref. 17).

would require a free carrier density $N \simeq 2 \cdot 10^{21}$ cm^{-3}. This is an
order of magnitude higher than the maximum concentration of free
carriers that one can estimate for Ge on the basis of its band
structure. In other words, if we write the real part of the
dielectric constant as

$$\varepsilon(\omega) \; = \; \varepsilon_0 + \varepsilon_f \tag{1}$$

where $\varepsilon_0 = 16$ is the dielectric constant of Ge at low excitation
intensity, and ε_f is the contribution arising from the polarizability of the free carriers, we can obtain $\varepsilon(\omega) = 0$, i.e., $\varepsilon_f =$
-16, only by approaching $N \simeq 2 \cdot 10^{21}$ cm^{-3}. On the other hand, at
$N = 2 \cdot 10^{20}$ cm^{-3}, $\varepsilon_f \simeq -2$.[10] However, as originally suggested by
Blinov et al.,[20] the direct free hole transitions can give a
negative contribution to $\varepsilon(\omega)$. This mechanism is not very effective in Ge, at 1.06 μm, and at low distribution temperatures, but
it is strongly temperature-dependent.[10] It is known that at high
power of excitation, when the number of created electron-hole
pairs saturates, very high distribution temperatures can be
achieved (several thousands °K).[7] An estimate shows that, at
$N \simeq 10^{20}$ cm^{-3}, the contribution ε_h due to direct free hole transitions becomes $\varepsilon_h \simeq -14$ at a distribution temperature slightly
higher than 4000°K. In other words, if we rewrite (1) as

$$\varepsilon(\omega) \; = \; \varepsilon_0 + \varepsilon_f + \varepsilon_h \tag{2}$$

the condition $\varepsilon(\omega) = 0$ is achieved at carrier densities that are easily reached in our experiment. We can therefore conclude that, under the influence of the direct free hole transitions, the plasma resonance condition is satisfied at a frequency close to the frequency of the radiation.

The relevance of the direct free hole transitions in increasing the reflectivity of the sample at excitation densities below the damage threshold is strongly suggestive of their important role also in the "plateau" observed, in the same excitation range, in the transmission experiment. Preliminary estimates show indeed that the absorption coefficient related to these transitions can be as high as 10^4 cm^{-1}. Moreover, the inclusion of the direct free hole transition mechanism in the ESSM model seems to substantially improve the agreement between theory and experimental results.[10]

We conclude this section with a few remarks concerning the relevance of the kind of research discussed above in terms of prospective applications. First, there are already many devices based on heavy doping and hot electron effects. With picosecond techniques, the physics of these devices can be studied in a "cleaner" and somewhat more controllable situation. For example, high density conditions are achieved here with negligible interference from impurity effects, contrary to what happens in heavily doped materials. Second, highly excited materials provide new systems that might be endowed with quite relevant properties. For example, it has been predicted that under certain conditions, a semiconductor in which population inversion is obtained, might become superconducting at relatively high temperatures.[21] Finally, to exploit the enormous potentialities of the picosecond techniques in a full range of technical applications (communications, etc.), one needs to develop ultrafast electronic switches. It is surely not a coincidence that Auston, one of the main contributors to the study of the optical properties of highly excited Ge, has also demonstrated the feasibility of ultrafast switches, directly exploiting the properties of the ultradense electron-hole plasma in semiconductors.[22]

III. PICOSECOND SPECTROSCOPY OF EXCITONIC PHASES IN SEMICONDUCTORS

The electron-hole plasma that we have considered in the case of Ge in the previous section displays a wealth of very interesting effects when very low temperature (4°K and less) are obtained. These effects are brought about by the excitonic interaction, which binds electrons and holes in exciton pairs, and by the exciton-exciton interaction. In particular, whereas the excitonic interaction is somewhat weakened by screening effects under high excitations, the exciton-exciton interaction becomes instead relevant, and leads to the formation of new kinds of transient excitations (excitonic phases). The simplest types of these excitations is the excitonic molecule (biexciton), consisting of two electrons and two holes bound together. Another alternative is the formation of a condensed phase of nonequilibrium carriers, which is expected to be an electron-hole liquid with metallic properties (electron-hole droplets), or a Bose-Einstein condensate of excitons, depending on what interaction prevails (exciton-exciton, or electron-hole, respectively). The literature on this subject is enormous,[23] and considerable controversy exists on the occurrence of the various phases. The most thorough investigation, to date, has possibly been carried out for Ge and Si, where the experimental results point to the formation of a condensed phase of electron-hole drops,[24] at sufficiently low temperatures and high concentrations. All the above phases are expected to give rise to specific recombination radiation, whose study is a main clue to the understanding of the processes involved. In general, the appearance of a new band at energy lower than the direct exciton recombination band, when the excitation increases, is a good evidence of the formation of a more stable phase, even if its detailed attribution is often controversial. Due to the transient nature of these phases, it is evident that time resolved studies of the associated luminescence bands are a powerful tool for studying the dynamics of their onset, growth, and decay. For example, it has been possible to establish that in Si the electron-hole liquid is formed directly from a hot dense plasma of electrons and holes, rather than through nucleation from an exciton gas.[25] Time resolved studies of this kind have been mostly performed, so far, on a nanosecond time scale (resolution >15 nsec). However, as the dynamics of several processes might involve shorter times, the importance of the higher time resolution afforded by picosecond techniques is clearly obvious. This point will be illustrated next by briefly reviewing the pioneering experiments of Kuroda et al.[26] on the time evolution of the exciton absorption and luminescence bands of CdSe, culminating in the possible observation of a Bose-Einstein condensation effect.

The luminescence spectrum of CdSe at high excitation and very low temperatures is characterized by several bands on the low energy side of the absorption peak for the A_1 exciton ($\hbar\omega_{A_1}$ = 1.825 eV). These are the M band, at $\hbar\omega_M$ = 1.819 eV, the P_M band at $\hbar\omega_{PM}$ = 1.816 eV, and the P band, at $\hbar\omega_P$ = 1.812 eV. According to Kuroda and coworkers, the M band is due to the decay of an excitonic molecule (possibly giving rise to a free exciton and a photon), the P_M band to the collision of two excitonic molecules, and the P band to the collision of two single excitons. A time dependence study of the luminescence from the various bands, excited by the second harmonic from a mode-locked Nd:glass laser, gives the following results.

The M band rises first, reaches a maximum after ∿100 psec from the excitation, and decays with a time constant of ∿100 psec. The P_M and P bands have their maxima after ∿200 psec and ∿200 to 350 psec respectively, and decay constants of ∿100 and ∿300 psec respectively. As for the absorption, the position of the A_1 exciton, probed with white picosecond pulses, was found to shift toward higher energy (blue shift) when the sample is simultaneously excited by intense picosecond pulses at 1.06 μm that generate a high concentration of excitons by two-photon absorption. The time dependence of the shift was measured by delaying the white pulses with respect to the exciting pulses at 1.06 μm. For an exciton concentration of ∿$5 \cdot 10^{16}$ cm^{-3} the shift reaches a maximum of ∿4 meV after ∿20 psec, then it decreases and disappears after ∿50 psec. A sketch of the interpretation of the above data is as follows. The blue shift indicates the formation of A_1 excitons from the electron-hole plasma, taking place in ∿20 psec after the interband excitation. This process is followed quickly by the formation of excitonic molecules, in ∿50 to 100 psec (disappearance of the blue shift, and rise of the M band). Molecule-molecule collisions contribute to the P_M band, slightly delayed with respect to the M band, and since every collision gives rise to a photon and three excitons, they cause an increase in the exciton population reflected, at longer delays, by the P band. The formation of excitonic molecules at high concentrations of excitons is seen to "beat" in speed the various dissipation processes. If one creates a sufficiently high concentration of excitonic molecules before the dissipation processes heat the exciton system above the Bose-Einstein transition temperature, one can hope to see their Bose-Einstein condensation. This is obviously impossible with nanosecond pulses, but it might be obtained with picosecond pulses. In fact, using two photon excitation with picosecond pulses at 1.06 μm, at temperatures between 1.8 and 4.2°K, Kuroda et al. found, for a restricted range of excitation intensity, a very sharp peak at 1.8195 eV superimposed on a much broader continuum, which according to them indicates Bose-Einstein condensation. Excitation with nanosecond pulses from a nitrogen laser did not give rise, instead, to any spike, as expected from the above discussion. Though the interpretation proposed by the Japanese group is somewhat controversial, and alternative explanations have been put forward,[27] the above experimental results provide nonetheless a good illustration of the wealth of information that picosecond techniques can provide in the study of the transient excitonic phases in semiconductors.

IV. TWO-PHOTON ABSORPTION OF PICOSECOND PULSES IN SEMICONDUCTORS

In this section the application of picosecond techniques to nonlinear optics will be briefly considered. The field has recently been thoroughly reviewed,[28] and we will simply discuss, as an example, the measurement of two-photon absorption (TPA) in semiconductors, using picosecond pulses. In particular, we will look at the case of GaAs, which is relevant also for the discussion in Section V.

Picosecond techniques are very interesting in nonlinear optics for several reasons. The ultrahigh power available from a mode-locked laser exceeds that obtainable from a Q-switched laser by several orders of magnitude. This allows the study of very weak processes and even the observation of phenomena that could not be detected by other means. Moreover, for the same power, one has much less energy in a picosecond pulse than in a longer one. This helps to discriminate against unwanted effects that are related to the energy (or the number of excited carriers) and/or that have relatively long time constants. Such effects include, for example, heating of the sample and, as it will be seen later, free carrier absorption (FCA) in a TPA experiment.

Let us now go directly to the case of GaAs. The absorption of Nd laser radiation in GaAs has been investigated several times.[29] Since the spectral energy $h\omega$ of the radiation is less than the energy gap E_g, whereas $2h\omega > E_g$, the absorption is unanimously discussed as a TPA process. A considerable disagreement, however, exists among the experimental values reported in the literature for the TPA coefficient β (cm/MW), which describes phenomenologically the effect according to the relation

$$\frac{dI}{dz} = -\alpha I - \beta I^2 \qquad (3)$$

where α is the one-photon absorption coefficient, I is the pulse irradiance (MW/cm^2) and z is the direction of propagation of the beam: values of β ranging from 0.02 cm/MW to 9.0 cm/MW have been reported.[29]

In view of this situation, it was interesting to measure β again, both at room and liquid nitrogen temperature, using the ultrashort pulses (in 8 psec) of our mode-locked Nd:glass laser. With the exception of the works of Bechtel and Smith,[30] and of Grasyuk et al.,[31] TPA measurements have been performed so far with pulses in the nanosecond region. Yet, there are considerable advantages in using picosecond techniques:

(1) The lower the duration τ of the pulses, the less contribution one expects from two-photon excited free carriers. This is on account of two effects:
(a) At the same power, the number of excited free carriers is much less for picosecond then for nanosecond pulses (roughly in the ratio of the

respective durations). Indeed, one can show[30] that
the critical irradiance I_{cr} at which FCA becomes
comparable to TPA is given by

$$I_{cr} = \frac{2\hbar\omega(1-R)}{\sigma\tau} \qquad (4)$$

where τ is the duration of the pulse, R is the
reflectivity, and σ is the FCA cross section.

(b)   As discussed in Section II, the rate constant for
      the FCA process, which requires the intervention
      of phonons to conserve the momentum, ranges pre-
      sumably from 10^{10} to 10^{12} sec^{-1}. Picosecond
      pulses might not have "enough time" to trigger
      this process, so one can hope to have a TPA result
      reasonably unaffected by FCA. In formula (4) this
      effect would imply a decrease of σ as τ decreases,
      and therefore, a further increase of I_{cr} as the
      pulses become shorter and shorter.

(2)   The higher power that can be reached allows the study of
      the curve transition vs irradiance over a more extended
      region, and consequently a more precise and stringent
      fit of the experimental results to the theoretical
      expression.

(3)   Due to multiple reflections within the sample, the total
      irradiance "seen" by the beam is actually increased.
      (At any point within the sample, one has to consider
      the overlapping of the electric fields arising from the
      reflected beams.[32]) This effect might lead to an over-
      estimation of β. Since it depends roughly on τ/D, where
      D is the thickness of the sample, the use of picosecond
      pulses allows it to be minimized.

Our results, at room temperature, are displayed in Fig. 6.[33]
(The results at liquid nitrogen temperature are not significantly
different.) The transmission is seen to decrease from a value of
~ 0.50 at low irradiance (corresponding to pure reflectivity
losses) down to $\sim 10\%$ at ~ 30 GW/cm^2, when damage of the sample
occurs. The curve has been fit with a formula[34] that presumes
that the pulses have Gaussian shape in space and time and keeps
into account multiple reflections for what concerns the contribution to the transmission, but not for the effect discussed at
point 3. The best fit gives $\beta \simeq 0.015$ cm/MW in good agreement
with Bechtel and Smith.[30] This value falls at the bottom of the
spectrum of values reported so far, as expected from the above
discussion, and seems to show that the high values of β reported
in the literature arise, at least in part, from free carrier
absorption effects. Evidence for the important role of FCA processes in multiple photon experiments has also been provided by

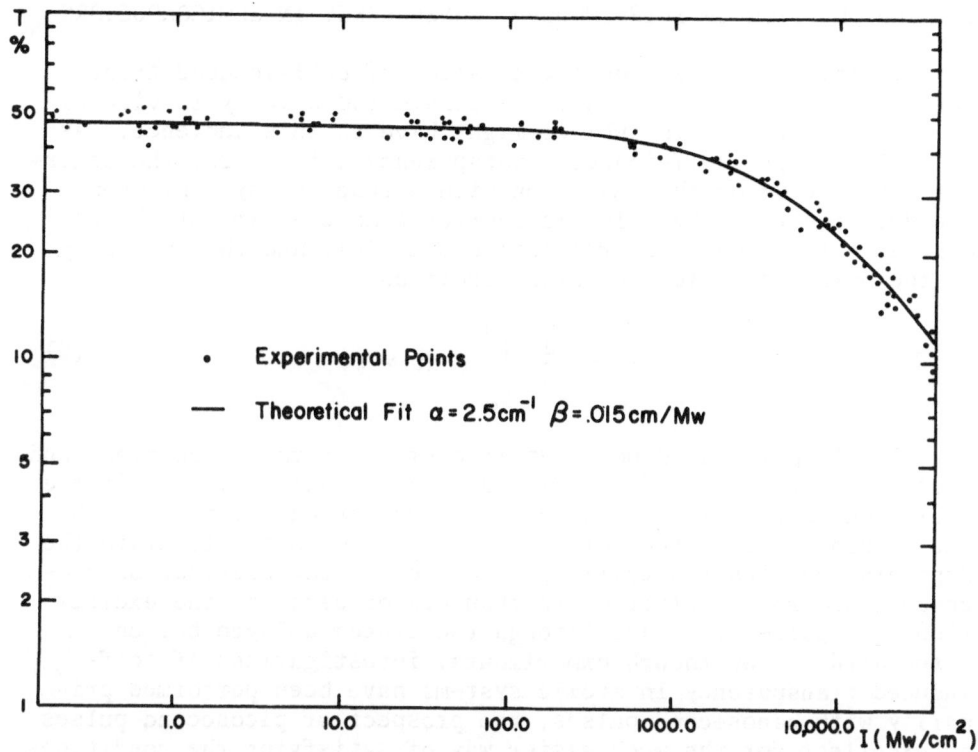

Fig. 6.   Transmission of a single pulse, as a function of its intensity, in a sample of GaAs (from ref. 33).

Penzkofer et al.,[35] in the case of CdS. For GaAs, Bechtel and
Smith[30] observe a possible onset of FCA at irradiances $\gtrsim$300 MW/cm.
In our case, no evidence for this process is found: the transmission curve is well fit by the theoretical expression over all
the irradiance range, up to the damage threshold. This difference
can be explained with the shorter duration of our pulses, according to the discussion given above (points 1a and 1b). A confirmation for this seems to come from some preliminary results,[36]
which show that, in a two-pulse experiment (excite-probe type,
but with both pulses having equal intensity), the ratio between
the transmission of the second and that of the first pulse is
equal to $\sim$1 at 0 delay and keeps decreasing as the delay is increased. Finally, we conclude this section by noting that our
transmission results are in disagreement with those of Grasyuk
et al.[37] They find a sharp difference between the transmission
at room temperature and that at liquid nitrogen temperature and
interpret this fact as evidence for coherent two-photon interaction. This point will be discussed further in the next section.

V.   THE QUESTION OF SELF-INDUCED TRANSPARENCY IN SEMICONDUCTORS

In this final section the question of self-induced transparency in semiconductors will be considered briefly in view of
the interest that this subject might have to this audience. As
is well known,[38] self-induced transparency arises from the coherent interaction of the radiation with a resonant system, when
the duration τ of the pulse is shorter than the time of phase
coherence (or transverse relaxation time T_2), and the intensity
of the pulse satisfies the area condition

$$\theta = \frac{\mu}{\hbar} \int_{-\infty}^{\infty} E\, dt > \pi. \tag{5}$$

In (1) μ is the dipole matrix element of the transition, and
E is the electric field of the radiation. Qualitatively, if the
radiation is intense enough to cause population inversion with
the leading edge of the pulse, and short enough to stimulate the
depopulation with the trailing edge, before the dissipation processes can provide alternative channels of decay to the excitation, the pulse will pass through the system delayed but un-
attenuated. Even though experimental investigations of self-
induced transparency in atomic systems have been performed pri-
marily with nanosecond pulses, the prospect of picosecond pulses
is excellent for the much easier way of satisfying the conditions
$\tau < T_2$.

Self-induced transparency for interband excitation in a semiconductor was first considered by Poluektov and Popov,[39] who
showed that the effect can indeed take place, provided that $\theta > \pi$
and $\tau << T_2$. The last condition put some limitations on the
wavelength of the radiation: one must have

$$h\nu < E_g + \hbar\omega_0 \tag{6}$$

where $\hbar\omega_0$ is the energy of optical phonons, which would cause a
strong reduction of T_2 due to the high rate of the optical phonon
relaxation processes. The above theory was subsequently generalized to cover the case of two-photon interband excitation.[40] In
spite of these works, however, the question of the possible
existence of coherent effects in interband excitations is still
controversial. If one limits his considerations to electron-
phonon interaction as the only mechanism that competes with
stimulated emission to remove carriers from the states coupled
by the radiation, then there is indeed some hope to "beat" in
speed this interaction with pulses that are fast enough. However, as we have seen in Section II, carrier-carrier interaction
is effectively at work in solid state systems. In particular,
Coulomb thermalization does not seem to allow much hope for the

existence of any self-induced transparency effect. These considerations perhaps might not apply to the case of self-induced transparency in an exciton system, for which there is but one crystal state having a given wave vector in a finite energy interval.[41] Self-induced transparency of excitons has also been proposed[42,43] and has given rise to a lively controversy.[44]

To date experimental results that have been interpreted as evidence for self-induced transparency effects in semiconductors have been reported for CdS_xSe_{1-x}[45] and for GaAs.[31,37] In the case of CdS_xSe_{1-x} the crystal composition was "tuned" to bring the second harmonic from a mode-locked Nd:glass laser in resonance either with interband transitions satisfying condition (6), or with the exciton transition. The measurements, performed at relatively high temperatures (130°K) show effects (enhanced transmission, pulse delay) that the authors interpret as evidence of coherent interaction effects. As noted in ref. 28, however, alternative explanations are possible.

In the case of GaAs, in which the Nd:laser radiation causes two-photon excitation, the self-induced transparency explanation is even less convincing, since condition (6) is apparently not satisfied. Moreover, the experimental results of refs. 31 and 37 are not in complete agreement with the data we have discussed in Section IV. For example, we do not find any substantial difference between the transmissions at liquid nitrogen temperature and at room temperature.[37] In addition, the transmission dependence on the pulse duration[31] possibly can be ascribed to free carrier absorption for the long pulses, as discussed in Section IV. Finally, the transmission over the range up to $\sim$20 MW/cm^2 reported in ref. 31 can be very simply explained in terms of interband two photon absorption, with $\beta = 0.015$ cm/MW, without any need of invoking self-induced transparency effects.

ACKNOWLEDGMENTS

The author is very grateful to Prof. M. O. Scully for the opportunity to work on picosecond spectroscopy in his group. Thanks are also given to M. O. Scully, C. Y. Leung, and J. S. Bessey for useful and extensive discussions. This research was supported in part by the Air Force Office of Scientific Research (AFSC), the United States Air Force, and the Army Research Office, United States Army.

REFERENCES

1. B. Bosacchi, this volume, p.
2. C. J. Kennedy, J. C. Matter, A. L. Smirl, H. Weichel, F. A. Hopf, S. V. Pappu, and M. O. Scully, Phys. Rev. Lett. 32, 419 (1974).

3. A. L. Smirl, J. C. Matter, A. Elci, and M. O. Scully, Opt. Commun. 16, 118 (1976).
4. D. H. Auston and C. V. Shank, Phys. Rev. Lett. 32, 1120 (1974).
5. C. V. Shank and D. H. Auston, Phys. Rev. Lett. 34, 479 (1975).
6. D. H. Auston, C. V. Shank, and P. LeFur, Phys. Rev. Lett. 35, 1022 (1975).
7. A. Elci, M. O. Scully, A. L. Smirl, and J. C. Matter, Phys. Rev. B 16, 191 (1977).
8. P. A. Wolff, Phys. Rev. Lett. 24, 266 (1970).
9. R. Newman and W. W. Tyler, Phys. Rev. 105, 885 (1957).
10. B. Bosacchi, C. Y. Leung, and M. O. Scully, to be published.
11. C. Haas, Phys. Rev. 125, 1965 (1962).
12. Private communication by S. McAfee and D. H. Auston to M. O. Scully.
13. H. M. van Driel, J. S. Bessey, and R. C. Hanson, Opt. Commun. 22, 346 (1977).
14. H. M. van Driel, A. Elci, J. S. Bessey, and M. O. Scully, Solid State Commun. 20, 837 (1976).
15. This point has been discussed in a paper by A. Elci, C. Y. Leung, M. O. Scully, and A. L. Smirl, presented at the International Conference on Hot Electrons in Semiconductors (Denton, Texas, July 1977), to be published.
16. J. S. Bessey, B. Bosacchi, K. Al-Katheeb, M. O. Scully, and F. C. Jain, presented at the International Conference on Hot Electrons in Semiconductors (Denton, Texas, July 1977), to be published.
17. J. S. Bessey, B. Bosacchi, H. M. van Driel, and A. L. Smirl, submitted for publication to Phys. Rev. B.
18. A. L. Smirl, P. Latham, A. Elci, and J. S. Bessey, presented at the International Conference on Hot Electrons in Semi-conductors (Denton, Texas, July 1977), to be published.
19. B. Bosacchi, C. Y. Leung, and M. O. Scully, submitted for publication to Opt. Commun.; and to be published.
20. L. M. Blinov, V. S. Vavilov, and G. N. Galkin, Sov. Phys. Solid State 9, 666 (1967).
21. D. A. Kirzhnits and Yu. V. Kopaev, JETP Lett. 17, 270 (1977).
22. D. H. Auston, Appl. Phys. Lett. 26, 101 (1975).
23. See, for example, *Excitons at High Density*, ed. by H. Haken and S. Nikitine (Springer-Verlag, 1975).
24. See, for example, Y. Pokrovskii, Phys. Stat. Sol. (a) 11, 85 (1972), and W. F. Brinkmann and T. M. Rice, Phys. Rev. B 7, 1508 (1973).
25. J. Shah and A. H. Dayem, Phys. Rev. Lett. 37, 861 (1976).
26. H. Kuroda and S. Shionoya, J. Phys. Soc. Japan 31, 476 (1974) and H. Kuroda, S. Shionoya, H. Saito, and E. Hanamura, J. Phys. Soc. Japan 35, 534 (1973).
27. D. von der Linde, in *Ultrashort Light Pulses*, ed. by S. L. Shapiro (Springer Verlag, 1977), p. 204.
28. D. H. Auston, in *Ultrashort Light Pulses*, ed. by S. L. Shapiro (Springer Verlag, 1977), p. 123.

29.   See refs. 30 and 33 for a review.
30.   J. H. Bechtel and W. L. Smith, Phys. Rev. B 13, 3515 (1976).
31.   A. Z. Grasyuk, I. G. Zubarev, V. V. Lobko, Yu. A. Matveets, A. B. Mirnov, and O. B. Shatberashvili, JETP Lett. 17, 416 (1973).
32.   R. A. Baltrameyunas, Yu. V. Vaitkus, Yu. K. Vishchaskas, and V. I. Gavriushin, Opt. Spectr. 36, 714 (1974).
33.   B. Bosacchi, J. S. Bessey, M. O. Scully and F. C. Jain, Bull. Am. Phys. Soc. (1977), and submitted for publication to Appl. Phys. Lett.
34.   B. Bosacchi, to be published.
35.   A. Penzkofer and W. Falkenstein, Opt. Commun. 16, 247 (1976).
36.   B. Bosacchi, C. Y. Leung, and M. O. Scully, to be published.
37.   T. L. Gvardzhaladze, A. Z. Grasyuk, and V. A. Kovalenko, Sov. Phys. JETP 37, 227 (1973).
38.   See for example, I. M. Slusher in *Progress in Optics*, Vol. 12, ed. by E. Wolf (North-Holland, 1974), p. 55.
39.   I. A. Poluektov and Yu. M. Popov, JETP Lett. 9, 330 (1969).
40.   I. A. Poluektov, Yu. M. Popov, and V. S. Roitberg, Sov. J. Quantum Electron. 2, 385 (1973).
41.   J. J. Hopfield, Phys. Rev. 112, 1555 (1958).
42.   A. Schenzle and H. Haken, Opt. Commun. 6, 96 (1972).
43.   V. V. Samartsev, A. I. Siraziev, and Yu. E. Sheibut, Proc. Acad. Sci. USSR 37, 140 (1973).
44.   E. Hanamura, J. Phys. Soc. Japan 37, 1553 (1974).
45.   F. Brückner, Ya. T. Vasilev, V. S. Dneprovskii, D. G. Koshchug, E. K. Silina, and V. V. Khattatov, Sov. Phys. JETP 40, 1101 (1975).

COHERENT EFFECTS IN PICOSECOND SPECTROSCOPY

A. Laubereau and W. Kaiser

Physik Department der Technischen Universität

München, Germany

The advance in laser technology and the progress in our understanding of nonlinear optical phenomena have opened up the possibility of generating intense picosecond light pulses in the spectral region extending from the near uv to the infrared /1,2/. Investigators in the field were quickly fascinated by the exceptional time resolution which can be achieved by these pulses allowing direct studies of ultrafast molecular processes in condensed phases. In fact, numerous investigations were conducted by several laboratories on a wide variety of problems during the past years /3/.

On the other hand, there are only a few papers concerned with the coherent interaction between light and matter on the picosecond time scale. This results from the stringent requirements on the pulse properties in experimental investigation of coherent effects where the dephasing times are 10^{-12} sec or even shorter. In fact, bandwidth limited pulses of several psec duration are necessary. In many experiments pulses close to the resonant frequency have to be applied, i.e. tunability of the pulse frequency is required. For gases at low pressure the situation is quite different. It is most fortunate that in these systems the relevant time constants are in general longer than 10^{-8} sec allowing the application of readily available laser pulses.

In this article we are interested in vibrational modes in the electronic ground state. Several experimental

techniques were developed to investigate the coherent
properties and measure interesting material parameters.
In our investigations, the molecular vibrations are excited by an intense laser pulse via stimulated Raman
scattering or by a resonant infrared absorption. After
the passage of the first pulse the excitation process
rapidly terminates and free precession decay of the
vibrational system occurs. The instantaneous state of the
excited system is monitored by a second interrogating
pulse of variable time delay.

The following experiments will be discussed: a) Coherent Raman probe scattering of a homogeneously broadened vibrational transition determining the dephasing
time T_2 from the observed signal decay; b) Coherent probe
scattering from a Raman line with discrete substructure
due to the isotope effect; a collective beating effect
on the picosecond time scale is observed under suitable
k-matching conditions; c) Coherent probe scattering of
an inhomogeneously broadened vibrational transition; we
have developed a selective k-matching geometry which
allows us to study the dephasing time of a molecular sub-
ensemble in a small frequency interval. In this way, we
are able,for the first time, to measure a homogeneous dephasing time of an inhomogeneous vibrational band in the
liquid state; (d) the propagation of a small area pulse
of several psec tuned to the resonance frequency of an
infrared active vibrational mode in liquid solution is
studied.

It is convenient to treat the vibrational system as
a two-level model; transitions involving higher excited
states are negligible on account of the anharmonic frequency shift. In this picture the coherent interaction
of a vibrational mode with a nonresonant light pulse via
stimulated Raman scattering (and also the interaction
with a resonant light pulse via electric dipole coupling)
is quite analoguous to the well-known case of magnetic
dipole transitions of a spin system. For the subsequent
discussion it is advantageous to consider the more general
case of an ensemble of two-level systems with an inhomogeneous distribution of vibrational transition frequencies.
Molecules in a small frequency interval of this distribution are grouped together to vibrational components j
with number density Nf_j, where N denotes the total number
density; $(\Sigma f_j = 1)$. Transient stimulated Raman scattering is considered first as an excitation process of the
vibrational system. This scattering mechanism generates

a coherent excitation represented by the amplitude <q>, the expectation value of the normal mode operators of the individual vibrational components j. Introducing the absolute values Q_j and phase factors ϕ_j of the amplitudes <q_j> and also the field amplitudes E_L and E_S of the laser and Stokes scattered light, the stimulated process is described by the following set of equations /4,5/:

$$(\frac{\partial}{\partial t} + \frac{1}{v_S} \frac{\partial}{\partial t}) \ E_S = \kappa_1 \ E_L \ \sum_j \ f_j \ Q_j \ \cos \ (\Delta\omega_j t + \phi_j) \quad (1)$$

$$(\frac{\partial}{\partial t} + \frac{1}{T_2}) \ Q_j = \kappa_2 \ E_L \ E_S \ \cos \ (\Delta\omega_j t + \phi_j) \quad (2)$$

$$\frac{\partial}{\partial t} \ (\Delta\omega_j t + \phi_j) + \frac{\kappa_2 E_L E_S}{Q_j} \ \sin \ (\Delta\omega_j t + \phi_j) = \Delta\omega_j \quad (3)$$

where

$$\kappa_1 = \frac{\pi\omega_S^2 N}{c^2 k_S} \frac{\partial\alpha}{\partial q} \ ; \qquad \kappa_2 = \frac{1}{4m(\omega_L - \omega_S)} \frac{\partial\alpha}{\partial q} \quad (4)$$

and

$$\Delta\omega_j = \omega_L - \omega_S - \omega_j \quad (5)$$

The coupling coefficients κ_1 and κ_2 collect various material parameters (Eq. 4). We assume for all components j equal values of the reduced molecular mass m, the Raman polarizibility $\partial\alpha/\partial q$ and the dephasing time T_2. v_S is the group velocity of the Stokes pulse.

We have made a detailed study of Eqs. 1 to 4 using material parameters relevant for the following experiments. Some of the findings are briefly summarized:

The excitation and subsequent free precession decay is illustrated by numerical data depicted in Figs. 1 to 3 for three different physical situations. Fig. 1 considers the simple case of a homogeneous line. The exponential decay of Q^2 with time constant $T_2/2$ should be noted. In Fig. 2, a vibrational system with isotopic substructure is considered. The total excitation $|Q_{tot}|^2 = |\sum f_j <q_j>|^2$ is plotted. The free precession decay of the substructured system shows a beating effect which is a direct consequence of the phase relation between the individual components established in the excitation process. The beating time reflects the constant frequency spacing

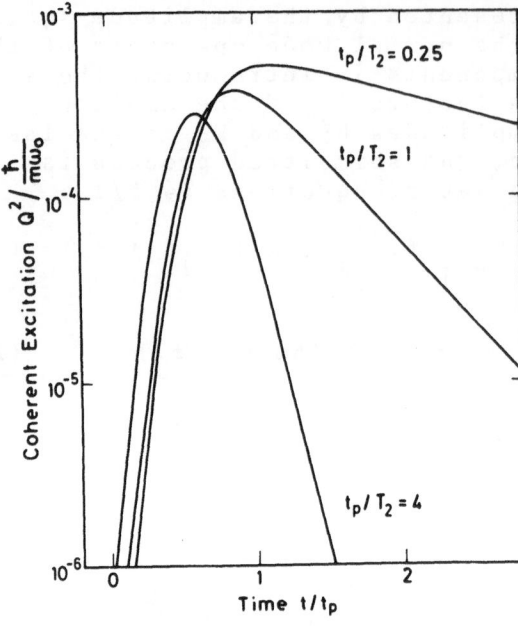

Fig. 1 Coherent vibrational excitation in units of $\hbar/m\omega_0$ versus time for three values of the parameter t_p/T_2 (duration of pump pulse t_p to dephasing time T_2). Note the exponential decay of the freely relaxing system with time constant $T_2/2$ (Stokes conversion of the excitation process 1%, interaction length 1 cm).

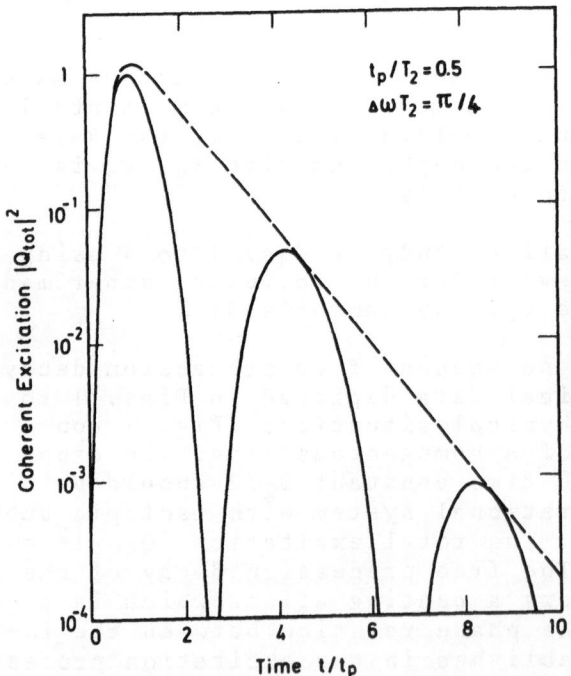

Fig. 2 Calculated coherent vibrational excitation of three molecular components versus time t/t_p (equal frequency spacing $\Delta\omega$; relative abundance 1:0.5:0.5). The solid curve represents the beating due to the superposition of the vibrational excitations. The broken line indicates the vibrational excitation of one molecular component.

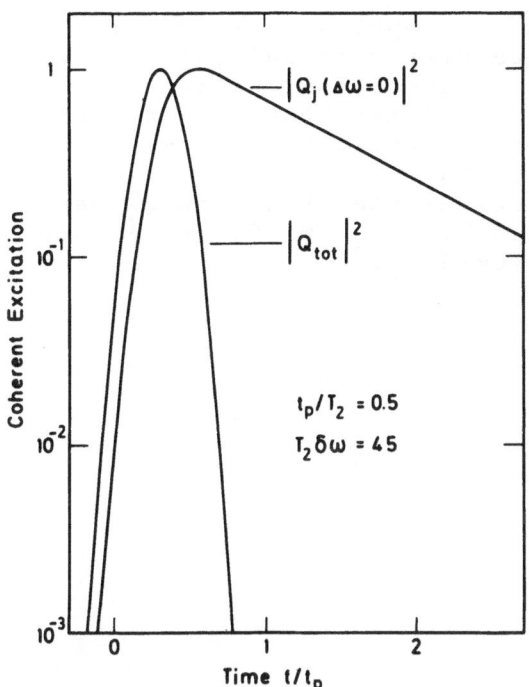

Fig. 3 Calculated coherent excitation of a vibrational system with a Gaussian distribution of transition frequencies (bandwidth $\delta\omega$). The solid curve represents the total vibrational decay due to destructive interference of the various components. The broken line shows the time behavior of a molecular sub-ensemble with negligible spread of transition frequencies; the exponential decay with time constant $T_2/2$ should be noted.

$\Delta\omega$ of the vibrational components. Quite different is the result shown in Fig. 3 for a quasi-continuous distribution of transition frequencies of width $\delta\omega_{inh}T_2 = 45$. The total coherent excitation $|Q_{tot}|^2$ disappears very rapidly on account of the broad frequency distribution while the individual components decay more slowly with time constant $T_2/2$.

Our calculations indicate that the stimulated Stokes field is highly coherent; i.e., a Stokes field of constant phase is generated. Stokes components with frequencies differing from ω_S disappear during the high gain excitation process. As a result, E_S in Eqs. 1 to 3 is taken to be real. The high degree of monochromaticity of the Stokes (and also of the laser) field is important for the generation of well defined wave vectors of the vibrational excitation. These results are substantiated by experimental findings. In Fig. 4, we present three spectra taken with an optical multichannel analyser and a spectrometer with a resolution of $\Delta\nu = 0.5$ cm^{-1}. The spectral width of the laser pulse entering the Raman sample (see Fig. 2a) is $\Delta\nu_L \simeq 5$ cm^{-1}, the Fourier transform of the laser pulse of $t_p \simeq 3.5$ psec. Figs. 4 b and c show stimulated Stokes spectra generated in methanol and ethylene glycol, re-

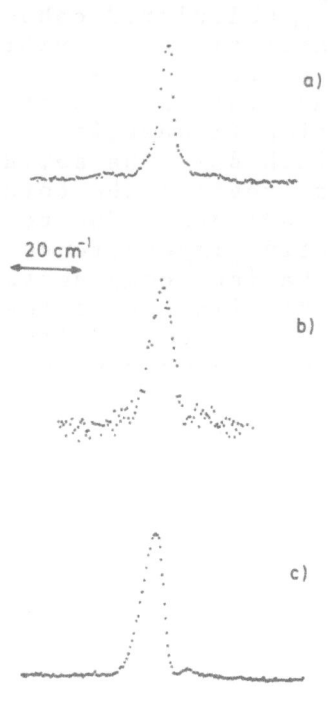

Fig. 4   Optical spectra of
three different pulses:
a) incident laser pulse,
b) stimulated Stokes pulse gene-
rated by the $\tilde{\nu}$ = 2835 cm^{-1} mode
in CH$_3$OH,
c) stimulated Stokes pulse gene-
rated by the $\tilde{\nu}$ = 2935 cm^{-1} mode
in (CH$_2$OH)$_2$.

spectively. The Stokes emission of the two samples has
the same bandwidth of $\Delta\nu_S \approx$ 8 cm^{-1}, in spite of the substantial difference in spontaneous Raman bandwidth of
$\Delta\nu_{spon}$ = 20 cm^{-1} and 60 cm^{-1} of the excited modes in
CH$_3$OH and (CH$_2$OH)$_2$, respectively. The value of 8 cm^{-1} is
determined by the duration of the Stokes pulse; we recall that the Raman Stokes pulses are somewhat shorter
than the laser pulse due to the highly nonlinear generation process.

     The coherent vibrational excitation is monitored
by coherent Raman scattering of a delayed probe pulse.
The coherent amplitudes <q> give rise to a macroscopic
polarization $P = N\frac{\partial\alpha}{\partial q} E \Sigma_j f_j <q_j>$, where E denotes the electromagnetic field of the probe pulse /5,6/. The induced po-
larization generates the scattering emission, which is
shifted by the vibrational frequency to larger (anti-
Stokes) or smaller (Stokes) frequencies. The molecules
vibrate with a defined spatial phase relation described
by a wave vector $\vec{k}_o = \vec{k}_L - \vec{k}_S$, prepared in the excitation
process by the laser ($\vec{k}_L$) and stimulated Stokes pulse
($\vec{k}_S$). The coherently excited subensembles j behave like
oscillating three-dimensional phase gratings. Scattering-
off these gratings occurs when the k-matching condition

for the relevant wave vectors of the incident probe pulse (k_{L2}), the scattered light and the vibrational components is fulfilled /6/. For Stokes scattering we have the condition:

$$|\Delta k_j| \Delta \ell = |k_{L2} - k_{Sj} - k_o| \Delta \ell \leq 0.5 \qquad (5)$$

Correspondingly, efficient anti-Stokes scattering occurs if

$$|\Delta k_j| \Delta \ell = |k_{Aj} - k_{L2} - k_j| \Delta \ell \leq 0.5 \qquad (6)$$

Here, k_{Sj} and k_{Aj}, respectively, denote the Stokes and anti-Stokes scattering components of the subensembles j with $k_{AS,j} = (\omega_{L2} \pm \omega_j)n/c$. $\Delta \ell$ is the interaction length of the probing process. Vibrational components with mismatch $\Delta k_j \Delta \ell$, considerably larger than indicated by Eqs. 5 or 6, give only negligible contribution to the probe scattering. For large values of $\Delta \ell$ (≥ 1 cm) a highly selective k-vector geometry may be adjusted in order to observe probe scattering of a molecular subensemble with negligible spread of transition frequencies. For a small interaction length ($\Delta \ell \ll 1$ cm), on the other hand, the k-matching is not sensitive to the distribution of transition frequencies and coherent superposition of neighboring frequency components with interesting beating effects may be observed; i.e., the total coherent excitation, discussed above in context with Figs. 1 to 3, may be investigated. Different values of $\Delta \ell$ may be adjusted varying the experimental geometry (sample length etc.).

Our experimental system for coherent Stokes probe scattering with collinear wave vector geometry is depicted schematically in Fig. 5 /4/. For the generation of the ultrashort light pulses we use a Nd:glass laser system consisting of a mode-locked laser oscillator, an electro-optic switch to select a single pulse, and a laser amplifier /7/. We emphasize the need to work with single bandwidth limited pulses of well defined duration, pulse shape, frequency width, and peak intensity. Bandwidth limited pulses /8/ are required in our investigations to achieve high k-vector resolution. After frequency doubling in a KDP crystal we work with pulses of frequency $\nu_L = 18,910$ cm^{-1}, width $\delta\nu_L \simeq 5$ cm^{-1}, pulse duration $t_p \simeq 3.5$ psec and approximately Gaussian shape. The probing pulse is generated by a first beam splitter and is properly delayed in a variable delay system. A second beam splitter allows the probe pulse to travel collinearly

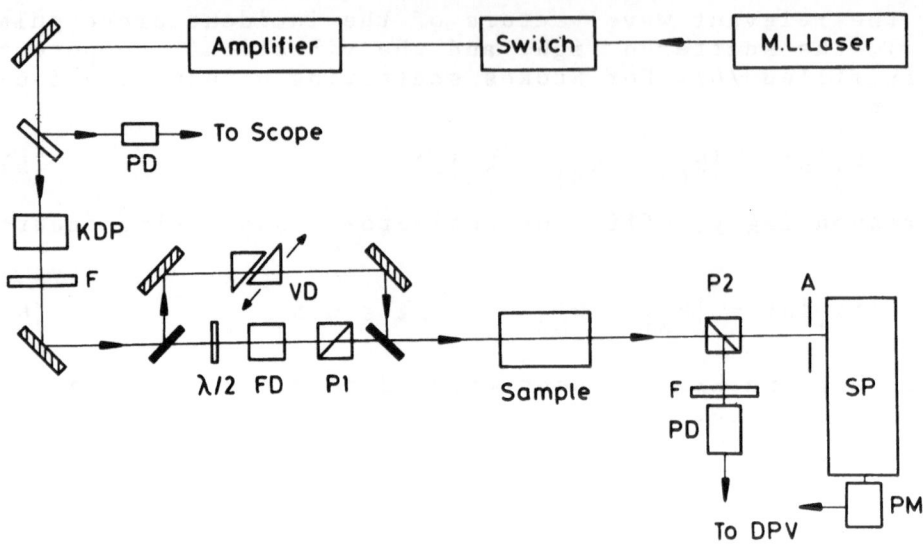

Fig. 5   Experimental set-up to measure coherent probe
scattering in a collinear geometry. A beam splitter generates the probe pulse which is properly delayed before
travelling collinearly with the exciting pulse. The
scattered signals of the two pulses are separated by two
polarizers P1 and P2.

with the exciting pulse through the medium. The intensity
of the probe pulse is less than 1% of the excitation
pulse. Coherent probe scattering in the forward direction
is measured. The pump pulse passed a $\lambda/2$ plate and polarizer P1 defining the plane of polarization of the excitation process. The stimulated Stokes emission of the
pump pulse is effectively blocked by a factor of $\geqq 10^5$
with the help of the second polarizer P2. The probe pulse
and the corresponding Stokes scattering signal, on the
other hand, pass P2 without significant attenuation. An
aperture, A, determines the angle of acceptance γ of the
detection system consisting of a spectrometer and a photomultiplier.

   a) As a first example of free precession decay in a
liquid we discuss the symmetrical CH_3-stretching mode of
CH_3CCl_3. Fig. 6 presents our results on the dephasing
time of the molecular vibration. The coherent probe scattering signal is plotted as a function of delay time between excitation and probe pulse. From the exponential
decay of the signal curve, we deduce a dephasing time of

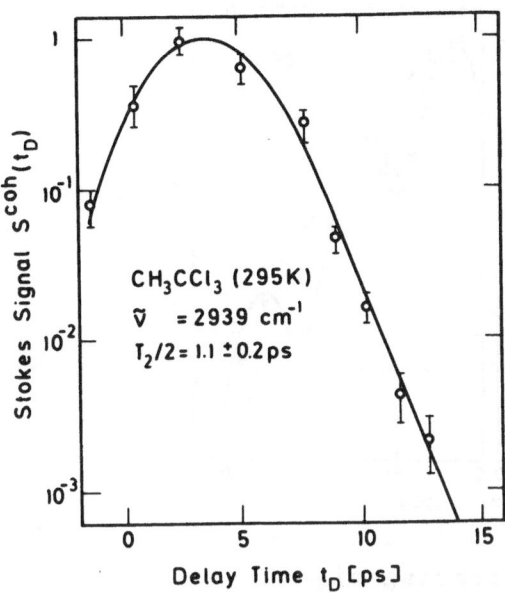

Fig. 6 Coherent Raman probe scattering signal $S^{coh}(t_D)$ versus delay time between pump and probe pulse for the symmetric CH_3-strechting mode of CH_3CCl_3.

$T_2/2 = 1.1$ psec, which corresponds to a homogeneous line broadening of ~ 5 cm^{-1}. This number is in excellent agreement with the linewidth of the isotropic scattering component observed in spontaneous Raman spectroscopy; i.e., the dynamic processes described by the dephasing time T_2 fully account for the spectroscopic line broadening. It is interesting to note that for the population lifetime T_1 of the first excited level of the same vibrational mode a value of $T_1 = 5.2$ psec was found. The difference between the time constants $T_2/2$ and T_1 indicates that the coherent excitation of the CH_3CCl_3 molecules decays under (partial) conservation of the excited state population. Theoretical estimates of T_2 from a quasi-elastic collision model satisfactorily agree with the experimental results within a factor of approximately two /9/.

b) For a homogeneously broadened mode, e.g. CH_3CCl_3 discussed above, the time constant T_2 is directly deduced from the free precession decay of the probe scattering signal. The situation is more complex for molecular vibrations with isotopic substructure. For these cases the observed time dependence of the coherent probe scattering depends on the k-matching situation. CCl_4 is discussed here as an example. The two isotopes of ^{35}Cl and ^{37}Cl give rise to vibrational multiplicity of the totally

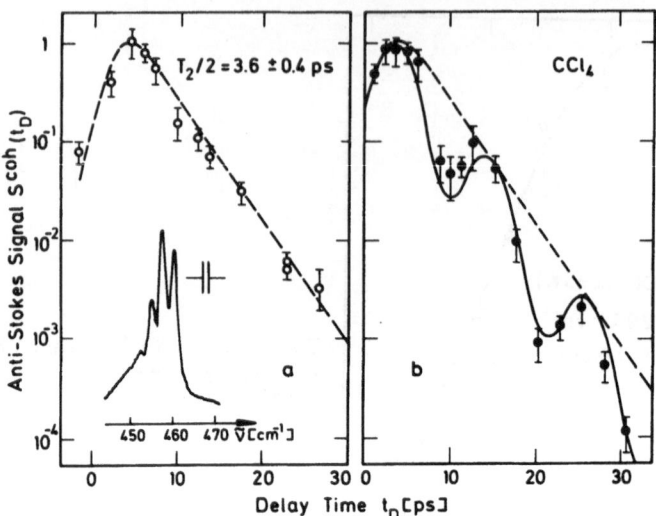

Fig. 7   Coherent probe scattering signal versus delay
time for CCl_4 with natural isotope abundance; a) selective
k-matching observing free precession decay of a single
isotope component; b) non-selective k-matching with coherent superposition of vibrational states; curves are
calculated. Inset: spontaneous Raman band.

symmetric tetrahedron vibration around 460 cm^{-1}. The isotopic structure is clearly seen in the spontaneous Raman
spectrum with a frequency spacing of approximately 3 cm^{-1}
of the various lines (see inset of Fig. 7a). The results
for coherent probe scattering with a selective k-vector
geometry are depicted in Fig. 7a /6/. The decaying part
of the signal curve represents free precession decay of
a single isotope component with time constant $T_2/2$ = 3.6
psec. The measured dephasing time fully accounts for the
Raman linewidth of one isotope species of 1.4 cm^{-1}.
Additional information is obtained by coherent probing
with non-selective k-matching. In these experiments, a
coherent superposition of the different isotopic species
is observed (Fig. 7b) /6/. The various isotope components
are first excited with approximately equal phases (rising
part of the signal curves). The excitation process then
terminates and free relaxation of the coherent excitation
is observed. The frequency differences of the individual
species lead to a striking interference phenomenon with
a beating period of ∿ 12 psec.

The beating phenomenon originates from the coherent superposition of the vibrational amplitudes $<q>_j$ of the excited quantum states. There are differences to the quantum beats previously observed in gases /10/: a) The different excited states belong to various molecular molecular species;   b) the vibrational states have no optical allowed transitions, i.e. there is no emission to be observed;   c) The beat frequency is very high, of the order of 10^{11} sec^{-1}, and the dephasing times are very short of the order of several 10^{-12} sec. Ultrashort pump and probe pulses are required to study this coherent phenomenon in the condensed phase.

c) There are vibrational modes where experimental evidence exists of a distribution of transition frequencies. For instance, liquids with strong hydrogen bonding show extended inhomogeneously broaded OH-bands. Spontaneous spectroscopic techniques do not allow a ready

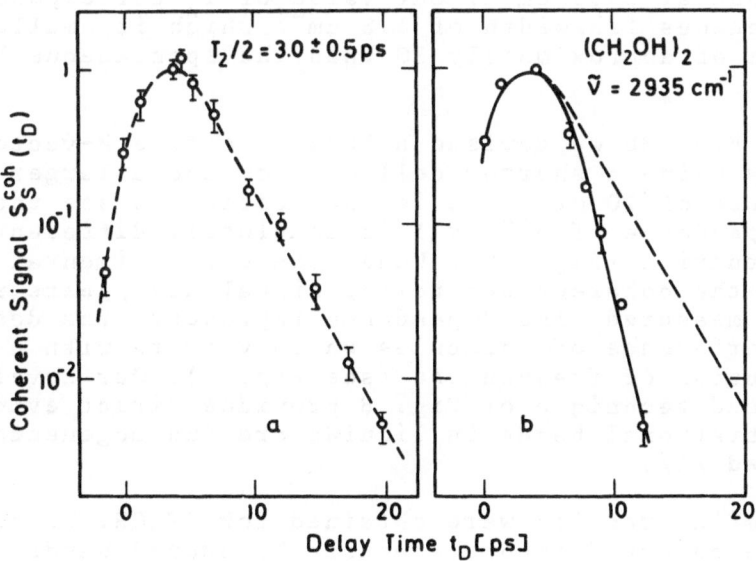

Fig. 8   a) Coherent Stokes scattered probe signals versus delay time measured in a highly selective collinear k-matching geometry in $(CH_2OH)_2$. From the observed free precession decay a dephasing time of $T_2/2 = 3.0$ psec is obtained corresponding to a homogeneous linewidth of 1.8 cm^{-1}.   b) Coherent Stokes signals for a less selective k-matching geometry. The signal decays rapidly on account of destructive interference of neighboring molecules oscillating with different transition frequencies.

separation between homogeneous and inhomogeneous line broadening factors. The investigation discussed now is aimed to tackle this problem. The vibrational system we discuss here is the CH-stretching mode of pure $(CH_2OH)_2$ at 2935 cm^{-1}. The spontaneous Raman band of this mode is broad with a linewidth of $\sim$60 cm^{-1}.

We present experimental data of two widely differing experimental situations /4/. In Fig. 8a, a highly select-ive k-matching geometry is achieved with a sample length of 10 cm and with a small Stokes divergence of $\gamma \simeq 3$ mrad. After the maximum of the probe signal we find an expo-nential decay with a time constant of $T_2/2 = 3.0\pm0.5$psec. For an interpretation of this time constant we estimate the frequency spread of the molecules, monitored by the selective k-matching, to be smaller than 0.5 cm^{-1}; i.e., dephasing by a distribution of frequencies is neglible. The time constant of 3 psec represents the homogeneous dephasing of a small group of molecules, the frequency of which is close to the center frequency of the broad Raman band at 2935 cm^{-1}. Our value of T_2 corresponds to a homogeneous linewidth of 1.8 cm^{-1} which is smaller by a factor of approximately 30 than the spontaneous Raman band.

In Fig. 8b we devised a less selective k-vector geo-metry by using a shorter cell of 1 cm and a larger Stokes divergence of 10 mrad. It is interesting to see that the time dependence of $S^{coh}(t_D)$ is completely different for this situation (Fig. 8b). Under these experimental con-ditions the coherent scattering signal disappears rapid-ly. The measured time dependence represents the destruct-ive interference of molecules which vibrate with a wide distribution of frequencies (see Fig. 3). Our novel picosecond technique of Fig. 8 provides direct evidence that vibrational bands in liquids are inhomogeneously broadened /4/.

Similar results were obtained for CH_3OH. We believe that the observed inhomogeneously broadened bands in al-cohols are connected with the local order in the liquid, in particular with the hydrogen bonding of the individual molecules /11/.

d) In the previous examples the coherent excitation of a physical system by the non-resonant Raman process was discussed. We now consider the resonant electric di-pole interaction of a picosecond pulse with a molecular

Parametric Generator

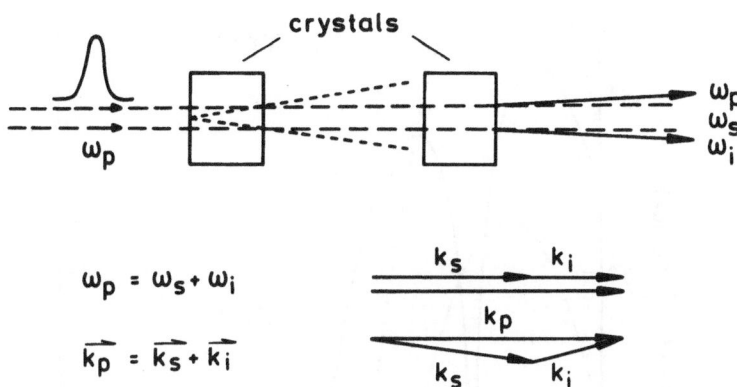

$$\omega_p = \omega_s + \omega_i$$

$$\vec{k_p} = \vec{k_s} + \vec{k_i}$$

Fig. 9   Schematic of the experimental set-up for a parametric generator consisting of two nonlinear crystals
(e.g. $LiNbO_3$). The parametric emission of the first
crystal is used as input for further amplification in the
second specimen. The conservation laws of the nonlinear
process are indicated.

vibration. Experimental investigations of this process
necessitate tunable pulses in the infrared. Generation
of the pulses is achieved via three-photon parametric
amplification (see schematic of Fig. 9) /2/. A coherent
picosecond laser pulse at fixed wavelength passes through
two properly oriented nonlinear crystals of $LiNbO_3$. The
signal and idler emission of the first crystal serve as
input radiation for further parametric amplification in
the second specimen. In this way, conversion efficiencies
of several per cent and a small divergence of the parametric light of $\sim$ 3 mrad is achieved. Pulses of $\sim$ 4 psec
duration and 6 cm^{-1} frequency width in a tuning range of
$\sim$ 2500 cm^{-1} to $\sim$ 7000 cm^{-1} are generated, close to the
theoretical bandwidth limit /12/.

We have studied theoretically the propagation of a
resonantly interacting light pulse at moderate intensity
level (small area pulses /13/). The rapid rotational motion
of the molecules, which proceeds in many liquids at comparable speed to vibrational dephasing, was included in
our calculations. Some numerical results are shown in
Fig. 10. The intensity of the incident (dotted line) and
of the transmitted (solid curve) pulse is depicted for a

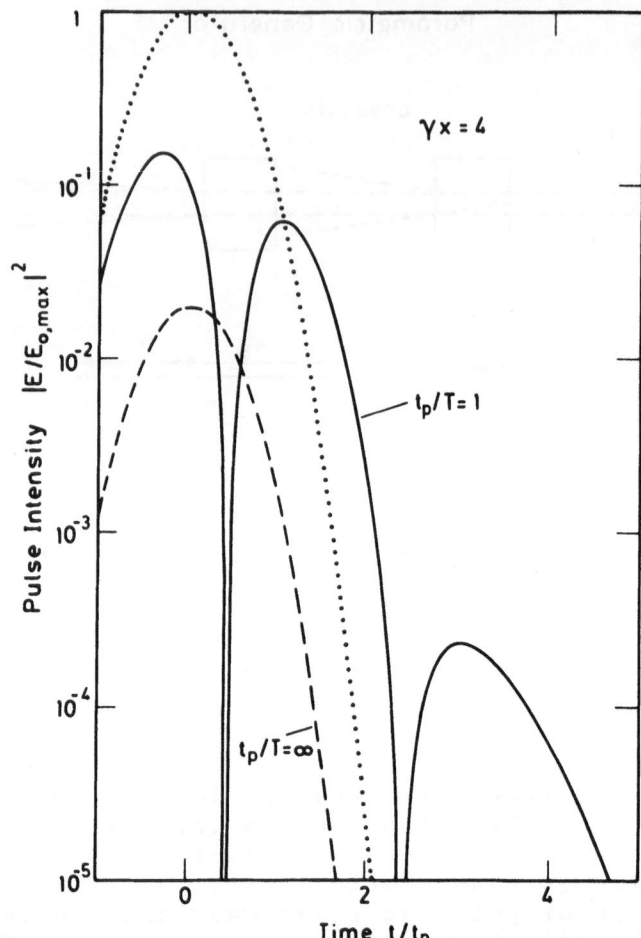

Fig. 10   Intensity versus time of a light pulse after
propagation of a distance $x = 4\ \gamma^{-1}$ through an absorbing
medium of absorption coefficient γ. Solid and broken
curves correspond to $t_p/T = 1$ and ∞, respectively, where
t_p denotes the pulse duration and T is the effective dephasing time. The dotted curve represents the Gaussian
input pulse.

ratio of pulse duration to effective dephasing time
$t_p/T = 1$. (T is determined by the contributions of the
vibrational and rotational motion). For the propagation
length we assume in Fig. 10 $\gamma x = 4$ where γ is the familiar
peak absorption coefficient. The break-up of the propagating pulse should be noted. It is a consequence of destructive interference via the 180° phase shift between

the incident field and the reemitted light of the molecular ensemble. Of special interest is the peak intensity
of the pulse (solid curve) which is a factor of ~ 8 larger
than Beer's law for steady-state conditions (broken curve).
The transmission of the pulse is notably enlarged for the
transient interaction of the short pulse with the vibrational transition. The transient transparancy discussed
here is connected with negligible population changes.
This is different from the general case of self-induced
transparency where large population changes are involved
in the light-matter interaction.

    We have experimentally studied the transparency phe-
nomenon in liquids on the picosecond time scale with our
parametric pulses discussed above. Results are presented
in Fig. 11 /14/. The energy transmission of picosecond
pulses resonantly tuned to CH-modes around 3000 cm^{-1} is
plotted versus propagation length in units of γ^{-1}. The
dotted straight line illustrates the exponential law
valid for stationary conditions. The full points denote
our data for ethanol (2928 cm^{-1}) where we expect a large
value of $t_p/T > 10$ close to the steady-state limit /15/.

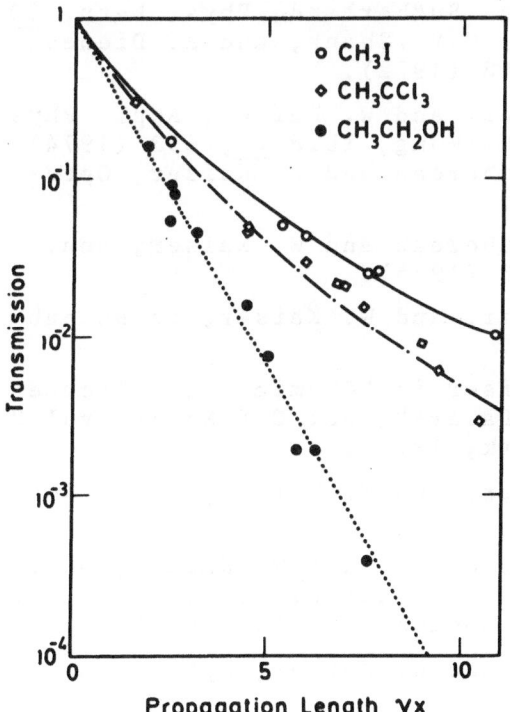

Fig. 11   Energy trans-
mission of an infrared
picosecond pulse versus
propagation length x in
units of γ^{-1}. Experimental
points for CH_3I, CH_3CCl_3,
and ethanol dissolved in
CCl_4 (mole fraction 5%).
The curves are calculated
for the transient cases
$t_p/T = 2.5$ (solid line)
and 3.1 (dash-dotted line)
and for stationary con-
ditions ($t_p/T = \infty$, dotted
line). The increased
transmission values in
the transient case should
be noted.

The good agreement with the theoretical line is readily
seen in the Figure. Different is the situation for CH_3I
(2950 cm^{-1}) and CH_3CCl_3 (2940 cm^{-1}) where transmission
factors more than two orders of ten above the steady-state
results are observed.The plotted curves in Fig. 11 were
calculated using the values of $t_p/T = 2.5$ and 3.1 for CH_3I
and CH_3CCl_3, respectively. These numbers are fully con-
sistent with estimates of T from available information on
vibrational dephasing times and on rotational relaxation
times /15-17/. The data of Fig. 11 represent the first observation of coherent interaction of a resonant infrared
pulse with a molecular vibration in a liquid on the picosecond time scale. It is felt that the present investigations advance our understanding of dephasing processes
of condensed systems.

References:

/1/   A.J. DeMaria, D.A. Stetser and J.Heyman, Appl.Phys.
       Lett. 8,174 (1966); W. Schmidt and F.P. Schäfer,
       Phys. Lett. A26, 558 (1968); D.J. Bradley, A.J.F.
       Durant, F.O'Neill and B. Sutherland, Phys. Lett. 30/
       535 (1969); E.P. Ippen, C.V. Shank, and A. Dienes,
       Appl.Phys. Lett. 21, 348 (1972).

/2/   A. Laubereau, L. Greiter, and W. Kaiser, Appl. Phys
       Lett. 25,87 (1974); A.H. Kung, ibid 25, 653 (1974);
       for a review see A. Laubereau and W. Kaiser, Opto-
       electronics 6,1 (1974).

/3/   For a review see A. Laubereau and W. Kaiser, Ann.
       Rev. Phys. Chem. 26, 83 (1975).

/4/   A. Laubereau, G. Wochner, and W. Kaiser, to be pub-
       lished.

/5/   A. Laubereau and W. Kaiser in "Chemical and Bioche-
       mical Applications of Lasers", ed. C.B.Moore, vol.2
       Academic Press (New York, 1977).

/6/   A. Laubereau, G. Wochner, and W. Kaiser, Phys. Rev.
       A13, 2212 (1976).

/7/   D.von der Linde, O. Bernecker, and W. Kaiser, Opt.
       Commun. 2, 149 (1970); D.von der Linde, O.Bernecker
       and A. Laubereau, Opt. Commun. 2, 215 (1970).

/8/   W. Zinth, A. Laubereau and W. Kaiser, Opt. Commun.
       22, 161 (1977).

/9/   S.F. Fischer and A. Laubereau, Chem. Phys. Lett.
      35, 6 (1975);   P.A. Madden and R.M. Lynden-Bell,
      Chem. Phys. Lett. 38, 163 (1976).

/10/ R.L. Schoemaker and R. Brewer, Phys. Rev. Lett. 28,
      1430 (1972);   W. Gornik, D. Kaiser, W. Lange,
      J. Luther, and H.H. Shuk, Opt. Commun. 6, 327 (1972);
      S. Haroche, F.A. Paisner, and W.L. Schawlow, Phys.
      Rev. Lett. 30, 948 (1973);   J.P. Heritage, T.K. Gus-
      tafson, and C.H. Lin, Phys. Rev. Lett. 34, (1975).

/11/ G.C. Pimentel and A.L. McClellan, "The Hydrogen
      Bond", Freeman (San Francisco, 1960);   Ann. Rev.
      Phys. Chem. 22, 347 (1971).

/12/ A. Seilmeier, K. Spanner, A. Laubereau, and W. Kaiser,
      to be published.

/13/ M.D. Crisp, Phys. Rev. A1, 1604 (1970);   A2, 2172
      (1970);   H.P. Grieneisen, J. Goldhar, N.A. Kurnit,
      A. Javan, and H.R. Schlossberg, Appl. Phys. Lett.
      21, 559 (1972);   S.M. Hamadani, J. Goldhar, N.A.
      Kurnit, and A. Javan, Appl. Phys. Lett. 25, 160
      (1974).

/14/ K. Spanner, A. Fendt, and A. Laubereau, to be pub-
      lished.

/15/ A. Laubereau, D. von der Linde, and W. Kaiser, Phys.
      Rev. Lett. 28, 1162 (1972).

/16/ K. Spanner, A. Laubereau, and W. Kaiser, Chem. Phys.
      Lett. 44, 88 (1976); see also our data of Fig. 6.

/17/ F.J. Bartoli and T.A. Litovitz, J. Chem. Phys. 56,
      413 (1972).

NONTHERMAL MICROWAVE RESONANCES IN LIVING CELLS

Fritz Keilmann

Max-Planck-Institut für Festkörperforschung

7 Stuttgart 80 – Fed. Rep. Germany

1. INTRODUCTION

In this contribution I would like to make you acquainted with a fascinanting new topic in spectroscopy: microwaves at extremely small power levels are able to induce drastic changes in living systems, and these effects are highly resonant in respect to frequency. I cannot give you any proven explanation of these effects, but I will just try to describe to you a theoretical concept due to H. Fröhlich who has predicted such effects in the right frequency region. The basis of this model is a very high single-mode vibrational excitation within, for example, a protein molecule switching on its enzymatic activity. Any speculation on the nature of such an oscillator has to account for the narrow resonances observed: It has to explain how a vibrational linewidth in a condensed system could be as small as in a gas with a pressure of a few Torr.

At this stage I consider it important to give you in some detail the experimental results obtained by us together with those of previous workers in this field. Note that in addition to the spectroscopic and physical interest in this subject there may be fundamental consequences for our understanding of biology. The way microwaves interfere directly in biological processes suggests that microwaves are actively used by living systems. Nonthermal microwave effects might also be applied in many related areas like medicine or agriculture. Microwaves are already used for many purposes ("microwave pollution" [1]). Economical and political consequences seem likely when a severe safety limit e. g. of 10 $\mu W/cm^2$ is esta-

blished internationally as has been in existence in the USSR since 1960 [1].

2. EXPERIMENTS ON NONTHERMAL BIOLOGICAL EFFECTS OF MICROWAVES

A great number of investigations were undertaken to study possible health hazards of microwaves to living beings. From the beginning of the era of high frequency radio half a century ago it
seems that researchers in the USSR paid much more attention to
these question than in other countries [1]. An up-to-date bibliography compiled in [2] lists about 4000 relevant publications. The ma-
jority of these deal with thermal effects. For example, detrimen-
tal heating of the low thermal-conducting fluid of the eyeball causing protein coagulation may occur at relatively low intensities.
Such hazards have led to the definition of a safety limit of $10mW/cm^2$
since 1958 in the western countries.

On the other side there are many reports about microwaves in-
fluencing nervous action or blood pressure, to cite only two,
where a thermal explanation is not satisfactory. Even quasisen-
sorial perception of microwaves ("hearing the radar pulse") has
been observed. These experiments deal with very complex systems.
The data obtained are rather incomplete, e. g. , frequency or power dependences are generally not available, excluding explanations in most cases. I shall not comment further on these experiments apart from stating that they seem to provide sufficient evidence to motivate large scale investigations, from which a coherent picture of their mechanism might evolve.

2.1 Resonances at Millimeter Wavelengths

The first spectral study of a nonthermal biological microwave
effect, to my knowledge, was performed by Webb and Booth [3].
They irradiated aqueous suspensions of living Escherichia coli
bacteria and measured the growth rate as well as the metabolic
intake of radioactive substances. Both measurements gave a clear
indication that (i) the growth rate could either be accelerated or
decelerated while no significant temperature change occured, and
(ii) the effect strongly depend on frequency which was set in 1 GHz
steps from 65 to 75 GHz (fig. 1). Note here that water absorption
is very strong but varies smoothly from 45 to 90 cm^{-1} in the frequency range from 40 to 120 GHz (at 30°C) [4]. No spectral feature

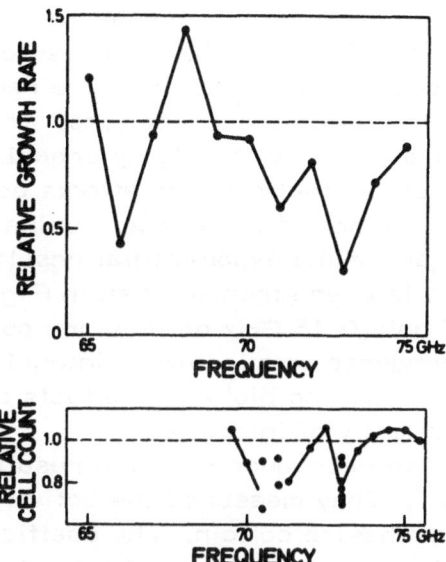

Fig. 1 – Growth Response of <u>Escherichia coli</u> to weak microwave irradiation after Webb and Booth [3] (upper diagram) and Berteaud et al. [5] (lower diagram). In both
cases the parameter measuring the growth is normalized to the value without irradiation.

could be related to water absorption as is visible in fig. 1.
    The results of Webb and Booth were corroborated by a repeti-
tion of the experiment by Berteaud et al. [5]   They also used a liquid suspension which this time was kept constant in temperature.
Also the microwave intensity was kept constant at about 10mW/cm.[2]
Adjusting the frequency in steps of 0.5 GHz they found a similar
frequency dependence (Fig. 1). Their results are compatible with
resonance structures with a width of 1 GHz or less. In both experiments [3,5] it was found that the death rate was not increased by

the irradiation. In ref. [5] a microwave influence on the mutagenic rate was not observed.

Independently a great effort had been conducted in the USSR on nonthermal biological effects of millimeter microwaves in the frequency range from 39 to 60 GHz. In a session of the USSR Academy of Sciences in 1973 [6] results were reported on a dozen different experiments at different locations. The objects investigated ranged from enzymes to the highly organized animal level. The overall tenor seems that striking effects occurred where one looked for them. More specifically, two fascinating aspects are common to a greater part of the experimental results: (i) The frequency dependence is even stronger than in Fig. 1. Typically, resonance widths of only 0.15 GHz or less are compatible with the data. (ii) The dependence on microwave intensity exhibits a threshold region below which no biological effects occur and above which they occur independently of intensity.

As an example we reproduce here the results by Smolyanskaya and Vilenskaya in [6]. They measured the activity of Escherichia coli bacteria to synthesize colicin. The coefficient K (Fig. 2) measures the number of cells that synthesized colicin in a given period, during which radiation was applied, divided by the respective number in an unirradiated control. Fig. 2 shows that after a certain period of the order of one hour K has been increased by up to about 300% by the irradiation. While no decrease in K was reported, increases were seen for widely different wavelengths, 5.8, 6.5 and 7.1 mm. The spectrum in Fig. 2 is compatible with a resonance bandwith of about 0.1 GHz. At first sight the periodic spectrum in Fig. 2 points to a thermal effect because interference of transmitted and reflected waves propagating between oscillator and sample cuvette could change the net power delivered to the sample in a frequency periodic fashion. With a monomode transmission line one would expect the observed periodicity of 200 MHz for a totale length around 1 m. However, the saturation evident at intensities above 10^{-5} W/cm^2 certainly contradicts a thermal explanation for the observed effects.

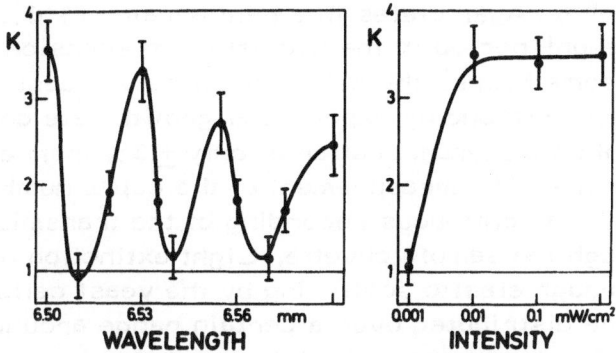

Fig. 2 – Relative number K of <u>Escherichia coli</u> bacteria syn-
thesizing colicin after microwave irradiation (K=1
for unirradiated control). Left diagram shows frequency dependence, right diagram intensity dependence at 6.50 mm (vacuum) wavelength, after
Smolyanskaya and Vilenskaya [6].

2.2 Yeast Experiment

We have recently begun an experiment which is designed to (i)
test one of the experiments mentioned in Sect. 2.1 and to (ii) possibly have a system usable for further experimental work and te-
sting of pertinent theoretical models. Entering this field from the
laser spectroscopy side required first of all to find a biologist
who knew his material and who shared a mutually understandable
language. Old ties and an incidental discussion on a winterly Al-
pine summit worked favourable in our case. So the expertise of
Werner Grundler's laboratory on yeast of the type <u>Saccharomyces
cerevisiae</u> quickly put us in the position of having a reproducible
sample to work on [7]. We cultivated "fresh" yeast cells daily in
stirred aqueous suspension. They were taken from a stock supply

kept "sleeping" on Agar plates in a refrigerator (4°C). After a
certain "wake-up" period in the nourishing suspension (at 32°C)
which lasted some hours, the cells reduplicated about every 75
minutes. The corresponding exponential growth rate could be re-
produced within a maximum scatter of only $\pm$ 3% from day to day
over many months. The measurement of the cell's number density
was achieved by a continuous recording of the transmission of visible light through the sample cuvette. Light extinction occurred pre-
dominantly through elastic scattering by the yeast cells. Their
diameters were distributed over a certain range around 3 um, but
this distribution stayed approximately constant.

The (aqueous) suspension of the yeast cells was chosen in order to avoid any overheating of the cells by absorption of microwaves. We estimated from simple heat conduction arguments a
characteristic time of 50 μs for the thermal equilibration of an
overheated cell. A steady-state overtemperature of the cell above
its surrounding by 1°C would therefore require a heating power
of 10 μW per cell. Let us discuss whether one could imagine such
a value in our experiment. As seen in Fig. 3 the microwave entered
into   the   aqueous   suspension   through a teflon interface approx.
10 cm^2 in area. The penetration depth in water is 200 μm at 42
GHz [4], so the intensity decays rapidly with the distance from the
interface. With a total absorbed microwave power of 30mW (ref.
below to measurement technique) the average intensity was 3mW/cm^2
just inside the aqueous suspension. Locally this value might have
been enhanced up to a few times, i.e. to about 10mW/cm^2, because
of standing wave interference in the teflon structure.   Under
these circumstances an absorption of 10 μW per cell would indeed
require a very high hypothetical absorption cross section per cell:
with the typycal cell density of 2. 10^5 cm^{-3} in our experiments,
this cross section would have to be very large (1 mm^2), so that
it would have reduced the penetration depth to 4 μm. Such a strong
absorption seems totally unrealistic. Since the partial volume occupied by the cells was only about 10^{-5} this would mean, roughly,
that the absorption in the cells was something like 10^6 times larger than in water, for which we cannot conceive any mechanism.

Overheating might in principle occur in moderately absorbing
but thermally isolated regions. Nothing is known of their existence
within a living cell.   Let us however, imaging, for a moment ,
a two-level system resonantly absorbing at ν = 42 GHz with a

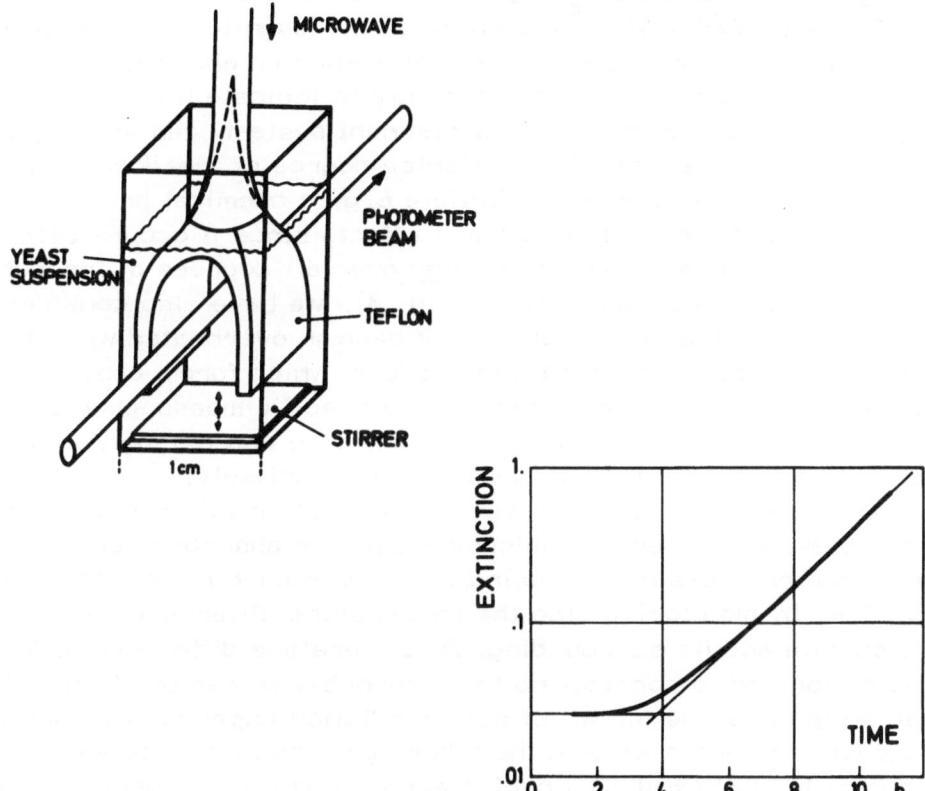

Fig. 3 – Schematic drawing of microwave irradiation experiment on yeast [7]. The extinction of the photometer
beam is used to observe the growth behaviour of the
stirred aqueous cell suspension. Millimeter microwaves enter through a metallic hollow waveguide from
above into a teflon antenna structure, from where
they penetrate into the aqueous suspension over an
interface area of about 10 cm^2. Also shown is a typical growth curve (without irradiation); note that
the freshly prepared culture acquires exponential
growth after a few hours.

homogeneous linewidth ν/Q. Such a system can have a minimum energy relaxation time of Q/ν which would be 24 ns for a Q-factor of 1000; $h\nu \cdot \nu/Q = 10^{-15}$ W then gives the magnitude of the absorbed power which would suffice to drive such an absorber into saturation. This remark is given here only to indicate how small powers, if applied coherently to a resonant system, can actually lead to a significant heating of a particular degree of freedom. If this heating were able to directly initiate e. g. a chemical process we would however call it a nonthermal effect, since it occurs before equilibration of the absorbed energy over all degrees of freedom. Turning back to our experiment (Fig. 3), we have also considered the possibility that some cells might become overheated when they escape from the suspension into the foam which forms above the surface due to the stirring motion. No specific measurements have been conducted up till now, but both the size and the sign of our results below make such an influence very unlikely.

The sample temperature was continuously monitored using thermocouples. Moreover, the microwave power absorbed could also be absolutely measured by using the sample cuvette as a calorimeter, i. e. by monitoring also the temperature difference between the cuvette and its surrounding. A temperature difference of 0.5°C was calibrated to correspond to an absorbed power of 24 mW. This was a typical value for all of our irradiation experiments. Due to slow temperature cycles in the laboratory the sample temperature did vary by up to $\pm$ 0.5°C during each growth experiment. A typical growth curve is shown in Fig. 3. When microwave irradiation was to be applied, it commenced from the beginning of the estinction recording, about 1 hour after preparation.

The resulting exponential growth rate without irradiation, as a function of sample temperature, is shown in Fig. 4. The small scatter of up to $\pm$ 3% about a mean curve is clearly evident. In Fig. 5 we show the results of the irradiation experiments. Here the normalized growth rate is shown with an error margin of $\pm 3\%$, demonstrating the reproducibility we obtained for the growth rate without irradiation. As for the irradiation frequency, we had both a short- and long-term stability of $\pm$ 1 MHz, but a resettability of our measuring cavity of only $\pm$ 3 MHz. Fig. 5 clearly shows irradiation-induced positive and negative changes in the growth rate. In addition to the considerations above, the nonthermal origin of the effect can be demonstrated by examining Fig. 4. This indicates that overheating could lead to only a modest increase of the growth rate by up to 10%, while further overheating beyond 37°C is known to again decrease it.

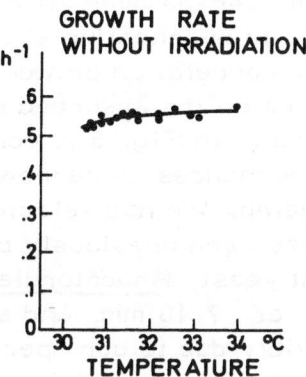

Fig. 4 – Dependence of exponential growth rate of yeast culture[7] on temperature (solid curve). The points result from measurements without irradiation taken in between the irradiation experiments (Fig. 5).

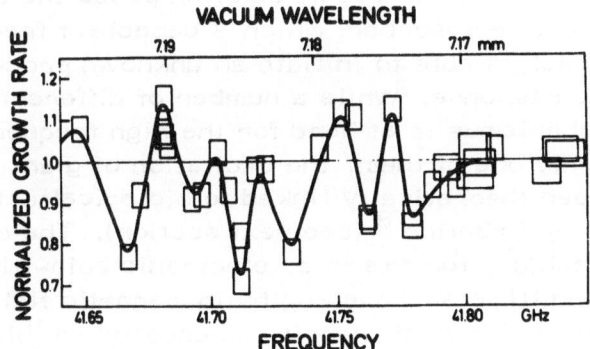

Fig. 5 – Exponential growth rate of irradiated yeast culture[7] vs. frequency. Normalization to correct for different temperatures is done by dividing by the unirradiated growth rate at the same temperature, as taken from the solid line in Fig. 4.

Since at any instance only a small part of all cells in the cuvette
is either near the teflon surface or in the foam, the observed
growth rate, which is an average over all cells, could not be increased by e. g. 15% due to overheating at these locations. Also,
our data do not show any correlation between the growth effects
and either the temperature or the absorbed microwave power.

A smooth curve is drawn in Fig. 5 to connect all data rectan-
gles. It indicates that resonances as narrow as 0.01 GHz exist
in the range we investigated. We had selected this region for our
experiment because it was used previously by Devyatkov in [6].
He irradiated a different yeast, Rhodotorula rubra, and found a
growth increase effect,   at   7.18 mm, and a decrease at 7.16,
7.17 and 7.19 mm. Possibly due to our spectral resolution our
experiments reveal a much finer structure. Clearly we were not
prepared to find such narrow resonances. A better frequency
measurement is necessary to fully resolve the structure. Furthermore this will allow an observation of the intensity dependence of
the microwave irradiation effect.

3. THEORETICAL CONSIDERATIONS

The mechanism leading to the experimental observations above
is unknown. The general result of highly resonant, nonthermal response to millimeter microwave irradiation poses the question of
an unknown sensitive absorber, which is capable of frequency discri-
mination and which is able to initiate an unknown process leading
to the observed response. While a number of different narrow-band
absorption mechanismus is at hand for the high frequency microwave region, only one of them, the excitation of giant dipole vibrations has been theoretically lihked to biological activity, as
early as 1968, by Fröhlich [8] (see next section). The other types
of absprbers include, for example, electronic spin-flip transitions
which occur at millimeter wavelengths in magnetic fields of a few
Tesla. While these types of transitions usually exhibit narrow resonances, they are not likely to play a role here due to the absence
of the required field strength in biological specimens. Other
possibilities like transitions due to ionic reorientation in weak
potentials or vibrations would require an unusuale high degree of
isolation from disturbing interactions e. g. in a cage-like environment, in order to exhibit the narrow resonance widths observed.
For example, guided acoustic waves have been suggested travelling along the inner hydrophobic parts of membranes [9].

3.1 Fröhlich Theory

Since his original conjecture [8] Fröhlich has developed a general concept of long range coherence in biology [10]. We shall re-
peat here some of his thoughts because they possibly contain the
relevant basic physics needed to explain our experimental results.
On the other hand we cannot, at this point, expect any disclosure
from his theory on the identity of the absorber we are looking for,
not on the biochemical reaction which was influenced in our experiment.

Biological systems have the interesting property – when they
are alive – that they are not in thermal equilibrium since energy
is continuously pumped through them. Fröhlich has put forward
a close analogy to lasers [12] where now the source of "biological"
pumping is chemical energy ultimately coming from food or sunlight
(in photosynthesis). He has then conceived the existence of a threshold pump power at which the system would react and undergo a
Bose condensation, so that instead of the many modes available all
the power would be chancelled into a single mode or a small number of modes [11]. The excitations themselves are hereby conceived as the relatively low-frequency vibrations which exist in biomolecules (for example proteins), membranes or even more complex structures and which are coupled to electric dipoles. The
latter prerequisite does not impose any restriction at all since
these structures show extraordinary dielectric properties [10] due
to strong interaction of their charged groups with the surrounding
water. For example a strong dipolar layer of 10^{-6} cm thickness
exists across membranes giving rise to electric field strengths of
10^5 V/cm. In fact one may ask which task such a field must fulfill,
if one accepts the principle of biologival evolution on a molecular
scale [10]. Here Fröhlich has pointed out that strong electric polarization might be induced e.g. in biomolecules (e.g. embedded in
membranes) which would be accompanied by a mechanical deforma-
tion, i.e. a conformational change. The new state could be long-
lived because of stabilization through rearrangement of counterions
etc. It is this metastable polar conformation of a biomolecule which
is considered capable of high-amplitude vibration (nh$\nu \gg$ T) in a
single mode under a strong biological pumping conditions.

Such a strong vibration of a polar molecule has been linked
by Fröhlich [8] to biological activity. One knows that e.g. enzyme
molecules possess relatively small "active" sites which are directly involved in the chemical reaction. On the other hand their enzymatic activity does depend critically on changes at quite distant

locations in the big protein body (typ. 50 Å in size). The reason for such stringent requirement of the structural order of the molecule is not clear. In answer and in addition to this Fröhlich has set the requirement of motional order which could activate e. g. an enzyme molecule. The mechanism of this is based on long-range dipole forces which occur between systems capable of giant dipole vibrations. It is shown in ref. 3 that these forces are selective, i. e. they lead to attraction of systems having equal vibration frequencies, and that they either occur when the background medium shows strong dispersion at the relevant frequency ("dispersion forces") or when the vibration is strongly excited. In the latter case thus a vibrational excitation (e. g. through biological pumping) could switch on a long-range attraction force which can be envisaged to bring reaction pertners together more quickly and thus enhance the overall activity. In extension these forces could be the basis for coupling larger entities and provide a control, for example, of homologous pairs of chromosomes or cell arrangements.

Let me conclude at this point with two remarks on our experiments. Here we were concerned with microwave frequencies near 10^{11} sec^{-1}. It was just this frequency range which Fröhlich originally proposed to be of relevance[8] before any of the experiments in refs. 3, 5, 6, 7 were performed. He had found this number by simply dividing a sound velocity of 10^5 cm/sec, corresponding to a typical soft material, by the diameter of a big biomolecule, 100 Å, in order to find its lowest mechanical vibration frequency. Note that with a molecular weight of 10^5 about 10^4 more vibrational frequencies exist up to about 10^{14} sec^{-1}. Now at 10^{11} sec^{-1}, the vibrational quantum energy is only 1/60 th of the thermal energy, so that pumping to a very high quantum number is necessary to surpass a purely thermal excitation. Indeed one can imagine that optical pumping in the cited experiments (at intensities between 10^{-5} and 10^{-2} W/cm^2) has sufficed[14] to activate specific enzymes which happened to be on resonance. These would have influenced specific biochemical processes which in the case of our experiment[7] would have led to the resulting change of the overall growth rate. Intuitively more often a slowing than an acceleration of the growth is expected by this external influence, as is observed. Finally, note that the saturation observed in some of the experiments (e. g. Fig. 2) strongly supports the view of microwave pumping of a "Fröhlich state".

REFERENCES

1) P. Brodeur, "Microwaves", The New Yorker, Dec. 13(1976) pp. 50–110 and Dec. 20 (1976) pp. 43–83.

2) Z. R. Glaser and P. F. Brown, "Bibliography of Reported Biological Phenomena ("Effects") and Clinical Manifestations Attributed to Microwave and Radio Frequency Radiation", Naval Medical Research Institute Report, Dahlgren, VA 22448, USA (1977).

3) S. J. Webb and A. D. Booth, "Absorption of Microwaves by Microorganismus", Nature 222, 1199 (1969).

4) P. S. Ray, "Broadband Complex Refractive Indices of Ice and Water", Appl. Optics 11, 1836 (1972).

5) A. J. Berteaud, M. Dardalhon, N. Rebeyrotte et D. Averbeck, "Action d'un Rayonnement Electromagnétique à Longueur d'Onde Millimétrique sur le Croissance Bactérienne", C. R. Acad. Sc. Paris 281, D843 (1975).

6) K. I. Gringauz, I. A. Zhulin, A. D. Sytinskii, N. D. Devyatkov, E. B. Bazanova, A. K. Bryukhova, R. L. Vilenskaya, E. A. Gelvich, M. B. Golant, N. S. Landau, V. M. Melnikova, N. P. Mikaelyan, G. M. Okhokhonina, L. A. Sevastyanova, A. Z. Smolyanskaya, N. A. Sycheva, R. L. Vilenskaya, V. F. Konrateva, E. N. Christyakova, I. R. Shmakova, N. B. Ivanova, A. A. Treskunov, S. E. Manoilov, M. A. Strelkova, N. P. Zalyubovskaya, R. I. Kiselev, N. P. Zalyubovskaya, V. I. Gaiduk, Y. I. Khurgin and V. A. Kudryashova, Scientific Session of the Division of General Physics and Astronomy, USSR Academy of Sciences, Sov. Phys. –Usp. (Translation) 16, 568 (1974).

7) W. Grundler, F. Keilmann and H. Fröhlich, "Resonant Growth Rate Response of Yeast Cells Irradiated by Weak Microwaves", Phys. Lett., in print; W. Grundler and F. Keilmann, "Nonthermal Effects of Millimeter Microwaves on Yeast Growth", Z. Naturf. c, in print.

8) H. Fröhlich, "Long Range Coherence and Energy Storage in Biological Systems", Inter. J. Quantum Chem. 2, 641 (1968); H. Fröhlich, "Long Range Coherence and the Action of Enzymes", Nature 228, 1093 (1970).

9) K. Dransfeld, personal communication.

10) H. Fröhlich, "The Extraordinary Dielectric Properties of Biological Materials and the Action of Enzymes", Proc. Nat. Acad. Sci. USA 72, 4211 (1975); H. Fröhlich, "Long Range Coherence in Biological Systems", Il Nuovo Cimento, in print.

11) H. Fröhlich, "Bose Condensation of Strongly Excited Longitudinal Electric Modes", Phys. Lett. 26A, 402 (1968).

12) H Haken, "Synergetics", pp. 225 and 293 ff. Springer-Verlag, Berlin, Heidelberg and New York (1977).

13) H. Fröhlich, "Selective Long Range Dispersion Forces Between Large Systems" Phys. Lett. 39A, 153 (1972).

14) D. Bhaumik, K. Bhaumik and B. Dutta-Roy, "On the Possibility of "Bose Condensation" in the Excitation of Coherent Modes in Biological Systems", Phys. Lett. 56A, 145 (1976).

THE FREE ELECTRON LASER: A SINGLE PARTICLE CLASSICAL MODEL

A. Bambini [(+)] and A. Renieri

C.N.E.N. - Divisione Nuove Attività, Centro di Frascati

C.P. 65 - 00044 Frascati, Rome, Italy

INTRODUCTION

Stimulated processes are the basic phenomena in laser devices, operating either as amplifiers or as optical oscillators. Although the laser operation is best described in the framework of a semiclassical theory, in which the electromagnetic field is not treated quantum mechanically, the basic processes are better understood if we quantize the field. When there are no photons in the final modes of the electromagnetic field, vacuum fluctuations cause an atom or a molecule in an excited level to emit spontaneously a photon in one mode. All the modes with an energy close to the transition energy are equivalent. But when one of these modes in occupied by a large number n of photons, then the cross section for the emission in that mode is greatly enhanced (by a factor n): this is what is called stimulated emission.

Conventional lasers operate through stimulated emission of radiation from an ensemble of inverted atoms or molecules.

Recently the first operation of a laser based on stimulated scattering from free electrons has been reported by Madey and co-workers [1] . We describe in this lecture a simple model of the free electron laser (FEL)based on a single particle theory. Indeed, laser action is a single particle process as far as the electromagnetic (e.m.) field can be considered as an injected signal, grown up from noise, which drives each particle to emit or scatter coherently with it. The model is completely classical, i.e. both e.m. field and the motion of the electrons are treated classically. In the single particle model one can have a physical description of the behaviour of

[(+)]   Lab. Elettronica Quantistica del C.N.R., Via Panciatichi 56/30 Firenze, Italy.

each interacting electron, what is of great importance for the design
of electron storage rings [2] suitable for FEL operation. Using this
model, which is not perturbative, the small signal gain and the saturation behaviour can be evaluated. The results are in agreement
with the previously derived ones [3,4].

Before concluding these introductory remarks, we want to stress
that the FEL is a single particle process. Although the single particle picture shows little connection with current laser theories,
which describe the medium as a whole, in that framework we can better
understand the physics involved in the FEL.

We describe firstly the spontaneous scattering of radiation from
an electron beam passing through a static periodic magnetic field.
Then we discuss the model which includes the stimulated scattering.
The small signal gain formula is derived and a brief physical picture
of the saturation behaviour is given.

1. SPONTANEOUS RADIATION FROM RELATIVISTIC ELECTRON BEAMS IN WIGGLER
DEVICES

An electron with relativistic velocity passing through a static
magnetic field, spatially modulated along the electron beam (e.b.)
direction (wiggler), radiates e.m. waves of peculiar characteristics
[5]. Let us assume that the spatially varying magnetic field is circularly polarized

$$\underline{B} \equiv (B_o \cos(k_q z),\ B_o \sin(k_q z),\ 0),\hspace{2cm}(1)$$

where we have chosen a reference frame with the z-axis parallel to
the e.b. direction. This field, with B_O = const, cannot exist on
large sections, because it does not satisfy the conditions

$$\mathrm{div}\ \underline{B} = 0;\ \mathrm{curl}\ \underline{B} = 0,\hspace{2cm}(2)$$

which apply for a static, current free magnetic field. The field (1)
is experienced by the electrons if they do not displace from the
wiggler axis more than $r \sim 1/k_q$ [6].

It can be shown [7] that the electron *sees* the field (1) as a
running wave of a radiation field, travelling oppositely to its motion, and with a wave vector given by

$$k_{PRF} = \frac{k_q}{2}\ ,\hspace{2cm}(3)$$

we have denoted the wave vector with the suffix PRF, to indicate
that the effective field is not a true radiation field, but it is a
Pseudo Radiation Field. The validity of this approximation is confined to ultrarelativistic electrons, and we shall use it throughout
our lecture.

The frequency of the PRF is then given by

$$\omega_{PRF} = ck_{PRF}.\hspace{2cm}(4)$$

Owing to the Doppler effect, the frequency is shifted up to higher values in a reference frame where the longitudinal electron velocity is zero. For ultrarelativistic electrons the Doppler shifted frequency is

$$\omega'_{PRF} = 2\gamma \, \omega_{PRF} \, . \tag{5}$$

We denote with a prime all quantities evaluated in the electron frame. In Eq. (5) γ is given by

$$\gamma = 1/\sqrt{1 - (v_z/c)^2}, \tag{6}$$

where v_z is the electron longitudinal velocity in the lab frame. The spatial wavelength of the static magnetic field, $\lambda_q = 2\pi/k_q$, is of the order of several centimeters. In the electrons' frame, with $\gamma \sim 100$, the effective wavelength of the PRF is of the order of several tenths of millimeters. Therefore, in that frame, Thomson scattering of the PRF occurs. Let us focus our attention on the backscattered radiation.

In the lab frame the field scattered in this direction has a wavelength given by

$$\lambda_s = \frac{\lambda'}{2\gamma} = \frac{\lambda_q}{2\gamma^2} \, . \tag{7}$$

Therefore, with $\lambda_q \sim 20$ cm, $\gamma \sim 100 \div 1000$, we obtain radiation with a wavelength

$$\lambda_s \sim 10 \ \mu\text{m} \ \div \ 0.1 \ \mu\text{m}.$$

Wiggler devices have been proposed as a source of radiation, to be inserted in storage ring machines [5].

Let us look at the properties of the scattered radiation. Its bandwidth is limited by two factors. There is a *homogeneous* bandwidth, due to the finite length of the wiggler, that is to the finite flight time (interaction time) of the electron through the wiggler. This type of broadening is called homogeneous because it affects all the scattering processes, independently of the particles' velocity and position. The bandwidth is given by [8]

$$\frac{\Delta\omega}{\omega} = \frac{\lambda_q}{2L} \tag{8}$$

where L is the wiggler length.

There is also an inhomogeneous broadening of the scattered radiation. This comes from the distribution of the velocities of the scattering particles and their displacement from the wiggler axis. Indeed the scattered radiation depends on γ (Eqs (5) and (6)), then its spectrum reflects the distribution of the velocities. Moreover, the magnetic field experienced by the electrons depends on their transverse displacements, as we have noted earlier. A change in the magnetic field amplitude affects the longitudinal motion of the particles, i.e. it changes their γ. The detailed analysis of these broadening mechanisms is out of the scope of our lecture. Further information can be found in the literature [9].

Now, let us look at the intensity of the scattered radiation. All the elementary scattered waves maintain a definite phase relation with the incident PRF. If the particles were randomly distributed in space, all elementary wave amplitudes would add incohrently to zero in all directions other than the forward one, and no scattering would be observable. But fluctuations in the density distribution occur. These fluctuations determine a non vanishing scattered wave in all directions. This phenomenon is very much the same as the Rayleigh scattering of radiation from a gaseous medium [10]. As a consequence, the scattered radiation in the backward direction (i.e. the radiation whose wavelength is given by (7)), depends linearly on the number density n_e of the electrons.

There is a close analogy between the radiation scattered from a relativistic e.b. in a wiggler and the radiation emitted through spontaneous emission from a sample of gaseous matter excited to high energy levels. Even in the latter case, radiation has a homogeneous broadening, determined by the finite lifetimes of the excited levels or the dephasing times of the elementary dipoles of the medium. It has an inhomogeneous width, due to the velocity distribution of the emitting atoms or molecules (Doppler broadening). Furthermore, radiation intensity is proportional to the number density of atoms or molecules in the medium. In the atomic samples it is possible to produce an inversion in the populations of two levels. This in turn yields amplification of an e.m. wave propagating through the medium, if its frequency is tuned to the frequency of the transition between the two levels. Obviously such a population inversion cannot be induced in an electron beam. But, as we shall see later on, stimulated scattering from the PRF to some properly chosen laser field can occur. Before concluding these remarks on the *spontaneous* scattering from an e.b., we want to point out that spontaneous radiation can be enhanced if we induce in the e.b. a longitudinal density modulation with a spatial wavelength equal to the wavelength of the scattered radiation [11].

However, this method applies only in the millimeter wavelength range, where it is possible to prebunch the e.b.

2. THE SINGLE PARTICLE CLASSICAL MODEL

Let us consider now the stimulated Thomson scattering (STS). As we have noted earlier the STS occurs when one of the modes of the e.m. field is occupied by a large number n of photons. The cross section for scattering in this mode is enhanced by a factor n, while the cross sections for all other modes remain unchanged. Therefore almost all the photons are scattered in that mode. In each scattering process the electron momentum changes by $2\hbar k'_{PRF}$ ($k'_{PRF} = \omega'_{PRF}/c$, see Eq. (5)), and the number of these processes is very large with respect to the spontaneous ones. As a consequence a proper treatment of the STS must include the transfer of momentum in the interaction process. This can be accomplished by taking into account the self field of the electrons. To this end we use the Coulomb (or radiation) gauge [12] for the e.m. field.

Within this gauge the Hamiltonian of the system depends on a transverse vector potential $\underline{A}_\perp$ (div $\underline{A}_\perp$ = 0), and an instantaneous electrostatic potential φ , which can be ignored in so far we are describing a single particle process.

We neglect all the modes in which spontaneous scattering occurs, and consider only two modes of the e.m. field, i.e. the PRF, which is the incident one, and the stimulating field (laser field), which is propagating in the same direction of the e.b. This latter field stimulates the backscattering of the PRF.

We assume that both fields are strong enough to be treated classically [13]. We choose the frame where the two oppositely running waves have the same frequency and are propagating along the z axis. In this frame the electron is at rest or has a small, nonrelativistic, velocity.

The classical, non relativistic, Hamiltonian is [12]

$$H = \frac{1}{2m} (\underline{p} - \frac{e}{c} \underline{A}_\perp)^2 + \frac{1}{2} (\widetilde{\underline{P}}_L^2 + \omega_L^2 \widetilde{\underline{Q}}_L^2) + \frac{1}{2} (\widetilde{\underline{P}}_w^2 + \omega_w^2 \widetilde{\underline{Q}}_w^2) \qquad (9)$$

where $\widetilde{\underline{P}}_L, \widetilde{\underline{Q}}_L$ ($\widetilde{\underline{P}}_w, \widetilde{\underline{Q}}_w$) are the canonical coordinates of the laser field (PRF), $\underline{p}$ is electron canonical momentum and $\underline{A}_\perp$ is given by

$$\underline{A}_\perp = \frac{1}{k} \left[\frac{4\pi}{V}\right]^{\frac{1}{2}} (\omega(\widetilde{\underline{Q}}_L + \widetilde{\underline{Q}}_w) \cos (k \cdot z) - (\widetilde{\underline{P}}_L - \widetilde{\underline{P}}_w) \sin (k \cdot z)),$$
$$(10)$$

where V is the mode volume. From now on we evaluate all the physical quantities in that frame.

We assume that the fields are circularly polarized, so that we have (for both laser and PRF radiation):

$$\widetilde{\underline{P}} \equiv \left[\frac{P}{\sqrt{2}} , -\frac{\omega Q}{\sqrt{2}} , 0\right]; \qquad \widetilde{\underline{Q}} \equiv \left[\frac{Q}{\sqrt{2}} , \frac{1}{\omega} \frac{P}{\sqrt{2}} , 0\right]. \qquad (11)$$

Furthermore we assume that, in the chosen reference frame, the electron transverse canonical momentum is zero (p_x = p_y = 0). This assumption is allowed by the invariance of p_x and p_y (H does not depend on their conjugate variables x and y).

We write the Hamiltonian (9) in a more compact form, by using the contact transformation for the fields, generated by the function,

$$F(Q_L, \phi_L; Q_w, \phi_w | t) = \frac{\omega}{2} (Q_L^2 \cotg (\omega t + \phi_L) + Q_w^2 \cotg(\omega t + \phi_w)) \qquad (12)$$

We have,

$$\begin{cases} P = \dfrac{\partial F}{\partial Q} = \omega Q \cot g \,(\omega t + \phi) \\[4mm] I = -\dfrac{\partial F}{\partial \phi} = \dfrac{\omega}{2} \, Q^2 \, \dfrac{1}{\sin^2(\omega t + \phi)} \end{cases} \tag{13}$$

The contact transformation resolved with respect to one set of variables (old and new respectively) are

$$\begin{cases} P = \sqrt{2I\omega}\,\cos(\omega t + \phi) \\[3mm] Q = \sqrt{\dfrac{2I}{\omega}}\,\sin(\omega t + \phi) \end{cases} \quad ; \quad \begin{cases} I = \dfrac{1}{2\omega}\,(P^2 + \omega^2 Q^2) \\[3mm] \phi = \arctan\left(\dfrac{\omega Q}{P}\right) - \omega t \end{cases} \tag{14}$$

The new Hamiltonian is given by ($r_o = e^2/mc^2$ = classical electron radius)

$$H_{new} = H_{old} + \frac{\partial F}{\partial t} = H_{old} - \omega\,(I_L + I_w) =$$

$$= \frac{p_z^2}{2m} + \frac{2\pi r_o c^2}{\omega V}\,(I_L + I_w + 2(I_L I_w)^{\frac{1}{2}}\cos\,(\phi_L - \phi_w - 2\,\mathbf{k}\cdot\mathbf{z})), \tag{15}$$

We note that H does not change if we consider linearly polarized fields instead of circularly polarized ones, apart terms rapidly oscillating at frequency 2ω.

The new canonical variable ϕ and its conjugate I are respectively the phase and the action of the field. The energy density w is related to the action I by the equation

$$w = \frac{1}{8\pi}\,(E^2 + H^2) = \frac{\omega I}{V}\,. \tag{16}$$

From Eq. (15) we find immediately the motion invariants, which have a very simple physical meaning:

1) Energy conservation: the Hamiltonian itself does not depend on time, i.e. the whole system energy is conserved through the interaction.

2) Conservation of the total field intensity: the Hamiltonian depends on $\phi_L - \phi_w$, therefore the sum of the conjugate coordinates is conserved

$$I_L + I_W = I_o = \text{const.} \tag{17}$$

(note that I_L and I_W contain also the associate electron self field).

3) Momentum conservation: a simple derivation shows that

$$p_z + \bar{k}(I_L - I_W) = \mu = \text{const.} \tag{18}$$

From Eq. (17) we derive that (in the chosen frame) the energy is supplied to the laser beam (if there is a positive gain) by the PRF, while (from Eq. (18)) the electron matches the momentum.

We note that H (Eq. (15)) is obtained from (9) without any

approximation. We can derive from it all the features of the STS.

Using the motion integrals (17) and (18) we can single out a closed differential equation for p_z:

$$\overset{0}{p_z^2} = \left(\frac{4\pi r_o c}{\omega V}\right)^2 \cdot ((kI_o)^2 - (p_z - \mu)^2) - $$

$$ - 4k^2 \left[H - \left(\frac{p_z^2}{2m} + \frac{2\pi r_o c I_o}{\omega V}\right)\right]^2 . \qquad (19)$$

The r.h.s. of Eq. (19) is a polynomial of fourth degree in p_z. Therefore an analytical solution can be found in terms of Jacobi elliptic functions [14]. However, in this lecture we shall not deal with Eq. (19).

3. SMALL GAIN APPROXIMATION

We write the Hamiltonian (15) in a more compact form by using the invariants (17) and (18) and by defining

$$\psi = \phi_L - \phi_w - 2kz; \quad p = p_z . \qquad (20)$$

We obtain in the independent variables p and ψ,

$$H = \frac{k}{m} p^2 + mcR\cos\psi, \qquad (21)$$

where we have defined

$$R = \frac{8\pi r_o}{mV} (I_L I_w)^{\frac{1}{2}} \qquad (22)$$

Note that R is a function of p (see Eqs (17) and (18)).

We assume that the variation of the laser field during the interaction time is small (small amplification per pass). Then we may assume

$$R \sim R_o = \frac{8\pi r_o}{mV} (I_L(0) \, I_w(0))^{\frac{1}{2}} \qquad (23)$$

We shall use this approximation throughout the lecture. In this connection the Hamiltonian (21) describes a pendulum-like motion. The small oscillation frequency is given by

$$\Omega = (2\omega R_o)^{\frac{1}{2}} . \qquad (24)$$

Let us define the new adimensional variables

$$\delta = 2 \frac{\omega}{\Omega} \frac{p}{mc} ; \quad \tau = \Omega t \qquad (25)$$

The motion equations derived from (21) are

$$\begin{cases} \delta' = - \sin \psi \\ \psi' = - \delta \end{cases} \qquad \left(\delta' = \frac{d\delta}{d\tau} = \frac{1}{\Omega} \frac{d\delta}{dt} , \text{ etc.} \right) \qquad (26)$$

The gain for the laser field is proportional to the variation of the electron momentum. We have indeed (see Eqs (17), (18) and (25))

$$\omega \cdot \delta I_L = - \frac{\omega}{2k} \Delta p = - \frac{mc^2}{4} \left(\frac{\Omega}{\omega} \right) \Delta \delta . \qquad (27)$$

The total variation $\omega \Delta I_L$ due to the whole e.b. is given by

$$\omega \cdot \Delta I_L = - \frac{mc^2}{4} \cdot \frac{\Omega}{\omega} N \left(\frac{1}{N} \sum_{i=1}^{N} \Delta \delta_i \right), \qquad (28)$$

where N is the number of the electrons and $\Delta \delta_i$ refers to the variation of the i-th electron. The number of particles in the e.b. is generally very high ($\sim 10^{10} \div 10^{12}$). Then assuming a continuum distribution for the initial phases and the initial momentum δ_o we get

$$\omega \Delta I_L = - \frac{mc^2}{4} \cdot \frac{\Omega}{\omega} N \int \rho (\delta_o, \psi_o) \Delta \delta (\delta_o, \psi_o; \Delta \tau) d\delta_o d\psi_o, \qquad (29)$$

where we have put

$\rho (\delta_o, \psi_o) = $ initial electron distribution function

$$\left(\int \rho (\delta_o, \psi_o) d\delta_o d\psi_o = 1 \right)$$

$\Delta \tau = \Omega \Delta t$; $\Delta t =$ interaction time.

In the following we shall limit ourselves to the study of a monochromatic and uniformly distributed (in z) e.b. In this condition we have ($0 \leqslant \psi \leqslant 2\pi$)

$$\rho (\delta, \psi) = \delta(\delta - \delta_o) \frac{1}{2\pi} , \qquad (30)$$

where $\delta(\delta - \delta_o)$ is the Dirac δ-function.

The gain per pass of the laser field is then given by

$$g = \frac{\omega \Delta I_L}{\omega I_L} = - \frac{mc^2}{4} \cdot \frac{\Omega}{\omega} \cdot \frac{N}{\omega I_L} \Delta (\delta_o, \Delta \tau), \qquad (31)$$

where we have defined

$$\Delta(\delta_o, \Delta \tau) = \frac{1}{2\pi} \int_0^{2\pi} \Delta \delta (\delta_o, \psi_o; \Delta \tau) d\psi_o. \qquad (32)$$

In order to evaluate the gain we must investigate the behaviour of the average e.b. momentum variation $\Delta(\delta_o, \Delta \tau)$. For this purpose it is very useful to inspect the phase plane (δ, ψ). In this way we can

better understand the physical mechanism which leads to amplification
in a FEL device.

4. PHASE PLANE PICTURE

The motion equations (26) can be derived from the Hamiltonian

$$H = \frac{\delta^2}{2} + \cos \psi \ . \tag{33}$$

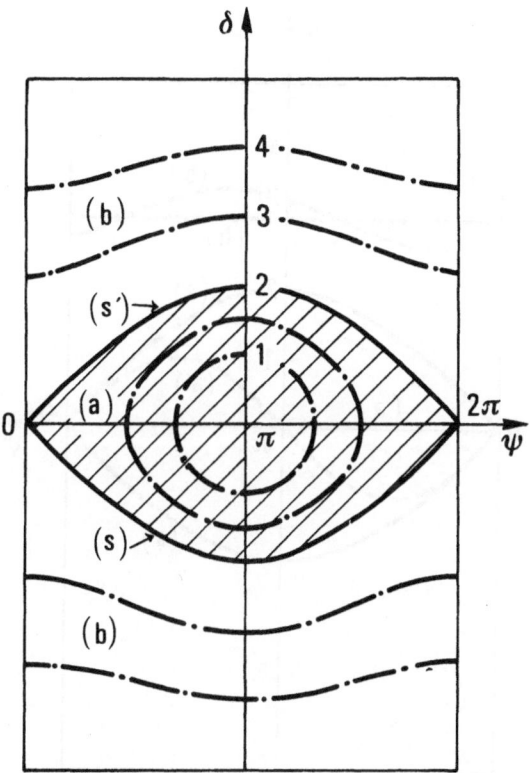

Fig. 1.   Phase plane trajectories relative to the Hamiltonian (26)
          (s)   separatrix
          (a)   periodic motion region
          (b)   aperiodic motion region

In Fig. 1 we have plotted the pendulum-like phase plane trajectories generated by (33). The phase plane is divided in two regions
by the separatrix

$$\delta = \pm\ 2 \sin \frac{\psi}{2} \tag{34}$$

Namely we have
a) Periodic motion around the stable equilibrium point $\delta = 0, \psi = \pi$

(dashed region inside the separatrix).
b) Aperiodic motion (outside the separatrix).
    We shall see that region b) corresponds to the small signal
regime and region a) to the saturation (see Sect. 6).
    The initial e.b. distribution is represented (see Eq. (30)) by
a straight line parallel to the ψ axis (see Fig. 2 curves a) and b)).

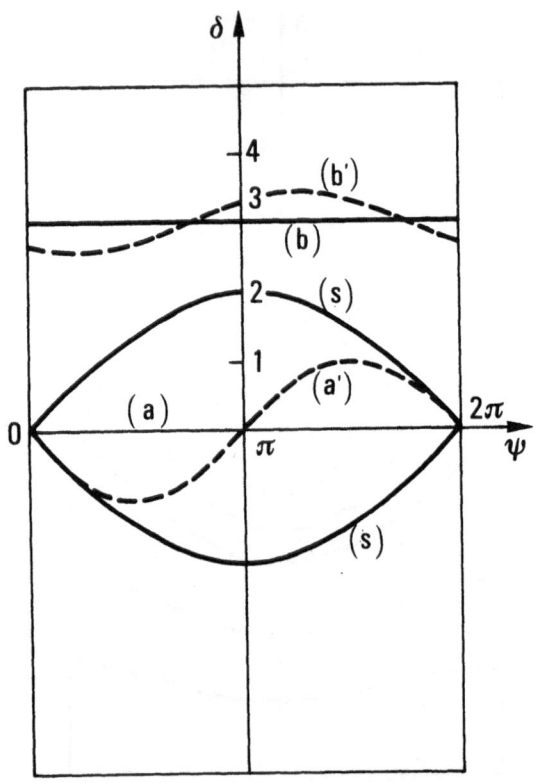

Fig. 2.   Monochromatic e.b. distribution function evolution.
          (s)   separatrix
          (a) δ_o = 0 ⎫
          (b) δ_o^o = 3 ⎬ initial distribution

          (a') δ_o = 0 ⎫
          (b') δ_o^o = 3 ⎬ final distribution

After the interaction the e.b. distribution is distorted (see Fig. 2
curves a')) and b')). The gain is due to the asymmetry of this modification which leads to a non vanishing Δ (δ_o ,Δτ).

When $\delta_o = 0$, Δ is identically zero at all times. This effect is due to the symmetry of the phase plane trajectories (see curve a') in Fig. 2. If $\delta_o \neq 0$ this symmetry is lost. (See curve b') in Fig. 2). Note that we have (again from the symmetry of the phase plane).

$$\Delta (\delta_o, \Delta\tau) = - \Delta(-\delta_o, \Delta\tau). \tag{35}$$

As a consequence the gain is antisymmetric with respect to δ_o. From the Eq. (25) we see that, for a fixed momentum p_o, δ_o increases as Ω/ω decreases. That is, for sufficiently small laser fields (see Eqs. (23) and (24)), the phase plane trajectories belong to the aperiodic region (region b) in Fig. 1).

5. SMALL SIGNAL REGIME

In this section we evaluate the function Δ in small signal regime, i.e. for $I_L \to 0$. We can express I_L as a function of Ω (see Eqs. (16) and (17))

$$I_L = \frac{V}{\omega} \cdot \left(\frac{m}{16\pi r_o}\right)^2 \cdot \frac{\Omega^4}{w_W}, \tag{36}$$

where we have defined (see Eq. (9))

$$w_W = \text{wiggler energy density} = \frac{\omega I_W}{V} \tag{37}$$

The gain, written in terms of Ω and p_o, is given by (see Eq. (25) and (36))

$$g = - \frac{(8\pi r_o c)^2}{m\omega} \cdot \frac{N}{V} w_W \frac{\Delta\left(\frac{2\omega}{\Omega} \cdot \frac{p_o}{mc}, \Omega\Delta t\right)}{\Omega^3}. \tag{38}$$

In the small signal regime we have

$$g_o = \lim_{I_L \to 0} g = \lim_{\Omega \to 0} g =$$

$$= - \frac{(8\pi r_o c)^2}{m\omega} \cdot \frac{N}{V} w_W \lim_{\Omega \to 0} \left\{ \frac{\Delta\left(\frac{2\omega}{\Omega} \cdot \frac{p_o}{mc}, \Omega\Delta t\right)}{\Omega^3} \right\}. \tag{39}$$

The behaviour of Δ for small Ω is given by (see App. A)

$$\Delta\left(\frac{2\omega}{\Omega} \cdot \frac{p_o}{mc}, \Omega\Delta t\right) = - \frac{1}{16} (\Delta t)^3 \Omega^3 S\left(\frac{2\omega\Delta t}{\pi} \cdot \frac{p_o}{mc}\right) + 0 (\Omega^5) \tag{40}$$

where we have defined

$$S(x) = -\frac{4}{\pi} \frac{d}{dx} \left[\frac{\sin\left(\frac{\pi}{2} x\right)}{\frac{\pi}{2} x} \right]^2 \tag{41}$$

The function $S(x)$ is plotted in Fig. 3.

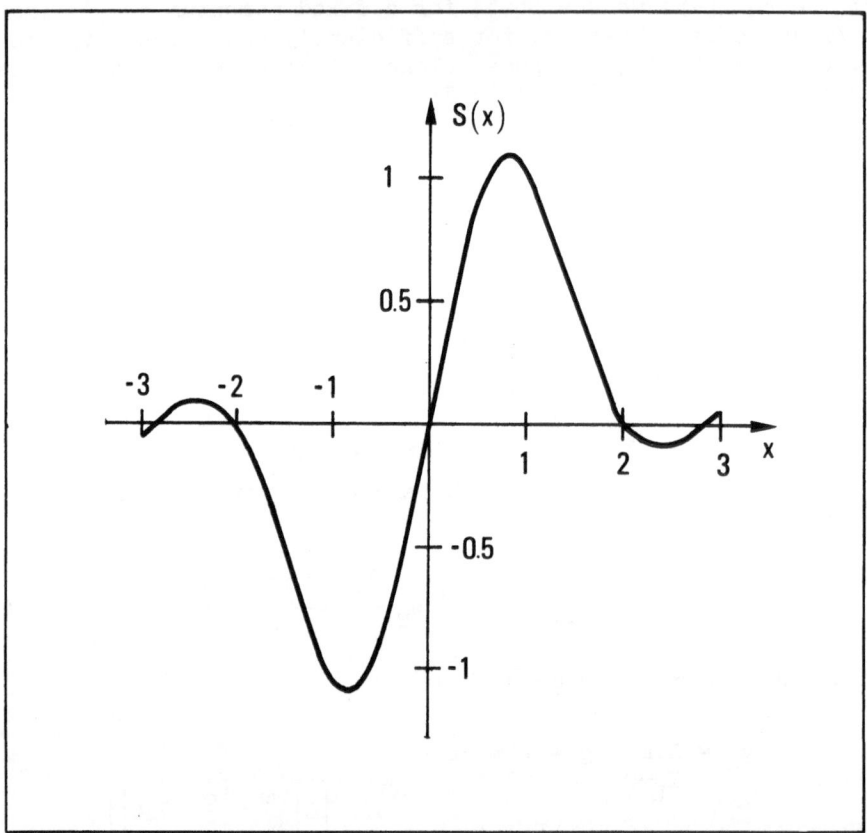

Fig. 3.   Gain profile in small signal regime.

The Eq. (39) becomes

$$g_o = \frac{(2\pi r_o c)^2}{m\omega} \cdot \frac{N}{V} (\Delta t)^3 w_W S \left(\frac{2\omega\Delta t}{\pi} \cdot \frac{P_o}{mc}\right). \tag{42}$$

It is straightforward to express g_o in terms of lab frame physical quantities. We have

$$\omega = \frac{2\pi c}{\lambda_q} \quad \gamma = \frac{\pi c}{\lambda \cdot \gamma} \qquad\qquad (\lambda = \text{laser wavelength in lab frame});$$

$$\frac{N}{V} = \frac{n_e F}{\gamma}$$

n_e = electron density,

F = filling factor = $\dfrac{\text{e.b. section}}{\text{laser beam section}}$,

w_W = $2\gamma^2 (w_W)_{LAB} = \dfrac{B^2}{4\pi} \gamma^2$,

B = peak wiggler magnetic field

Δt = $\dfrac{L}{\gamma c}$ (L = wiggler length in lab frame).

$$\frac{2\omega\Delta t}{\pi} = \frac{2L}{\lambda_q} = \left(\frac{\Delta\omega}{\omega}\right)^{-1} \qquad \text{(see Eq. (8))}. \qquad (43)$$

The electron intial momentum p_o has a simple physical meaning in terms of the physical quantities evaluated in the lab frame. We show in App. B that

$$\frac{p_o}{mc} = \frac{1}{2} \frac{\delta\omega}{\omega} , \qquad\qquad (44)$$

where $\delta\omega/\omega$ is the detuning between the backward spontaneous scattered radiation and the laser e.m. field.

The gain per unit lenght is then given by

$$\alpha_o = \frac{g_o}{L} = \frac{1}{2\sqrt{2}} \cdot \frac{r_o^2}{mc^2} \lambda^{3/2} \lambda_q^{3/2} B^2 n_e F \left(\frac{\Delta\omega}{\omega}\right)^{-2} S \left(\frac{\delta\omega/\omega}{\Delta\omega/\omega}\right) . \quad (45)$$

Equation (45) is in agreement with the previous gain derivations based on the Compton recoil [3] and the bunching mechanism [4].

6. STRONG SIGNAL REGIME

Let us consider a FEL device operating at the optimum detuning for the small signal regime (see Fig. 3).

$$\frac{\delta\omega}{\omega} \sim \frac{\Delta\omega}{\omega}$$

In this condition the electron initial momentum is given by (see Eqs (25) and (44))

$$\delta_o = \frac{\omega}{\Omega} \cdot \frac{\Delta\omega}{\omega} = \frac{\omega\lambda_q}{2L\Omega} = \frac{\pi}{\Omega\Delta t} . \qquad (46)$$

We assume that the initial electron distribution is given by the

Eq. (30), which corresponds to a non recirculated monochromatic e.b.
    For small fields, i.e. small Ω , δ_0 is very high. On the phase
plane this situation is represented by the curve a) of Fig. 4. All

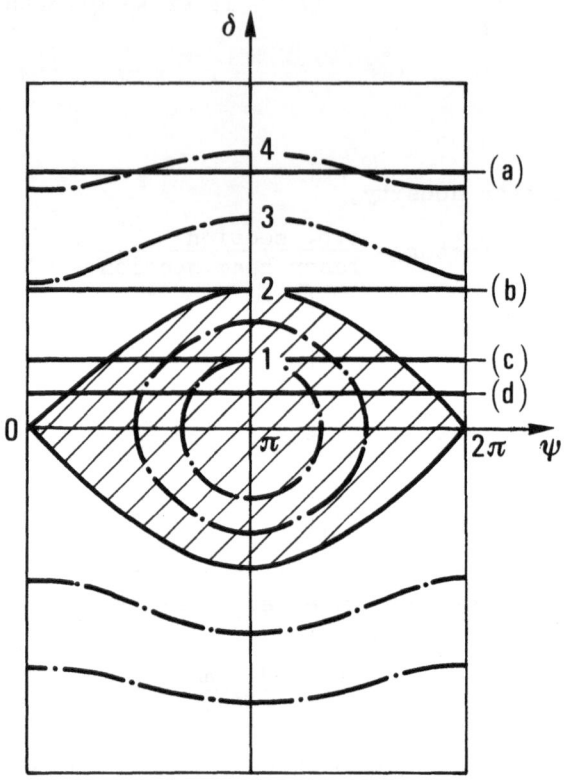

Fig. 4   Monochromatic e.b. initial distribution function in the
         phase plane ($\delta\omega/\omega = \Delta\omega/\omega$)
         (a) $\Omega\Delta t < \pi/2$, $\delta_0 > 2$ (small signal regime)
         (b) $\Omega\Delta t = \pi/2$, $\delta_0^0 = 2$
         (c) $\Omega\Delta t = \pi$, $\delta_0 = 1$
         (d) $\Omega\Delta t = 2\pi$, $\delta_0 = 0.5$

         — . —   phase plane trajectories.

the particles, as pointed out in the previous sections, are in the
aperiodic region. As Ω increases, δ_0 decreases. The e.b. enters the
periodic region when $\delta_0 \leqslant 2$ (see Fig. 4 curve b)). In this condition Ω is given by (see Eq. (46))

$$\Omega \cdot \Delta t > \frac{\pi}{2} \ . \tag{47}$$

In this situation a large fraction of the e.b. has a periodic motion
around the stable equilibrium point $\delta_0 = 0$, $\psi_0 = \pi$. The revolution

frequency is roughly given by the small oscillation frequency Ω. When Ω is big enough to satisfy the Eq.

$$\Omega \Delta t > \pi \qquad (\delta_o < 1) \qquad (48)$$

part of the e.b. moves back toward its initial configuration (see Fig. 4 curve c)).

   This effect leads to a net amplification decrease, which drives the FEL into the saturation regime. We have indeed that in one half of its closed trajectory the electron scatters radiation from PRF to the laser field ($0 \leqslant \psi \leqslant \pi$) while in the other half the opposite process takes place ($\pi \leqslant \psi \leqslant 2\pi$). After one period the electron has recovered its initial condition, i.e. its initial momentum is restored, then no net gain is available for the laser field. The situation is complicated by the fact that the periods of the electrons are different, because they depend on the initial phases. However we can say that, as order of magnitude, when the interaction time is equal to the small oscillation period, that is

$$\Omega \Delta t \sim 2\pi \qquad (\delta_o \sim 0.5)$$

a considerable fraction of the e.b. has roughly recovered its initial momentum. As a consequence the gain is practically reduced to zero.

   The saturation regime of the FEL has a close analogy with saturation in conventional lasers. In the latter case the amplifying medium consists of atoms or molecules with discrete energy levels. If an atom of the system is prepared in one of the two levels of the lasing transition, then, under the influence of the coherent laser field, it starts oscillating between the two levels. In the saturation regime, the oscillation period is shorter than the life time of the levels. So that several oscillations occur before the atom decays to its ground state. This oscillation is called Rabi flopping and its frequency is the Rabi frequency, which is given by

$$\Omega_R = \frac{\mu \cdot E_o}{\hbar} \quad , \qquad (49)$$

μ = dipole transition momentum,
E_o = resonant electric field.

   In one half of the oscillation period the atom absorbs energy from the field, while in the other one the atom is stimulated to emit energy in the field mode.

   Under this point of view we can call Ω the *Rabi frequency* of the FEL. In terms of the laser and wiggler electric fields (E_L, E_W), Ω is written in the form

$$\Omega = \frac{e\lambdabar_e}{\hbar} \sqrt{2E_L E_W}, \qquad (50)$$

where we have put,

$\lambdabar_e$ = reduced Compton wavelength of the electron.
   The corrispondence with the usual Rabi frequency is straightforward. We have (see Eq. (49))

$$\mu \;\rightarrow\; e \varkappa_e \;, \qquad E_o \;\rightarrow\; \sqrt{2 E_L E_W} \;.$$

Now we can derive easily the order of magnitude of the maximum energy which can be extracted from the e.b. Saturation is reached when the interaction time is about one half of the *Rabi* period, that is (see Eq. (48)).

$$\Omega \;\sim\; \frac{\pi}{\Delta t} \;, \quad \delta_o \;\sim\; 1 \;\Rightarrow\; \Delta \;\sim\; -2. \qquad (51)$$

The laser energy variation is given by (see Eq. (31)).

$$\omega I_L \;\cdot\; g = -\frac{mc^2 \Omega N}{4\omega} \; \Delta \;\sim\; \frac{mc^2 \pi N}{2\omega \Delta t} \;. \qquad (52)$$

From the Eq. (43) we obtain

$$\omega I_L \;\cdot\; g \;\sim\; \frac{mc^2 N}{2} \;\cdot\; \frac{\Delta \omega}{\omega} \;. \qquad (53)$$

In the lab frame we have

$\Delta \mathcal{E}_L$ = laser energy variation = $2\gamma(\omega I_L g) \sim$

$$\sim (mc^2 \gamma N) \left[\frac{\Delta \omega}{\omega} \right], \qquad (54)$$

which corresponds to the efficiency (e-beam → laser beam)

$$\eta = \frac{\Delta \mathcal{E}_L}{(mc^2 \gamma) N} \;\sim\; \frac{\Delta \omega}{\omega} \;. \qquad (55)$$

The Equation (55), derived in a qualitative way, is in agreement with previous analytical evaluations [4]. The use of a recirculated e.b. does not increase the efficiency of the FEL. Indeed the strong energy modulation destroys the monochromaticity of the e.b. This effect, in turn, lowers the amplification. It has been shown [15] that the maximum efficiency we can get in this situation is given again (roughly) by the Eq. (55). However the use of a storage ring for a FEL device is very actractive. It is possible to store high peak currents (many Ampères) for a long time (many hours), with high efficiency in the e.b. acceleration process ($\sim 25 \div 50\%$).

APPENDIX A

From the Eqs. (25), (26) and (32) we have

$$\Delta(\delta_o, \Delta \tau) = -\frac{1}{2\pi} \int_0^{\Delta \tau} d\tau \int_0^{2\pi} d\psi_o \; \sin \psi(\delta_o, \psi_o; \tau) =$$

$$= -\frac{\Omega}{2\pi} \int_0^{\Delta t} dt \int_0^{2\pi} d\psi_o \; \sin \psi (\delta_o, \psi_o; t). \qquad (A.1)$$

The evolution of ψ is given by (see Eqs (26) and (33))

$$\int_{\psi_0}^{\psi} \frac{d\overline{\psi}}{\sqrt{1 + \frac{2}{\delta_0^2} (\cos \psi_0 - \cos\overline{\psi})}} = - \delta_0 \tau = - 2\omega t \frac{P_0}{mc} . \qquad (A.2)$$

In the small signal regime we have $\delta_0 \gg 1$ (see Sect. 4). To the first order in $(1/\delta_0)^2$, the Eq. (A.2) becomes

$$\psi \cong \psi_0 - \pi x + \left(\frac{1}{\delta_0}\right)^2 (\sin \psi_0 (1 - \cos(\pi x)) + \cos\psi_0(\sin(\pi x)-\pi x)), \qquad (A.3)$$

where we have put

$$x = \frac{2\omega t}{\pi} \frac{P_0}{mc} . \qquad (A.4)$$

From Eq. (A.3) we have

$$\sin\psi \sim \sin (\psi_0 - \pi x) + \cos(\psi_0 - \pi x)\left(\frac{1}{\delta_0}\right)^2 (\sin \psi_0(1-\cos(\pi x)) +$$

$$+ \cos \psi_0 (\sin(\pi x) - \pi x)). \qquad (A.5)$$

The average of $\sin\psi$ over ψ_0 is given by

$$\frac{1}{2\pi} \int_0^{2\pi} d\psi_0 \sin \psi = \frac{1}{2\delta_0^2} (\sin (\pi x) - \pi x \cos (\pi x)). \qquad (A.6)$$

From Eqs. (A.1), (A.4) and (A.6) we obtain,

$$\Delta(\delta_0, \Delta\tau) = - \frac{\Omega}{2\delta_0^2} \int_0^{\Delta t} dt (\sin(\pi x) - \pi x \cos(\pi x)) =$$

$$= - \frac{2}{\delta_0^3} \left[\sin \left[\frac{\pi}{2} x\right] \left[\sin\left[\frac{\pi}{2} x\right] - \frac{\pi}{2} x \cos\left[\frac{\pi}{2} x\right]\right]\right] = \qquad (A.7)$$

$$= \frac{1}{16} (\Omega \cdot \Delta t)^3 \frac{d}{dx} \left[\frac{\sin \left[\frac{\pi}{2} x\right]}{\frac{\pi}{2} x}\right]^2 \Bigg|_{x = \frac{2\omega\Delta t}{\pi} \cdot \frac{P_0}{mc}}$$

The next term in the development of ψ (Eq. (A.3)) is of the order of $(1/\delta_0)^4$. As a consequence in the Eq. (A.7) we have neglect terms of the order $(1/\delta_0)^5$.

APPENDIX B

The laser frequency in the lab frame is given by

$$\omega_L = 2\gamma\omega = \frac{4\pi c}{\lambda_q} \gamma^2 , \qquad (B.1)$$

where we have put

$$\gamma = (1 - (v/c)^2)^{-\frac{1}{2}} ,$$

v = lab frame velocity with respect to the chosen reference frame (Σ_o).

λ_q = wiggler wavelength in lab frame

The spontaneous emission frequency is given by [8]

$$\omega_s = \left(\frac{E}{mc^2}\right)^2 \frac{4\pi c}{\lambda_q} \qquad (B.2)$$

where E is the electron energy in the lab frame. To the first order in p_o/mc we have (p_o = initial e.b. momentum in Σ_o)

$$E^2 \cong (mc^2)^2 \gamma^2 \left[1 + \frac{2p_o}{mc}\right] . \qquad (B.3)$$

From the Eqs (B.1), (B.2), (B.3) and (25), we obtain

$$\frac{\delta\omega}{\omega} = \frac{\omega_s - \omega_L}{\omega_L} = \frac{2p_o}{mc} . \qquad (B.4)$$

REFERENCES

[1]   L.R. ELIAS et al., Phys. Rev. Letts 36 (1976),717
      D.A.G. DEACON et al., Phys. Rev. Letts 38 (1977), 892
[2]   For a Comprehensive View of the Physics of the Electron
      Storage Rings see, M. Sands, SLAC Report n. 121 (1970)
      Stanford University (California)
[3]   J.M.J. MADEY, J. Appl. Phys. 42 (1971),1906
[4]   F.A. HOPF, P. MEYSTRE, M.O. SCULLY and W.L. LUISELL, Opt.
      Comm. 18, (1976) 413, and Phys. Rev. Letts 37 (1976),1342
[5]   See, for example, Wiggler Magnets,"Stanford Synchrotron Radia-
      tion Project (SSRP)", Report n. 77/05 (1977), Ed. by
      H. Winick and T. Knight, Stanford University (California)
[6]   J.P. BLEWETT and R. CHASMAN, J. Appl. Phys. 48 (1977),2692
[7]   W. HEITLER "The Quantum Theory of Radiation" Oxford 1960,p.414
[8]   B.M. KINCAID, J. Appl. Phys. 48 (1977),2684
[9]   See Ref. (5) p. VI - 104
[10] See, for example, L. LANDAU and E.LIFCHITZ "Electrodynamique
      des Milieux Continus" Edition Mir (1969) p. 496
[11] H. MOTZ, J. Appl. Phys. 22 (1951),527
      H. MOTZ, M. THON and R.N. WHITEHURST, J. Appl. Phys. 24 (1953),
      826
[12] J.J. SAKURAI, "Advanced Quantum Mechanics", Addison Wesley (1967)
      p.301

[13] See Ref. (5) p. II - 191
[14] See, for example, M. ABRAMOVITZ, I.A. STEGUN, "Handbook of
      Mathematical Functions", Dover Pub. (1970), p. 569
[15] A. Renieri, to be published

FREQUENCY TUNABILITY AND HIGH SPEED MODULATION OF FAR INFRARED

OPTICALLY PUMPED LASERS BY STARK EFFECT

Franco Strumia

Istituto di Fisica della Università

Piazza Torricelli, 2 - 56100 Pisa, Italy

By Far Infrared (FIR) we mean the portion of the electromagnetic spectrum between 25 μm and 1 mm wavelength. In the past this region was the less studied and used especially for the lack of suitable sources. In fact frequency multiplication from microwaves suffer from exceedingly high power losses while the difference frequency mixing from near infrared or visible show a poor frequency stability and a low efficiency as a consequence of the reduced transparency in the FIR of the electrooptic crystals.

A first significant improvement was obtained in 1963 with the discovery of the glow discharge molecular laser. The importance of this discovery was immediately recognized and extensive researches for new molecules and new lines were carried out. However only four molecules were found to lase in the FIR (H_2O, HCN, H_2S, and SO_2) (1,2). The key step toward a more general laser source for the FIR was made by T.Y. Chang and T.J. Bridges in 1970 with the discovery of the optically pumped molecular FIR laser (3). They observed that the 9-P(20) line of the CO_2 laser was strongly absorbed by the CH_3F molecules. This result was recognized as a near perfect coincidence with the $\nu_3 : \nu = 0 \leftrightarrow 1$ Q(12,2) vibrational transition of the CH_3F. By putting methylfluoride gas in a mirror resonator a strong laser emission at 496 μm was observed when the CO_2 radiation was also introduced in the resonator. In fig. 1 the scheme of the levels involved is given. The laser action is a consequence of the pumping of the CH_3F molecules into the rotational level of the excited vibrational state by the CO_2 radia-

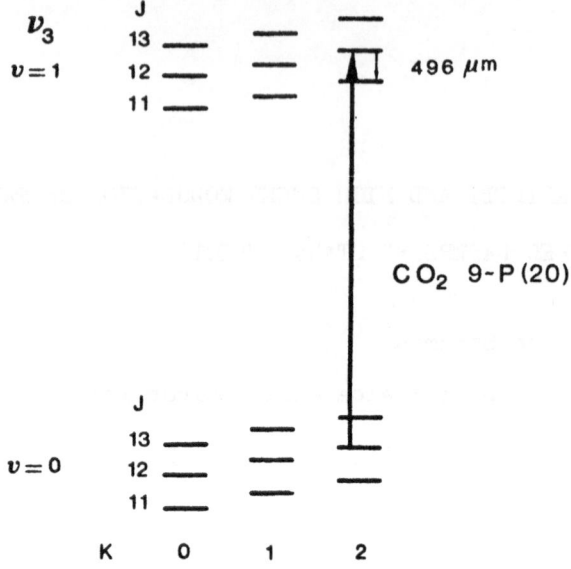

Fig. 1 – Partial scheme of the CH$_3$F levels.

tion. A good population inversion is possible because the energy of the pump radiation is larger than kT so that the other levels of the excited state are much less populated. It is obvious that this scheme is quite general and laser action is expected every time a molecular line is in resonance with a pumping radiation of sufficient power and with $h\nu > kT$. Since the FIR laser are generally associated to rotational transitions, molecules with a permanent electric dipole moment are also necessary. Up to the present more than 40 molecules were found to lase with more than 1000 lines covering the range from 12 μm to 2 mm. Many of this lines were found to lase in the CW mode (4,5). By changing the length of the resonator each of this lines can be frequency tuned by an amount that was found to be of the order of the Doppler width of the line wich is some 10^{-6} of their wavelength. So in spite of the large number of the discovered laser lines the FIR is covered only by discrete set of frequencies.

Another important feature to consider is the possibility of high speed frequency modulation which is necessary for many applications and in particular for phase-locking the FIR lasers to high order harmonics of the primary frequency standards. In this way the precision of frequency measurements in the FIR will be improved by orders of magnitude.

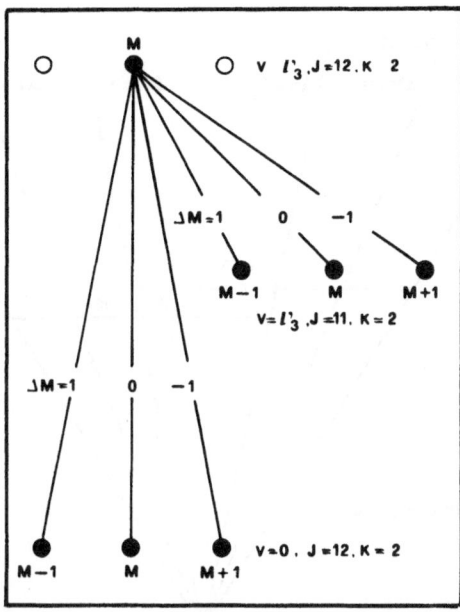

Fig. 2 - Labeling of the M sublevel adapted in the text.

Several methods can be considered for increasing the frequency tunability. However the collision broadening cannot be used because the collisions drastically reduce the population inversion between the active levels. The electrooptic crystals have two main draw-backs a) the transparency in the FIR is too low for intracavity operation, b) the electric field needed for obtaining a high modulation index for submillimeter wavelengths exceeds the breakdown limit. A more promising method for frequency tuning is the Stark effect on the lines (6). It is worth noting that an electric field applied to the lasing medium will not modify the pumping mechanism nor introduce lossy material into the laser. In addition the Stark shift is capable of high speed because the response time can be as short as the average time which photons spend inside the laser. The main limits for the use of the Stark effect are the breakdown threshold of the gas and the need for the absorbing vibrational line to remain in resonance with the pumping radiation. For each pumping coincidence, three FIR laser lines can be obtained in accordance with the $\Delta M = 0, \pm 1$ selection rule. These lines are tunable by an amount which is proportional to the width of the absorption line and to the tunability range of the pumping radiation. Moreover the Stark effect can also be used to obtain laser action from levels which are close to - but not in - resonance with the pumping radiation. In this way new lines from NH_3 were obtained (7).

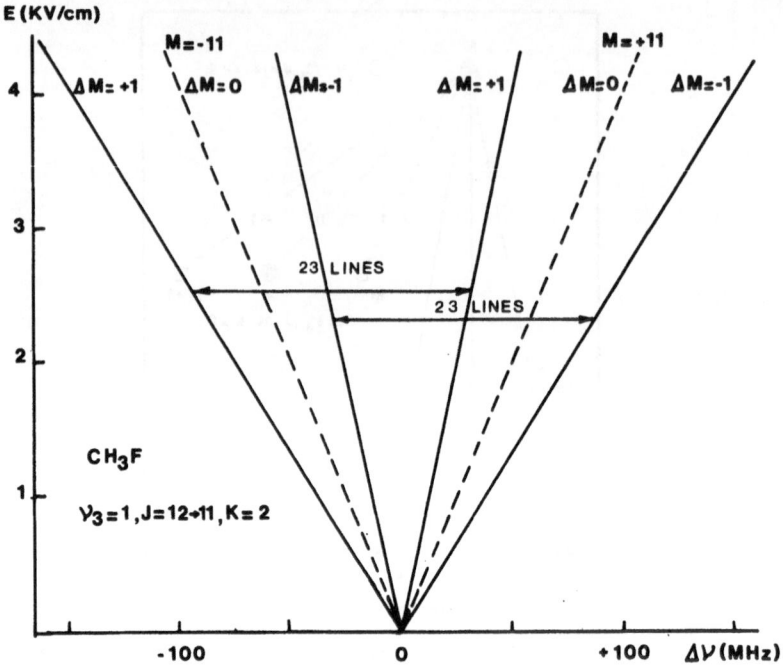

Fig. 3 – Positions of the Stark components of the J,K = 12,2 →11,2
transition as a function of the electric field.

The Stark effect tunability of the 496 μm line of CH_3F was
computed in ref. 6. The CH_3F was selected for two reasons a) the
496 μm line is one of the most powerful of the FIR lasers, b) CH_3F
is a well studied symmetric top molecule, the transitions are assigned and the spectroscopic constants are known with a precision sufficient to allow a detailed analysis. The 9-P(20) line of the CO_2
laser excite the J=12, K=2 transition of the V_3, Q branch, while
the J=12 → 11, K=2 rotational line is responsible for the laser
emission at 496 μm. In the presence of an electric field E the selection rules for the Q lines are

$$\Delta J=0, \quad \Delta K=0, \quad \Delta M=0, \pm 1$$

while for the rotational lines are

$$\Delta J=\pm 1, \quad \Delta K=0, \quad \Delta M=0, \pm 1$$

In the following we will label the level as shown in fig. 2.

The Stark frequency shift of the components of the rotational
transition is given by

$$\Delta V_R = \frac{\mu_e E K}{J(J^2-1)\hbar} \left[2M - (J+1)\Delta M\right] =$$

$$= (2.236\,M - 14.53\,\Delta M)\cdot E \qquad MHz/(kV/cm) \quad 1)$$

where μ_e is the permanent electric dipole moment of the excited vibrational state. Terms in E^2 are negligible up to a field of several kV/cm. For the vibrational transitions the Stark shifts are given by

$$\Delta V_v = \frac{E K}{J(J+1)\hbar} \left[(\mu_o - \mu_e)M - \mu_e \Delta M\right] =$$

$$= (-0.302\,M - 12.30\,\Delta M)\cdot E \qquad MHz/(kV/cm) \quad 2)$$

where μ_o is the electric dipole moment of the ground vibrational state. From eqs. 1 and 2 it follows that the shift in the vibrational lines is about one order of magnitude smaller as a consequence of the term in $(\mu_o - \mu_e)$. This result is important because, considering also that the Doppler broadening is larger for the vibrational transition, it permits to keep an efficient CO_2 pumping even in the presence of a high electric field.

In fig. 3 the positions of the Stark components of the rotational line are shown. As follows from eq. 1, 12 lines with $\Delta M = -1$ (from M = -12 to M = -1) have the same frequency as 12 lines with $\Delta M = +1$ (from M = +1 to M = +12). As a consequence for $\Delta M = 0$ we obtain 23 lines, while for $\Delta M = \pm 1$ we obtain 46 lines of which only 34 are resolved. The first evidence that the Stark effect induced some modifications on the FIR lasers was found by monitoring the output power of the laser as a function of the applied electric field (8,9,10,11,12). The lasers resonators were modified in order to allow the application of the electric field on the lasing medium. In refs. 8,9 as a resonator a dielectric circular waveguide with external Stark plates was used. As shown in fig. 4 in ref. 10-12 a rectangular metal-dielectric waveguide was used (13), which proved to be the best solution because, for field higher than about 500 V/cm, is avoided any screening effect due to deposition of electric charges on the inner side of the dielectric waveguides. The experimental evidence, that a frequency tuning of a optical pumped

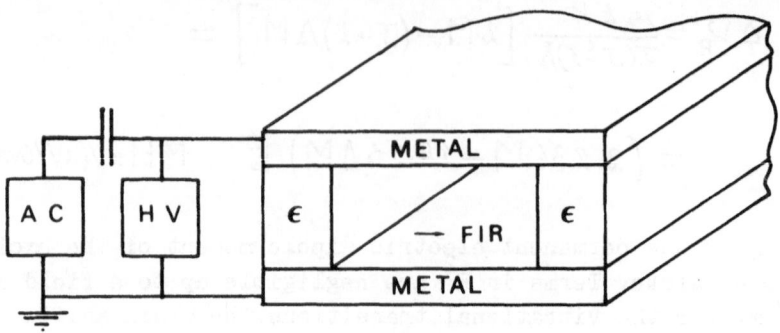

Fig. 4 – Structure of a metal–dielectric waveguide suitable for
Stark tuning of optically pumped FIR lasers.

FIR laser can be obtained by means of the Stark effect, is possible
only by directly measuring the frequency of the emitted radiation
as a function of the electric field intensity. In ref. 14 this
measurement was performed by mixing together the output ot two CH_3F
FIR lasers of which one was equipped with internal Stark electrodes
on the other was used only as a local oscillator. The experimental
apparatus is shown in fig. 5. A MOM open structure diode was used
for the mixing and the frequency resolving power was limited to a
few KHz by the instability of the free running reference laser.
Experimental results obtained in the case of a DC electric field
are shown in fig. 6. By applying an AC electric field a high speed
modulation was observed and, in suitable conditions, a nearly pure
frequency modulation was observed as shown in fig. 7. In refs. 15
and 16 the FIR laser output was mixed with a high order harmonic
of a frequency stabilized Klystron. In this case the resolution was
limited by the degradation of the spectral purity of the carrier as
a consequence of the very high multiplication order (17). The measured frequency tuning was in agreement with that reported in ref. 14.
In the above experiments the applied electric fields are too low
for a resolution of the single Stark components and the obtained
frequency tuning is a combined effect of a cavity pulling and Stark
effect. When the laser resonator is set for maximum output power,
i.e. the cavity resonance is set exactly in coincidence with the
CH_3F line, no Stark tuning is observed. In fact at low E fields the
Stark effect causes only a symmetric broadening of the unperturbed
gain profile and no shift direction is preferred. When the FIR resonator is slightly detuned, the application of an E field causes
a frequency shift because the cavity pulling is controlled by the

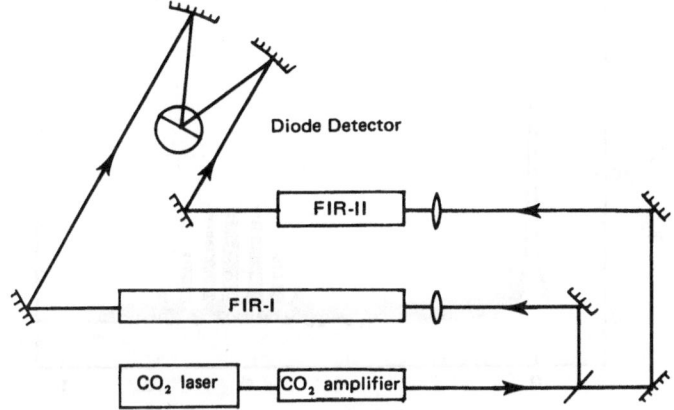

Fig. 5 – Schematic diagram of system for heterodyning two far infrared (FIR) lasers. The frequency of FIR-II may be tuned by applying an electric field (14).

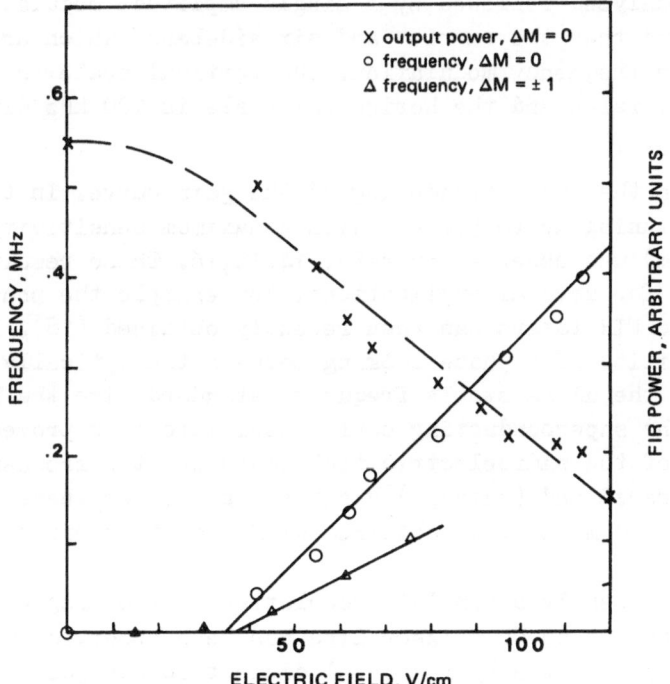

Fig. 6 – Frequency shift and output power of a far infrared (FIR) laser with static electric field applied parallel (Δ M=0) and perpendicular (Δ M=±1) to the oscillating electric field (14).

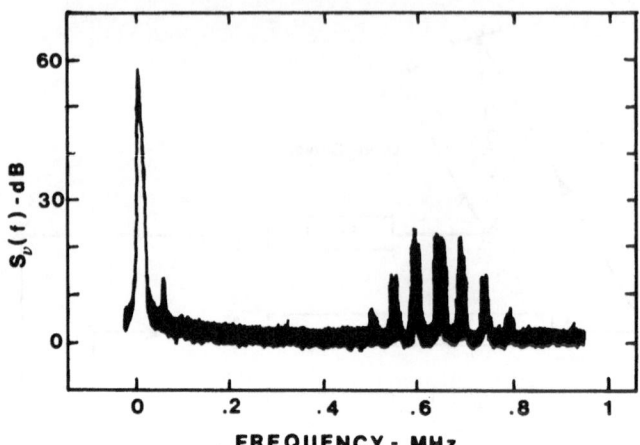

Fig. 7 – Frequency modulation (spectral density of voltage) of a
        far infrared laser modulated at 50 kHz with a modulation
        index greater than 1. From the left to the right are a
        large peak due to the local oscillator of the spectrum
        analyzed followed by a single amplitude modulation peak
        and then the carrier and six sidebands which are 90% due
        to frequency modulation. The vertical scale is 10 dB/
        division and the horizontal scale is 100 kHz/division(14).

E field via the Stark broadening of the gain curve. In this way a
frequency tuning up to 500 kHz with a maximum sensitivity of about
5 kHz/(V/cm) was observed in refs. 14,15,16. These results are just
sufficient for several applications. For example the phase locking
between two FIR lasers has been recently obtained (18). In this way
the feasibility of a phase locking between the optically pumped FIR
lasers and the ultra stable frequency standard like the hydrogen
maser or the superconducting cavity oscillators is proved with the
extension of the radioelectric techniques and the frequency synthesis
to the infrared and (perhaps) visible region. For these purposes
the high speed modulation obtained with the Stark effect is needed
(19).

    More recently a complete resolution of the single Stark compo-
nents of the 496 μm CH$_3$F laser line has been observed at NBS (20).
For this experiment a hybrid metal–dielectric waveguide was used
(see fig. 4) as a FIR resonator and the single components were observed by measuring the laser output power as a function of the resonator length. A frequency tuning of many MHz was observed. As a con-

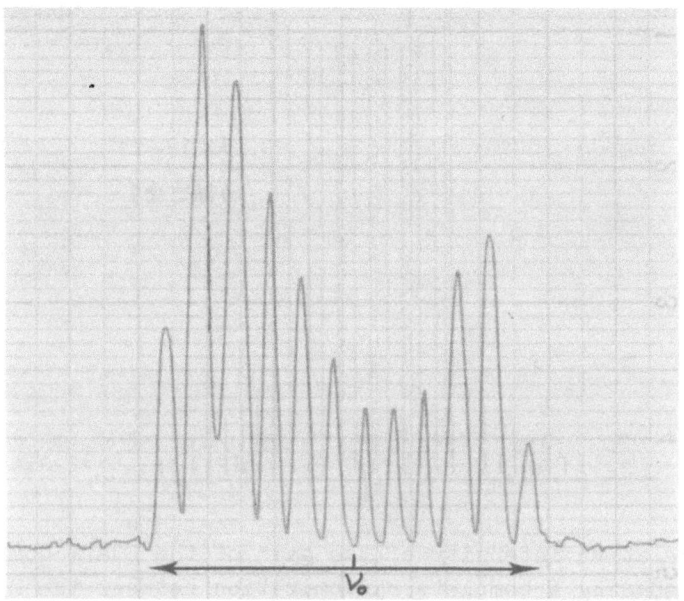

Fig. 8 — Emission from a CH_3F 496 μm laser when a field of
1.56 KV/cm is applied and the CO_2 pump is absorbed
with $\Delta M = 0$ selection rule. Total splitting 38.4 MHz
(25).

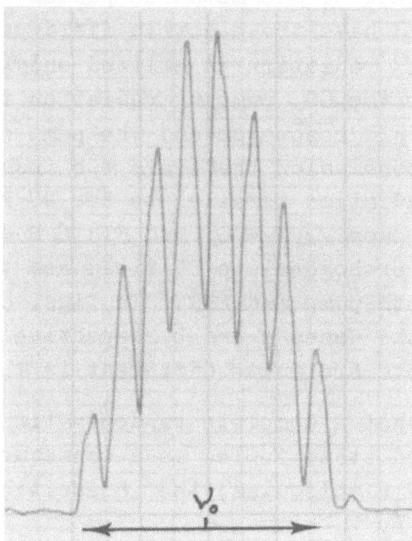

Fig. 9 — Emission from a CH_3F 496 μm laser when a field of
1.70 KV/cm is applied and the CO_2 pump is absorbed
with $\Delta M = \mp 1$ selection rule. Total splitting 26.6
MHz (25)

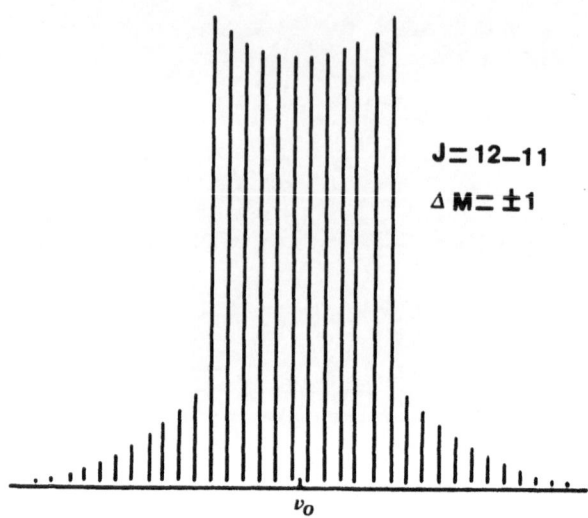

$$J = 12-11$$
$$\Delta M = \pm 1$$

Fig. 10 – Relative intensity of the gain of the Stark components
        assuming a complete randomization between the M state
        of the J,K=12,2 level.

sequence of the adopted waveguide structure the DC electric field
is obviously orthogonal to the metallic plates while all the low
loss modes in the FIR have the electric field parallel to the metallic plates and FIR radiation is emitted only with the selection
rules $\Delta M = \pm 1$. For the CO_2 pumping radiation these is no preferred
polarization in the FIR resonator and the pump selection rules are
$\Delta M = 0, \mp 1$. In conclusion there are two independent cases:
a) E field of CO_2 radiation parallel to the DC E field with the
selection rules for pump $\Delta M = 0$, and FIR $\Delta M = \mp 1$ respectively,
b) both CO_2 and FIR orthogonal to DC field and the selection rules
are $\Delta M = \pm 1$ for both pump and FIR. In figs. 8 and 9 the experimental results for the cases a and b respectively are shown. As
we can see the results are quite different in the two cases.

        If we assume that a complete randomization between the M sub-
levels of the excited state follow as a consequence of the collisions between the CH_3F molecules, the intensities of the different
components is given by

$$I(M \rightarrow M \mp 1) \propto (J \pm M)(J \pm M - 1) \qquad\qquad 3)$$

and an intensity pattern as shown in fig. 10 is expected irrespective of the polarization direction of the CO_2 pump radiation.

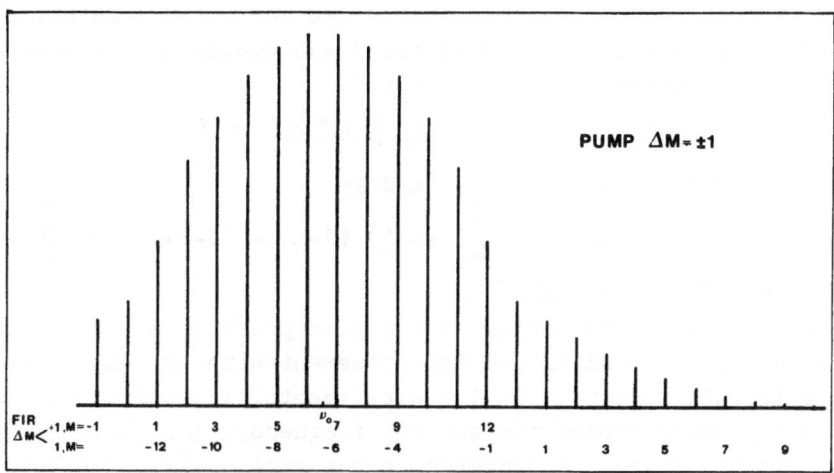

Fig. 11 – Relative intensity of the gain of the Stark components assuming no randomization in the upper state and absorption of the CO_2 pump radiation with the selection rule $\Delta M = \pm 1$. The structure is symmetric around ν_0 but the low intensity component with $M < 0$ are not reported for the sake of simplicity.

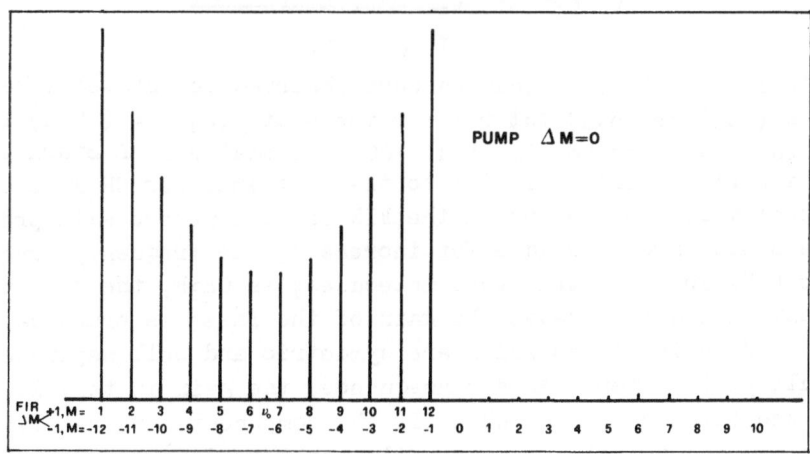

Fig. 12 – Relative intensity of the gain of the Stark components assuming no randomization in the upper state and absorption of the CO_2 pump radiation with the selection rule $\Delta M = 0$. Also in this figure some components with $M < 0$ are not shown.

Obviously this is not the case and we are forced to consider the case of no randomization between the M sublevels. As a consequence the relative intensities are given by

$$I_M \propto M^2 \; (\; J \overset{+}{\underset{-}{}} M) \; (\; J \overset{+}{\underset{-}{}} M - 1 \;) \qquad\qquad 4)$$

for the case a) $E(CO_2) \; // \; E_s$ and by

$$I_M \propto \frac{1}{4} (J^{+}_{-}M) (J^{+}_{-}M-1) \left[(J+M+1)(J-M) + (J-M+1)(J+M) \right] \quad 5)$$

for the case b) $E(CO_2) \perp E_s$.

The corresponding intensities of the gain profiles are given respectively in figs. 11 and 12 and the agreement with the experimental results is satisfactory. In the experimental intensities there is also an asymmetry around the central frequency ν_o. This asymmetry can be explained as a consequence of the small difference between the CO_2 pump frequency and ν_o . The experimental evidence of no mixing between the M sublevels is in agreement with the results obtained in collision–induced optical double resonance (21).

It was observed that for symmetric top molecules as CH_3F the dipole–dipole interaction of collision pairs dominates. The corresponding cross section depends strongly on the ratio between the K and J values being given by (22).

$$\sigma \propto \frac{K^2 \; (J \overset{+}{\underset{-}{}} M + 1) \; (J \overset{-}{\underset{+}{}} M)}{J^2 \; (J + 1)^2} \qquad\qquad 6)$$

In ref. 21 satellite resonances were observed for states like $(J,K) = (4,3)$ or $(5,3)$ but not for the state $(J,K) = (12,2)$ for which the cross section is about 100 time smaller. A Stark frequency tuning with results similar to that obtained for CH_3F is expected for other molecules lasing in the FIR and the method will probably become a standard technique for increasing the frequency coverage of the FIR. For symmetric topo molecules, as CH_3F, the case a) is the most favorable because the gain of the laser is practically restricted to few lines which are symmetric and well separated from the multiplet center. As a consequence, the gain in this lines is above the threshold, even when all the components are resolved.

In fig. 13 the computed gain profile for the $(J,K) = (12,2)$ line of CH_3F (23) is shown. As we can see the gain of the strongest components is about 1/8 of that of the unperturbed line and are symmetrically spaced of the amount

$$\Delta \nu = \; \overset{+}{\underset{-}{}} \; 12.3 \; \text{MHz}/(\text{kV}/\text{cm}) \qquad\qquad 7)$$

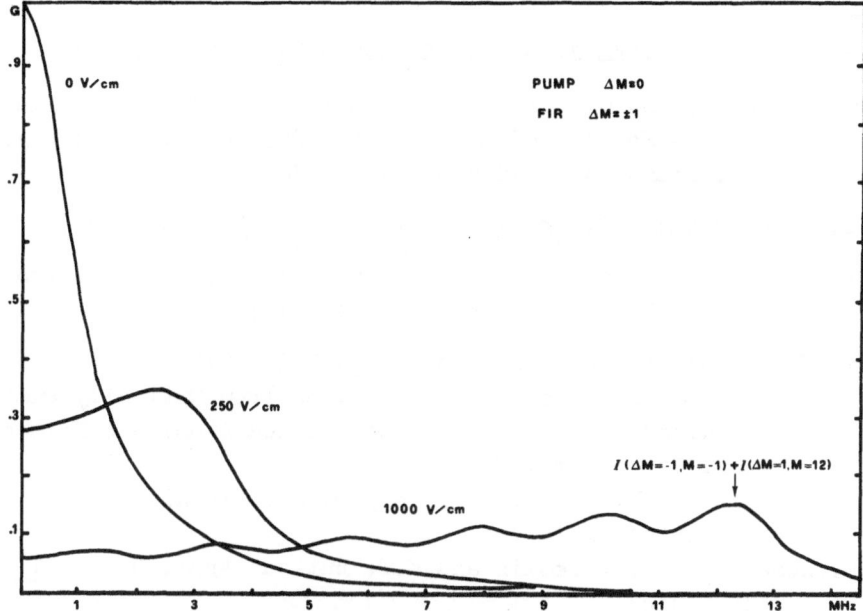

Fig. 13 – Gain profile of the 496 μm line in absence of electric
field and in presence of an electric field of 250 V/cm
and 1000 V/cm respectively. The curves are symmetric
around ν_0 and only one half is shown.

In conclusion we should expect that the frequency tunability of the
optically pumped FIR lasers will be increased by about two order of
magnitude by an appropriate use of the Stark effect. The Stark modulation has also been used to increase the output power of an NH_3
optically pumped FIR laser (24). In fact by the application of an
AC E field the frequency radiation sweeps across the whole profile
of the absorbing line reducing the saturation due to "hole burning"
and allowing the entire velocity distribution in the initial state
to be pumped. This method is interesting for the cases where the
inhomogeneous broadening is important as for molecules of low molecular weight. In the case of NH_3 a tenfold increase of the output
power was observed (24).

The author is particularly indebted to Drs. K.M. Evenson and D.A.
Jennings for the permission of reproducing unpuplished results and
to Dr. P. Minguzzi for critical reading of the manuscript.

References

1 — P.P. Coleman: IEEE J. Quant. Electron QE-9, 130 (1973)

2 — G.W. Chantry and G. Duxbury: "Molecular Lasing Systems" in
    Marton — Methods of Experimental Physics, vol. 3A pag.302,
    D. Williams ed. Academic Press 1974

3 — T.Y. Chang and T.J. Bridges: Optics Comm. 1, 423 (1970)

4 — T.Y. Chang: "Optical Pumping in Gases" in "Nonlinear Infrared
    Generator" Y.R. Shen ed. Springer 1977

5 — M. Yamanaka: Rev. of Laser Engin. 3, 253 (1976)
  — F. Strumia and M. Tonelli: "Laser Lines from Optically Pumped
    Molecules" Proc. Summer School Fisica Atomica e Molecola-
    re. F. Strumia ed. CNR 1976
  — F. Strumia and M. Tonelli: to be published in"Rivista del Nuo-
    vo Cimento" (1978)
  — M. Rosenbluh, R.J. Temkin and K.T. Button: Appl. Optics 15,
    2635 (1976)

6 — M. Inguscio, P. Minguzzi and F. Strumia: Conf. Digest Int. Conf.
    Infrared Phys. Zurich 1975 pag. C229 and Infrared Phys.
    16, 453 (1976)

7 — H.R. Fetterman, H. R. Schlossberg and C.D. Parker: Appl. Phys.
    Lett. 23, 684 (1973)

8 — F. Strumia and P. Minguzzi: Proc. 2nd Symp. on Freq. Standard
    and Metrology — Copper Mountain , 1976 pag. 65
    Available from Time & Frequency Standards Section NBS
    Boulder, Colorado

9 — F. Strumia, R. Benedetti, M. Inguscio, P. Minguzzi, M. Tonelli,
    E. Bava, A. De Marchi, A. Godone — Conf. Digest 2nd Int. Sub-
    mill. Waves — Puerto Rico 1976 pag. S9

10 — M.S. Tobin and R.E. Jensen: Appl. Opt. 15, 2023 (1976) and
    Conf. Digest, Puerto Rico 1976 pag. 167

11 — K.P. Koo and P.C. Claspy: Conf. Digest, Puerto Rico 1976,
    pag. 171

12 — M.S. Tobin and R.E. Jensen: IEEE J.Quant. Electr. QE-13, 481
    (1977)

13 — H. Steffen and F.K. Kneubühl: IEEE J.Quant.Electr. QE-4,992(1968)
    R. Adam and F. K. Kneubühl: Appl. Opt. 8, 281 (1975)

14 - S.R. Stein, A.S. Risley, H. Van de Stadt and F. Strumia: Appl.
     Optics 16, 1893 (1977)

15 - M. Inguscio, P. Minguzzi and M. Tonelli: Opt. Comm. 21, 208
     (1977)

16 - R. Benedetti, A. Di Lieto, M. Inguscio, P. Minguzzi, F. Strumia
     and M. Tonelli: Proc. 31th Symp. on Freq. Control -
     Atlantic City, 1977. Available from: Electr. Ind. Assoc.
     2001 I St. N.W., Washington D.C. 20006

17 - F.L. Walls and A De Marchi: IEEE Trans. on IM 24, 210 (1975);
     E. Bava, A. De Marchi and A. Godone: Proc. 31th Symp. on
     Freq. Control - Atlantic City, 1977

18 - S.R. Stein and H. Van de Stadt: Proc. 31th Symp. on Freq.
     Control - Atlantic City, 1977

19 - S.R. Stein: Proc. 2nd Freq. Standards and Metrology Symp.,
     Copper Mountain, 1976, p. 479

20 - K.M. Evenson et al.: to be published

21 - R.L. Shoemaker, S. Stenholm and R.G. Brewer: Phys. Rev. A10,
     2037 (1974)

22 - P.W. Anderson: Phys. Rev. 76, 647 (1949)

23 - M. Inguscio: Tesi di Perfezionamento, Pisa, 1978

24 - H.R. Fetterman, C.D. Parker and P.E. Tannenwald: Opt. Comm.
     18, 10 (1976)

25 - M. Inguscio, P. Minguzzi, A. Moretti, F. Strumia and M. Tonelli:
     to be published (In this paper will be reported also the
     observation of large Stark frequency tuning of the FIR
     laser lines of CH_3OH at 119 μm and 70.5 μm.

INDEX